Lecture Notes in Computer Science 15985

Founding Editors

Gerhard Goos
Juris Hartmanis

Editorial Board Members

Elisa Bertino, *Purdue University, West Lafayette, IN, USA*
Wen Gao, *Peking University, Beijing, China*
Bernhard Steffen, *TU Dortmund University, Dortmund, Germany*
Moti Yung, *Columbia University, New York, NY, USA*

The series Lecture Notes in Computer Science (LNCS), including its subseries Lecture Notes in Artificial Intelligence (LNAI) and Lecture Notes in Bioinformatics (LNBI), has established itself as a medium for the publication of new developments in computer science and information technology research, teaching, and education.

LNCS enjoys close cooperation with the computer science R & D community, the series counts many renowned academics among its volume editors and paper authors, and collaborates with prestigious societies. Its mission is to serve this international community by providing an invaluable service, mainly focused on the publication of conference and workshop proceedings and postproceedings. LNCS commenced publication in 1973.

Péter Balázs · Reneta P. Barneva ·
Valentin E. Brimkov · Antal Nagy
Editors

Combinatorial Image Analysis

23rd International Workshop, IWCIA 2025
Szeged, Hungary, September 24–26, 2025
Revised Selected Papers

 Springer

Editors
Péter Balázs
University of Szeged
Szeged, Hungary

Valentin E. Brimkov
SUNY Buffalo State University
Buffalo, NY, USA

Reneta P. Barneva
State University of New York at Fredonia
Fredonia, NY, USA

Antal Nagy
University of Szeged
Szeged, Hungary

ISSN 0302-9743 ISSN 1611-3349 (electronic)
Lecture Notes in Computer Science
ISBN 978-3-032-19346-9 ISBN 978-3-032-19347-6 (eBook)
https://doi.org/10.1007/978-3-032-19347-6

This Springer imprint is published by the registered company Springer Nature Switzerland AG
The registered company address is: Gewerbestrasse 11, 6330 Cham, Switzerland

Preface

This volume presents the Proceedings of the 23rd International Workshop on Combinatorial Image Analysis (IWCIA 2025), organized by the Department of Image Processing and Computer Graphics, University of Szeged, Hungary and convened on September 24–26, 2025.

Image analysis provides the theoretical foundations and methodological tools essential for addressing complex real-world problems across diverse areas, including medicine, robotics, defense, and security. As the majority of data encountered in practical applications are inherently discrete, combinatorial methods have emerged as a natural and increasingly prominent paradigm within the field. Such approaches frequently yield significant advantages in both efficiency and accuracy when compared with traditional continuous models.

The IWCIA workshop series provides a forum for researchers throughout the world to present cutting-edge results in combinatorial image analysis, to discuss recent advances and new challenges in this research area, and to promote interaction with researchers from other countries. IWCIA had successful prior meetings in Paris (France) 1991, Ube (Japan) 1992, Washington DC (USA) 1994, Lyon (France) 1995, Hiroshima (Japan) 1997, Madras (India) 1999, Caen (France) 2000, Philadelphia, PA (USA) 2001, Palermo (Italy) 2003, Auckland (New Zealand) 2004, Berlin (Germany) 2006, Buffalo, NY (USA) 2008, Playa del Carmen (Mexico) 2009, Madrid (Spain) 2011, Austin, TX (USA) 2012, Brno (Czech Republic) 2014, Kolkata (India) 2015, Plovdiv (Bulgaria) 2017, Porto (Portugal) 2018, Novi Sad (Serbia) 2020, Messina (Italy) 2022, and Fort Lauderdale, FL (USA) 2024.

The Workshop organized by the University of Szeged retained and enriched the international spirit of these previous editions. The IWCIA 2025 Program Committee members were renowned experts coming from 19 different countries from Asia, Europe, and North America, and the authors came from 9 different countries.

Each of the 20 submitted papers was sent to at least two reviewers for a double-blind review. The most important selection criterion for acceptance or rejection of a paper was the overall score received. Other criteria included: relevance to the workshop topics, correctness, originality, mathematical depth, clarity, and presentation quality. We believe that as a result, only 15 papers of high quality have been accepted for publication in this volume.

The Microsoft CMT service was used for managing the peer-reviewing process for this conference. This service was provided for free by Microsoft and they bore all expenses, including costs for Azure cloud services as well as for software development and support.

Excellent keynote talks were given by our invited speakers. Peter Gritzmann from the Technische Universität München, Germany presented recent results on anisotropic diagrams and their connection to constrained clustering for representing and analyzing polycrystalline materials and their dynamics. In particular, weight-constrained anisotropic clustering enable the computation of diagram representations of polycrystals using grain volume, center, and moment data obtained from tomographic measurements. He also introduced new coreset techniques, valuable in their own right, that substantially accelerate these computations. The improvements were demonstrated on real-world 3D data sets.

Lajos Hajdu from the University of Debrecen, Hungary presented applications of algebra and number theory to two problems in digital image processing. The first concerns discrete tomography, outlining an algebraic framework that clarifies the structure of ghosts – functions on finite subsets of Z^2 with zero line sums – and showing how this aids the characterization of integer-valued and binary solutions. The second focuses on detecting periodicity in medical images by finding grids that best fit point sets in R^n. He demonstrated how lattice theory and Diophantine approximation techniques address the problem and how the LLL algorithm efficiently produces well-approximating grids in practice.

This volume includes fifteen contributed papers spanning a broad spectrum of research in digital geometry, image analysis, shape representation, and discrete mathematical structures. The articles address topics such as bijective image transformations, multidimensional shape descriptors, thinning algorithms, edge detection on non-standard grids, affine and discrete geometric mappings, automata for picture languages, and structural properties of combinatorial sets. The collection also includes advances in deep-learning-based object detection, binary tomography, and geometric analysis on specialized grids, reflecting theoretical aspects and practical applications of contemporary combinatorial image analysis.

The success of IWCIA 2025 was made possible through the contributions of many dedicated individuals and supporting organizations. The Editors are indebted to IWCIA's Steering Committee for endorsing the candidacy of Szeged for the 23rd edition of the Workshop. We wish to thank everybody who submitted their work to IWCIA 2025. We are grateful to all participants and especially to the contributors to this volume. Our most sincere thanks go to the IWCIA 2025 Program Committee whose cooperation in carrying out high-quality reviews was essential in establishing a strong scientific program. We express our sincere gratitude to the keynote speakers – Peter Gritzmann and Lajos Hajdu – for their excellent talks and overall contribution to the Workshop program.

The organizers of IWCIA 2025 gratefully acknowledge the generous support of Tuxera and the Institute of Informatics at the University of Szeged. Their contribution and commitment were essential in making this event possible and in providing a stimulating environment for the exchange of ideas and recent research results.

The success of the Workshop was made possible through the dedicated efforts of the local Organizing Committee. We also extend our sincere appreciation to the host institution, the Department of Image Processing and Computer Graphics, for its valuable support. Finally, we wish to thank the team at Springer for their efficient and kind cooperation in the timely production of this book. We are confident that IWCIA 2025 offered an excellent environment for professional growth, and we anticipate that this volume will be of interest to a wide community of scholars and educators.

December 2025

Péter Balázs
Reneta P. Barneva
Valentin E. Brimkov
Antal Nagy

Organization

General Chairs

Péter Balázs   University of Szeged, Hungary
Antal Nagy   University of Szeged, Hungary

Co-chairs

Reneta P. Barneva   SUNY Fredonia, USA
Valentin E. Brimkov   SUNY Buffalo State, USA

Steering Committee

Gabor T. Herman   CUNY Graduate Center, USA
Valentin E. Brimkov   SUNY Buffalo State, USA
Tibor Lukić   University of Novi Sad, Serbia
Giorgio Nordo   University of Messina, Italy
Renato M. Natal Jorge   University of Porto, Portugal
João Manuel R. S. Tavares   University of Porto, Portugal

Keynote Speakers

Peter Gritzmann   Technische Universität München, Germany
Lajos Hajdu   University of Debrecen, Hungary

Program Committee

Michela Ascolese   University of Florence, Italy
Andreas Alpers   University of Liverpool, UK
Eric Andres   Université de Poitiers, France
Buda Bajić   University of Novi Sad, Serbia
Péter Balázs   Uranversity of Szeged, Hungary
George Bebis   University of Nevada at Reno, USA
Partha Bhowmick   Indian Institute of Technology Kharagpur, India

Contents

Fast Image Resizing Through Bijective Transformations: Applications in Classification Tasks

Dacian Goina[(✉)] and Marc Frincu

West University of Timisoara, 300223 Timisoara, Romania
`dacian.goina00@e-uvt.ro`

Abstract. Although Convolutional Neural Networks provided effective solutions for computer vision tasks across various domains in recent years, the high training times of the models remain a crucial aspect. A straightforward way to decrease the computational costs is to use smaller images during training. However, existing image resizing approaches are either computationally intensive or lossy methods. This paper introduces the usage of the Cantor pairing function for image resizing through pixel reduction. The pairing function is a fast bijective transformation that maps a pair of two integers into a single value. Combining two pixels into one leads to a decrease in the image size, while the bijective property ensures a lossless transformation. The experiments conducted on the MNIST datasets demonstrated that the proposed approach reduces resizing time at least six times, decreases training times significantly, and achieves similar performance with baseline methods.

Keywords: Cantor pairing function · Bijective transformation · Image resizing · CNN models

1 Introduction

In recent years, image classification tasks and computer vision methods have gained popularity with applications in fields like medicine [22], autonomous vehicles [9], and agriculture [13]. While Convolutional Neural Network (CNN) models are effective solutions for image classification tasks, they are costly considering the training processes. An effective approach for reducing CNN training times is to resize the images. However, scaling methods like Bicubic Interpolation and Nearest Neighbor [3] are computationally expensive for large datasets. These methods are lossy, causing the complete loss of certain parts of the original data during the transformation. This paper presents the usage of the **Cantor pairing function** [7] for image downscaling. The Cantor pairing function is a bijective transformation that maps two integer values into one. The key idea is to merge pairs of pixels into single values. The bijective property ensures that no data is lost during the transformation, allowing the original image to be entirely reconstructed later if needed. The proposed approach provides a transformation at least six times faster than the baseline methods used in the experiments.

© The Author(s), under exclusive license to Springer Nature Switzerland AG 2026
P. Balázs et al. (Eds.): IWCIA 2025, LNCS 15985, pp. 1–18, 2026.
https://doi.org/10.1007/978-3-032-19347-6_1

This research aims to verify that the model trained on resized images achieves similar performance to the model trained on the original data, while decreasing training time. The promising results obtained for MNIST datasets show the potential and efficiency of the proposed approach. The experiments show that the proposed method is at least as effective as, or even more effective in some cases compared with the Bicubic Interpolation and Nearest Neighbor techniques.

The key contributions of this paper are:

- Introducing the usage of the Cantor pairing function for image resizing in the context of image classification tasks;
- The usage of a quadratic-like downscaling transformation method applied on the resized image, for a better semantic preservation of the original image;
- Experiments on widely used datasets that validate the usability and performance of the proposed approach.

The rest of the paper is structured as follows: Sect. 2 reviews different approaches to the problem. Section 3 presents the proposed method and its theoretical elements. Section 4 describes the performed experiments and their results. Section 5 discusses the conclusions and future directions.

2   Related Work

The image resize problem is approached in multiple works using a variety of techniques. Amanatiadis and Andreadis [4] conducted a survey presenting interpolation-based techniques and their performances evaluated using various metrics.

Multiple studies analyzed content-aware image resizing methods, which reduce image size by removing less relevant content and preserve essential features. Nam et al. [16] presented a deep neural network architecture able to minimize the subtle local artifacts from images. The experimental results obtained for the proposed network outperformed those of the handcrafted feature-based method. Danon et al. [8] proposed a deep learning network used for image resizing from feature maps. The research concluded that the introduced method tends to preserve the important semantic regions of the original image.

Han et al. [10] proposed an image resizing method based on wavelet analysis that uses an energy map obtained by weighting multiscale subbands of the image. The experiments showed that the proposed method produces higher subjective results comparing with those obtained with other scaling methods. Other wavelet-based solutions [5] were used to resize images for different cases, such as medical image classification [14].

Several works studied the impact of using resized images on the models' performances for image classification tasks. Talebi and Milanfar [20] presented a CNN-based resizer architecture able to downscale the image and then to fit it directly to the classification model. The experiments demonstrated superior performances compared to the classification model trained on images resized with Bicubic Interpolation method. Hossain et al. [12] analyzed the effect of training

CNN models on images resized using bicubic and bilinear interpolation methods. The results showed only a small drop in accuracy comparing with the original model. Hirahara et al. [11] investigated the impact of using images resized with classic techniques like Nearest Neighbor and Lanczos Interpolation for training CNN models in chest X-ray image classification. The best performance was achieved by a model trained on images of size 64×64.

Various techniques have been proposed for the approached problem, each of them with its advantages and drawbacks. Traditional interpolation-based methods often provide good results, but they are lossy techniques. Wavelet-based approaches are reliable solutions but can be sensitive to noise in various situations. Content-aware image resizing methods present promising results but their reliance on deep neural networks makes them computationally expensive.

The usage of a bijective transformation such as the Cantor pairing function for image resizing was not explored before despite its great potential. Combining the values of two pixels into one reduces the number of pixels. A bijective function provides a one-to-one reversible transformation, ensuring data preservation, contrary to the lossy methods. The fast computational time is another benefit of the proposed approach.

3   Solution Methodology

3.1   Cantor Pairing Function

The Cantor pairing function [7] is a bijective transformation that maps each pair of two integers into a single unique integer. The function is shown in Eq. 1.

$$f : \mathbb{N} \times \mathbb{N} \rightarrow \mathbb{N}, \quad f(a,b) = \frac{(a+b) \cdot (a+b+1)}{2} + b \tag{1}$$

The reason for choosing the Cantor pairing function is mainly related to its simplicity and rarity. A bijective function that maps a domain D to a lower-dimensional space is not so common. There exist other pairing functions such as the Rosenberg-Strong function [17] which operates with negative integers. Considering the domain of the input data (image pixels), a function that operates on positive integers suffices.

3.2   Data Transformation

The data transformation flow is presented in Fig. 1. The transformation method takes as input an $m \times n$ matrix representing a grayscale image of size $m \times n$, a list h containing pairs of pixel indexes used for pairing, and the sizes u, v for the resulting image. The resized image size is $u \times v < m \times n$; values u, v must be *compatible* with k, i.e., $u \cdot v \mid (m \cdot n - k/2)$, with k being an even value that represents the number of pixels used for pairing (from list h). The pairing function combines two values into one (two pixels into one). In a base case, all pixels are combined two by two, leading to $u \cdot v = (m \cdot n)/2$ (i.e., the image size

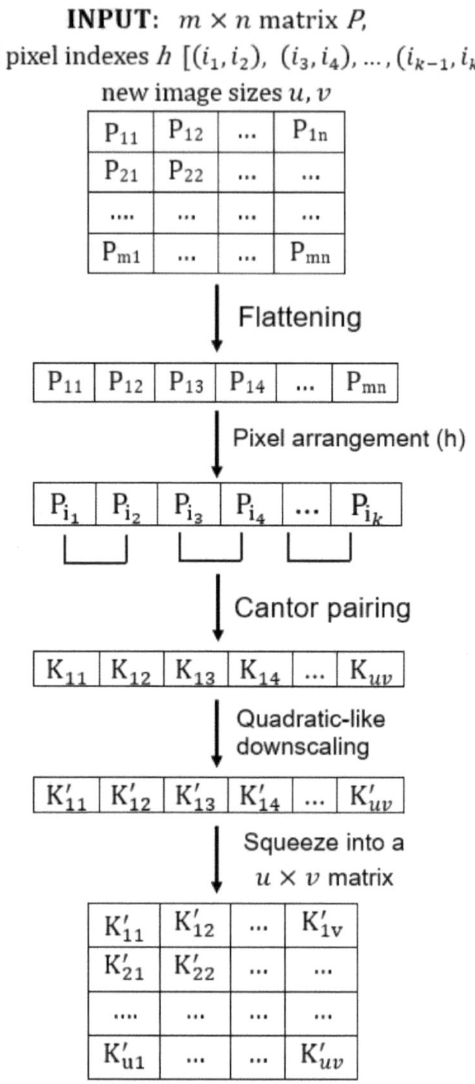

Fig. 1. Data transformation flow - usage of the Cantor pairing function for image resizing.

is halved). Similarly, just several pixels can be combined (instead of all of them), only to **choose a configuration h and values u, v** such that in the resized image, **the number of pixels $u \cdot v$ is a value that can be written as a product of two integers** (the height and width of the resized image). Next, we explain each transformation stage in detail.

Flattening Stage. The original matrix of size $m \times n$ is flattened into an $1 \times (m \cdot n)$ array, arranging the elements row by row. That is, the values from the first row are added to the one-dimensional array, followed by the values from the second row, and this process is repeated until all rows are included.

Pixel Arrangement Stage. The elements of the one-dimensional array obtained in the previous step are arranged according to the pairs of indexes from the list h. For pairing the pixels using the order from the original image, the **identity configuration** $h = [(1, 2), (3, 4), ..., (m \cdot n - 1, m \cdot n)]$ is used. This step allows observing the model's performances considering an unfavorable case of pixels pairing. Usually, when combining two neighboring pixels (e.g., P_{11} with P_{12}) the semantic meaning of the image is mostly preserved, i.e., the semantic context is maintained (humanly understandable). However, there are cases in which the semantic meaning is not preserved upon the combination of two pixels, e.g., pairing a pixel located at the end of one row with a pixel located at the start of the next row; in these situations, the semantic meaning can be heavily disturbed. The arrangement stage enables testing the model's performance in unfavorable cases where the pixels are arranged in random order. The same list of pairs h is used for transforming all images in the dataset.

Cantor Pairing Stage. Following the arrangement, consecutive pixels are paired two by two using the Cantor pairing function. The results are stored in a new $1 \times (u \cdot v)$ array. Compared with other image resizing methods, the proposed approach does not work directly with the size $u \times v$ of the new image in all cases. Obtaining a new custom size is not always possible by performing only one iteration of the proposed method. Pairing all the pixels once halves the image size; pairing all the pixels again leads to a size reduced by four times with respect to the size of the original image.

Quadratic-like Downscaling. After applying the Cantor transformation, many of the obtained values exceed the value 255, leading to an unusual situation for image representation. Shifting the values into the range [0, 1] or [0, 255] could be done using min-max normalization; this approach implies a linear transformation which is unsuitable in this context. Indeed, using a non-linear transformation like the Cantor pairing function changes the data distribution, but it is desirable to preserve the distribution at the grayscale level as much as possible. Before using min-max normalization to the obtained values, each pixel value v is updated according to Eq. 2.

$$v \leftarrow \frac{-1 + \sqrt{1 + 8 \cdot v}}{4} \tag{2}$$

Equation 2 shows a bijective transformation employed to align the grayscale value distribution of the obtained image with that of the original one - this process is explained in detail in Subsect. 3.3. After the usage of Eq. 2, the obtained values are moved to the range [0, 1] using min-max normalization.

Squeezing Stage. The obtained $1 \times (u \cdot v)$ array is reshaped into a matrix: each chunk of consecutive (nonoverlapping) u values forms a row. The obtained structure is a $u \times v$ matrix, thus it can be processed by a CNN model.

The steps presented and explained above are used to resize each image in the dataset to create a consistent new dataset.

3.3   Quadratic-Like Downscaling

A quadratic-like downscaling procedure is applied on the values obtained with the Cantor pairing function because usage of a non-linear transformation affects the data distribution, introducing non-linear discrepancies (distances) between the resulting values [6]. Before continuing with the explanations, let us consider an example. Starting with a set of values $l = [1, 2, 3, ..., 20]$, by applying an exponential transformation (e.g., $g(x) = 2^x$) the obtained values are largely distanced one from another. By applying min-max normalization (with range $[0, 1]$), the maximum value 2^{20} is mapped into 1 and the majority of the remaining values are mapped into values close to zero because they are far away from 2^{20}. The min-max normalization is a linear transformation and does not repair the discrepancies introduced by the exponential function. To address this issues, the logarithmic function (the inverse of the exponential function) must be used first. Indeed, in this scenario, the inverse function transforms the obtained values back to the original ones, but the idea of this explanation refers to how min-max normalization method is sensitive to largely distanced values.

Similarly, following the usage of the Cantor pairing function, the obtained values are largely distanced one from another. In our case, using the pairing function with maximum values as arguments (i.e., $f(255, 255)$), leads to 130,560 which is a large value compared with the other ones. Since the maximum possible value is so large, after the usage of min-max normalization, the majority of the rest of the values (they are quadratically smaller than the largest one) will gather close to the value zero. The previously described situation leads to a scenario with large distances between the values which were relatively close initially. Even if the pairing function changes the data distribution, it is desirable to preserve the distribution at the grayscale level as much as possible.

Since the Cantor pairing function is a bijective transformation, it is reversible, i.e., for a given output value obtained with f, the specific input values n_1, n_2 can be recovered. In the current context, since we aim to reduce the number of pixels, we do not want to unpair the result back to two values, but it is desirable to shrink the obtained value into a range $[0, 255]$. The Cantor pairing function performs a quadratic-like transformation, thus the inverse function is related to the square root function. The analysis and demonstration of the inverse of the Cantor pairing function is not the purpose of this paper, so several mathematical elements are skipped. Considering two values $n_1, n_2 \in \mathbb{N}$ and $z = f(n_1, n_2)$, according to the theory [19], to recover n_1, n_2 from z, the first step is to compute $\lfloor \phi(z) \rfloor$ where $\lfloor \cdot \rfloor$ is the floor function and ϕ is the function described in Eq. 3.

$$\phi : \mathbb{N} \to [0, \infty), \quad \phi(z) = \frac{-1 + \sqrt{1 + 8 \cdot z}}{2} \tag{3}$$

According to [19], the value $\lfloor \phi(z) \rfloor$ is the sum of the input values n_1, n_2. We discard the floor function component to retain more information from the value provided by the ϕ function, i.e., to not lose information from the resulting value after flooring. At this point, $\phi(z)$ is a value close to $n_1 + n_2$, but in some cases it is still higher than 255. Considering the case with maximum values for arguments, i.e., $n_1 = 255$, $n_2 = 255$, it leads to $\phi(z) \in [510, 511)$. Considering $v = \phi(z)/2$ for shifting the value v closer to the range $[0, 255]$ leads to Eq. 2. Further, min-max normalization is applied to the value v to bring it to the range $[0, 1]$. Even if there exist cases when v exceeds the value 255, i.e., $v \in [255, 256)$, this issue is solved by min-max normalization.

Summarizing, the quadratic-like downscaling transformation presented in Eq. 2 is required to reduce the quadratic-like distances between the values resulting from the usage of the pairing function. This processing step is recommended for obtaining an image with a histogram of grayscale values closer to the histogram of the original image. The effects following the usage of the Cantor pairing function and quadratic-like downscaling for image transformation are shown in Figs. 2, 3. As shown in Fig. 2c, *the usage of a quadratic-like downscaling method leads to an image with much more intense gray values, especially at the edges of the image*, comparing to Fig. 2b. Studying the histograms presented in Fig. 3, it can be observed that the histogram of the image obtained using the quadratic-like downscaling method (Fig. 3d) is much more similar to the histogram of the original image (Fig. 3a), compared to those from Figs. 3c, 3b.

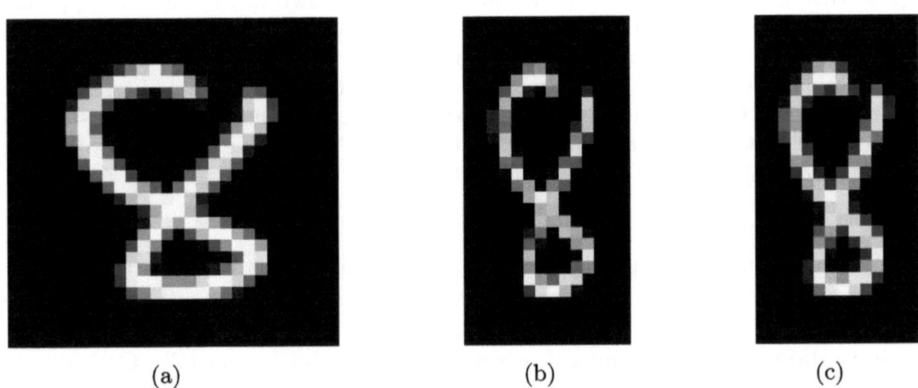

(a)   (b)   (c)

Fig. 2. An example of usage of the Cantor pairing function on an image. (2a) shows the original 28×28 image. (2b) shows the original image resized to 28×14 using the Cantor pairing function and normalized using min-max normalization. (2c) shows the original image resized to 28×14 using the Cantor pairing function and normalized using quadratic-like downscaling method.

3.4   Data Representation and Pairing Method Benefits

As a bijective mapping, the Cantor function provides a one-to-one transformation for a given input. After applying the Cantor method, even if the obtained image provides the same data meaning as the original one, they are different concerning the expressive power and representation. That is, the obtained image

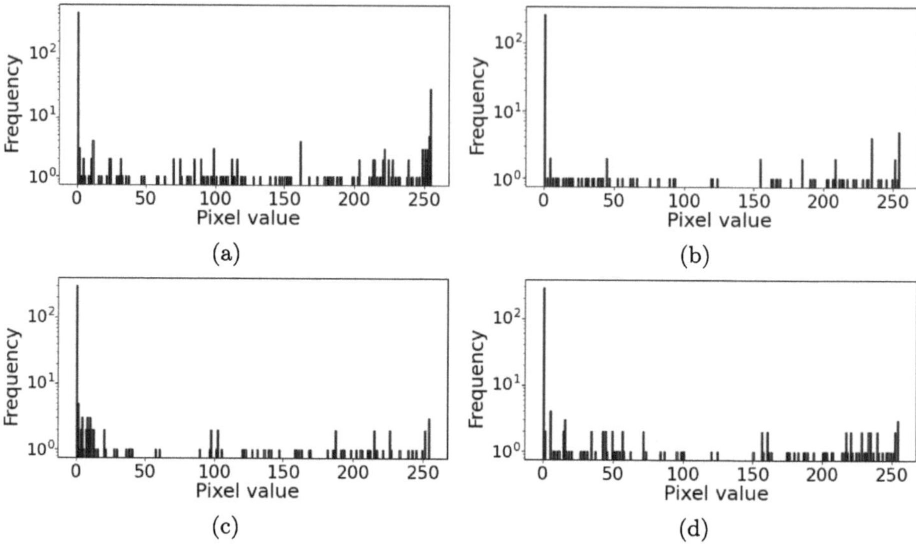

Fig. 3. Histograms of grayscale values for the image shown in Fig. 2a after applying different transformations on it. (3a): Histogram of the original 28 × 28 image. (3b): Histogram of the image obtained by resizing Fig. 2a to size 28 × 14 using Bicubic Interpolation method. (3c): Histogram of the image obtained by resizing Fig. 2a to size 28 × 14 using the Cantor pairing function and min-max normalization. (3d): Histogram of the image obtained by resizing Fig. 2a to size 28 × 14 using the Cantor pairing function and quadratic-like downscaling method.

has a smaller number of pixels than the original one, thus it is less representative; even if at the data level, those two images are the same (they 'say' the same thing), the transformed image exposes less information. These remarks are important because in general, deep learning models (especially CNNs) rely on data representation. A model trained on smaller images is likely to show a decrease in performance compared with the same model trained on the original data [18]. This situation is also encountered when applying other image-resizing methods.

The advantages of using the Cantor pairing function are its low computational time and the ability to provide a lossless transformation. The efficient computational time is presented in Sect. 4. The property of providing a lossless transformation is related to mappings performed during the **transformation chain**. The Cantor pairing function is a bijective transformation, the ϕ function used for quadratic-like downscaling provides a unique output for each unique input, and the min-max normalization method is a bijective mapping. Considering these facts, starting with the resized image, the original image can be recovered entirely. This property is important, as relevant details can be lost when reconstructing an image resized with a lossy technique.

4 Experiment

4.1 Experiment Methodology

Experimental Setup. All experiments were executed on a Google Colab notebook with default (free usage) specifications. All the code [1] was implemented in Python language; TensorFlow was used as the main framework.

Experiment Methodology. The MNIST Digits [2] and Fashion [21] datasets were used for the experiments. Each dataset contains 70,000 images, each of size 28×28. Different types of experiments were conducted to observe the evolution of model performance: some experiments used the entire dataset, while others utilized only 50% or 80% of the dataset samples, respectively. For data split, 80% of data were used for training and 20% for validation. Testing data were not considered since the model was evaluated on unseen (validation) data after each epoch, without influencing its configuration and training. The performance of the models was evaluated using the F1 score, accuracy (on validation data), and training time. Both the F1 score and accuracy metrics [15] take values in the range [0, 1], with higher values indicating better model performance. The experiments were executed five times (using different data split seeds) and the averaged resulted values are presented.

Separate models were trained on original 28×28 images, as well as on images resized to 28×14 and 14×14 pixels. After resizing, the pixel values were normalized to range [0, 1]. The proposed resizing approach was applied using the **identity configuration** (see Subsect. 3.2) to pair the pixels, but an experimental comparison with a random shuffle of pixels was also conducted. For comparison, Bicubic Interpolation and Nearest Neighbor methods were used. Deep learning resizing-based methods were excluded due to their high execution time, the aim being to reduce it. Each model was trained for 10 epochs without early stopping, as the experiments were conducted for demonstration purposes to observe the models' performance evolution. All other training parameters were kept at their default values from the TensorFlow implementation.

CNN Model Architecture. To reiterate, this research focuses on reducing the models' training time using smaller images while preserving the performances of the model trained on original data as much as possible. The model used for the experiments is a simple one chosen for demonstration (experimental) purposes and is described in Listing 1.1.

```
Convolutional2D(filters=8, kernel=(3, 3), activation='relu')
MaxPooling2D(pool_size=(2, 2), padding='same')
Convolutional2D(filters=16, kernel=(3, 3), activation='relu')
MaxPooling2D(pool_size=(2, 2), padding='same')
FlattenLayer
DropoutLayer(rate=0.5)
DenseLayer(units=10, activation='softmax')
```

Listing 1.1. CNN model layers.

4.2　Experiments Results and Analysis

Table 1 shows the averaged execution times (in seconds) over 10 runs for resizing all 70,000 28×28 images in MNIST Digits dataset to 28×14 size using each method. As observed, our approach is significantly efficient, 11 times faster than Bicubic Interpolation and 6 times faster than the Nearest Neighbor method. This behavior occurs likely because the proposed approach computes the value for each new pixel using a constant operation (using Eqs. 1, 2) leading to a linear time complexity. In comparison, the Bicubic Interpolation method also has linear time complexity, but it uses a 4×4 grid (it operates with 16 pixels) to compute each new pixel value.

Table 1. Averaged mean/std. execution time of methods over 10 runs.

Method	Mean execution time (s)	Std.
Proposed (Cantor pairing-based)	0.85	0.09
Bicubic Interpolation	9.54	1.03
Nearest Neighbor	5.65	0.48

Figures 4a, 4c shows several images resized to 28×14 with the proposed approach using the identity configuration for the arrangement of the pixels. Figures 4b, 4d shows the same images as in Figs. 4a, 4c but a random shuffle of pixel indexes was used for arrangement in the resizing procedure. Figure 5 shows the F1 scores for models trained on images resized with the proposed method, using identity configuration and random shuffle. This experiment was performed for demonstration purposes to examine the impact of disturbed image semantics on model performance, considering that the proposed resizing method could be applied in this manner, i.e., pairing randomly located pixels within an image. As anticipated, models trained on images resized using the identity configuration provided better results. CNN models learn patterns from images, thus when the pixels are spread randomly rather than organized semantically, it is harder for the model to recognize the patterns and deliver accurate classifications. However, the differences between models' performance are not significantly large.

In Figs. 6, 7, 8 and 9 the term *Original model* refers to a model trained on original images of size 28×28 pixels. This model was compared to models trained on images resized to 28×14 and 14×14 pixels using the specified methods. For the proposed method, the **identity configuration** was applied for pixels pairing.

Figures 6a, 6b show the training times for the models trained on images resized to 28×14, respectively 14×14 pixels, using entire MNIST Digits dataset. As expected, the models trained on smaller images provided lower training times for each epoch. Using smaller images for model training reduces training times since smaller images contain fewer pixels. As observed from the figures, the training time scales approximately linearly with the number of pixels.

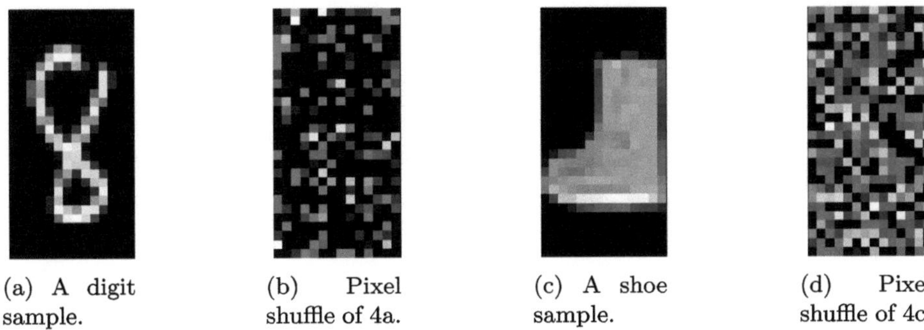

(a) A digit sample.

(b) Pixel shuffle of 4a.

(c) A shoe sample.

(d) Pixel shuffle of 4c.

Fig. 4. A digit (4a), respectively a shoe sample (4c), along with the images obtained after shuffling pixels order before resizing (4b, 4d).

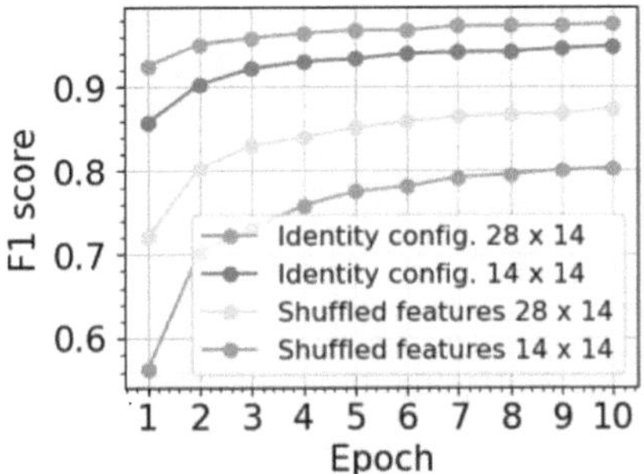

Fig. 5. F1 score values for models trained on images resized using identity configuration and random shuffle of pixels.

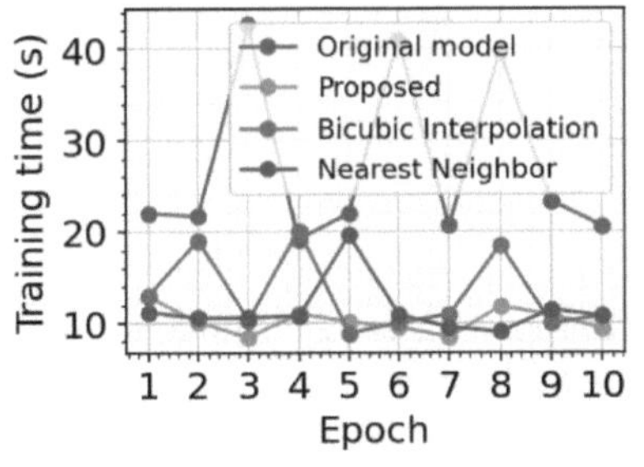

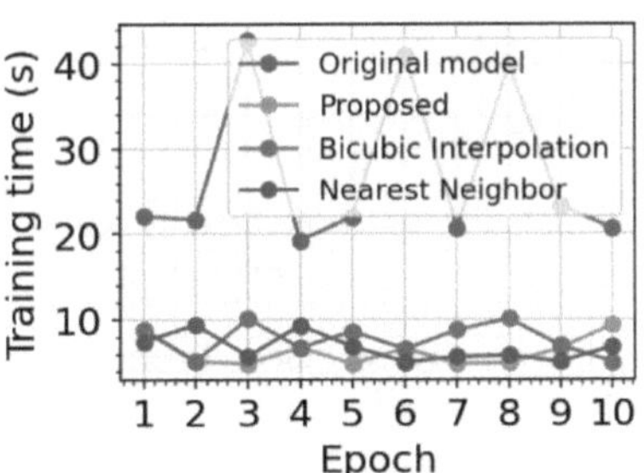

(a) Models trained on samples of size 28×14 pixels.

(b) Models trained on samples of size 14×14 pixels.

Fig. 6. Training times (in seconds) for models trained on MNIST Digits dataset, with samples resized to 28×14 (6a) and 14×14 pixels (6b), respectively.

Figures 7 shows the F1 score values obtained for the models trained and validated on subsets of MNIST Digits dataset with different sampling percentages. As observed, the models trained on larger subsets of data provided better results

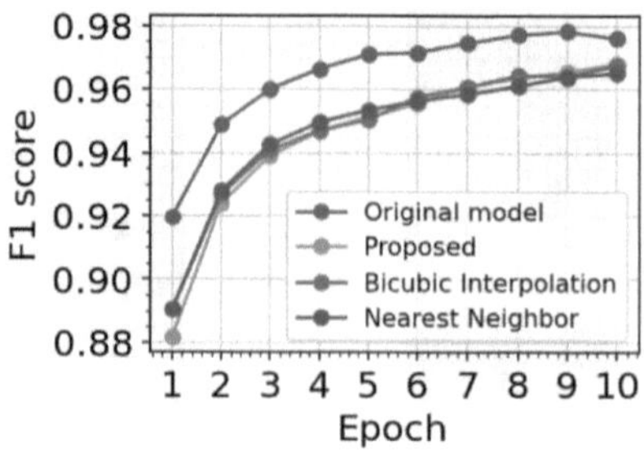

(a) Subset of 50% data used. Images resized to 28 × 14 pixels.

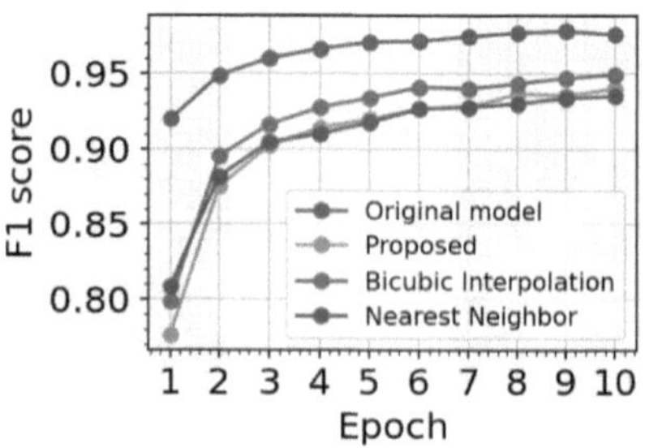

(b) Subset of 50% data used. Images resized to 14 × 14 pixels.

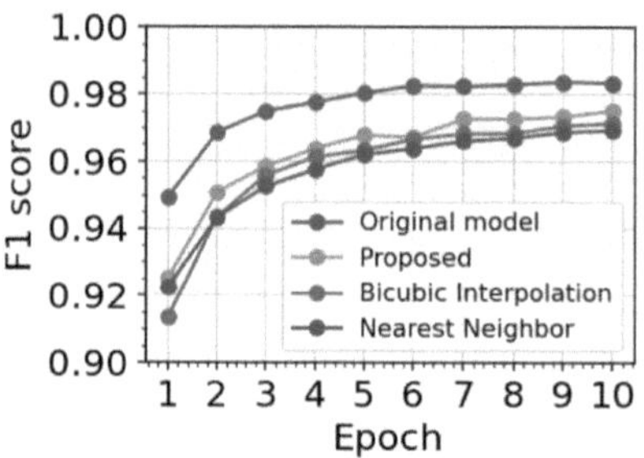

(c) Subset of 80% data used. Images resized to 28 × 14 pixels.

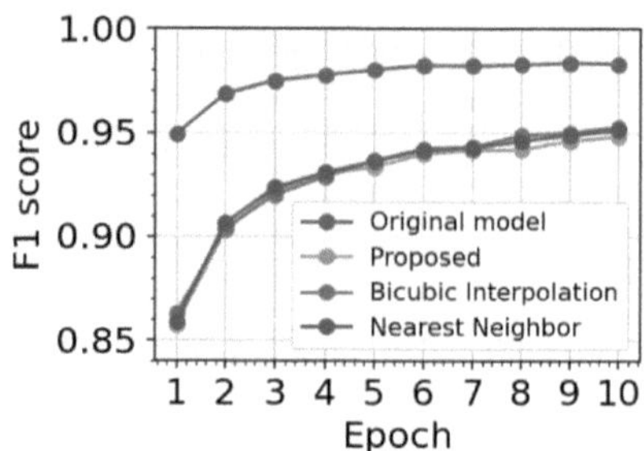

(d) Subset of 80% data used. Images resized to 14 × 14 pixels.

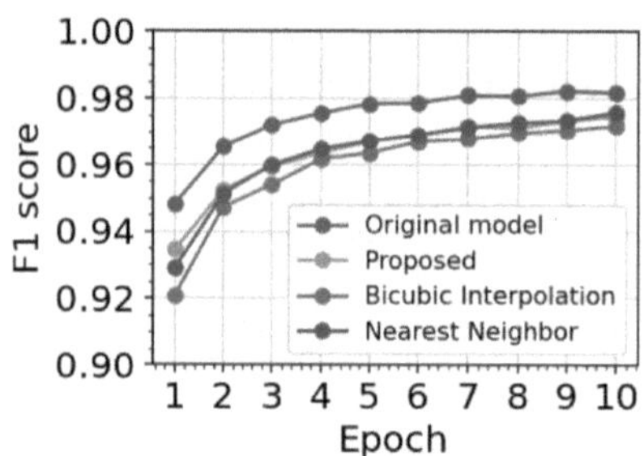

(e) Entire dataset used. Images resized to 28 × 14 pixels.

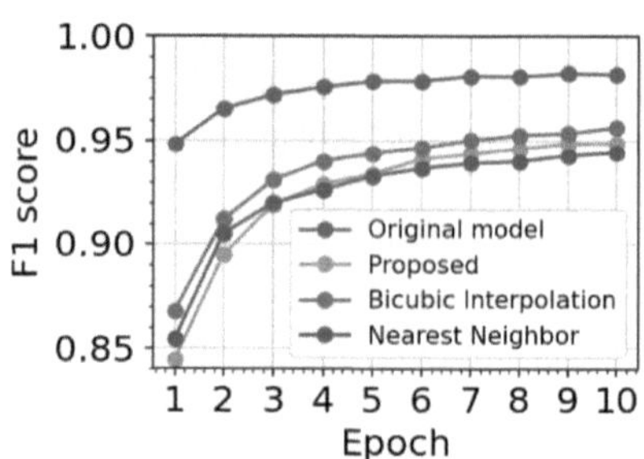

(f) Entire dataset used. Images resized to 14 × 14 pixels.

Fig. 7. F1 score values obtained on validation data for models trained on different subsets of MNIST Digits samples resized to different sizes.

compared with the models trained using smaller subset of data. This is a common behavior for deep learning models: an increase in the volume of training data leads to improved performance, as the model gains a better understanding of complex patterns in the data. Also, the models' performance declines as image size decreases. Being trained on data with fewer pixels, the models learn a weaker representation of the data compared to the case when the original data (all pixels) are used. However, the difference in the performance of the Original model and the models trained on smaller images was not significant. After 10 epochs of training, the maximum F1 score difference was 0.01 between the Original model and the models trained on images of size 28×14, respectively 0.03 for models trained on images of size 14×14. Considering the reduced training time, these slight decreases in F1 score values present a beneficial trade-off.

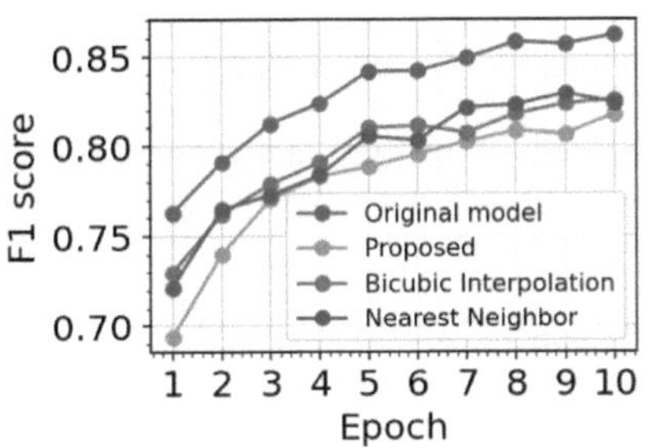

(a) Subset of 50% data used. Images resized to 28×14 pixels.

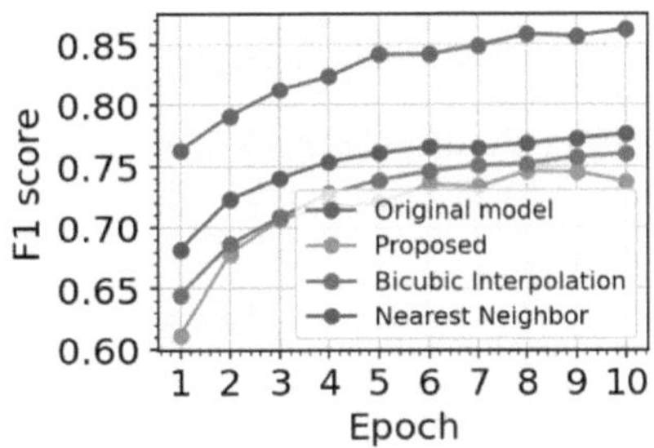

(b) Subset of 50% data used. Images resized to 14×14 pixels.

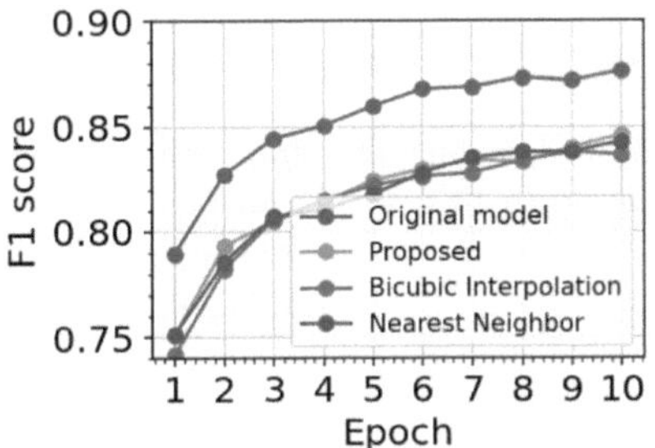

(c) Subset of 80% data used. Images resized to 28×14 pixels.

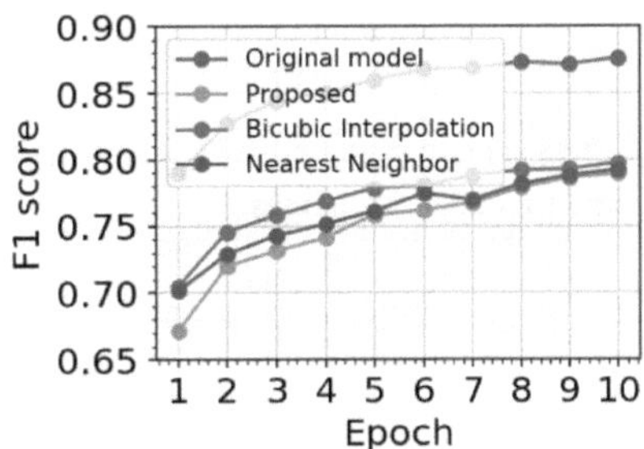

(d) Subset of 80% data used. Images resized to 14×14 pixels.

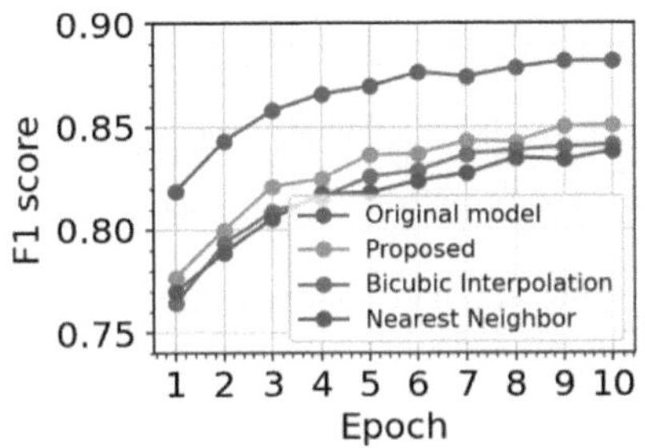

(e) Entire dataset used. Images resized to 28×14 pixels.

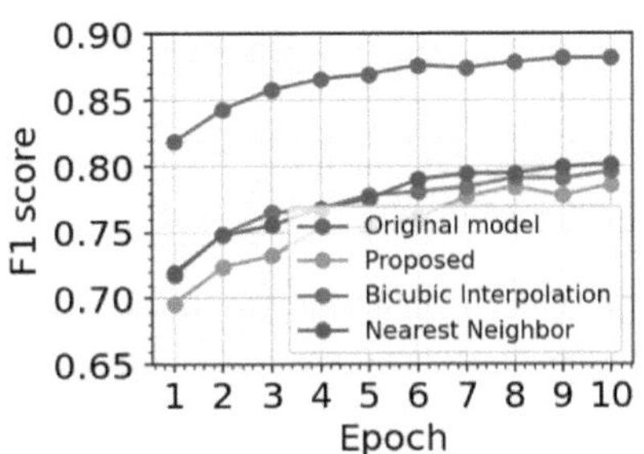

(f) Entire dataset used. Images resized to 14×14 pixels.

Fig. 8. F1 score values obtained on validation data for models trained on different subsets of MNIST Fashion samples resized to different sizes.

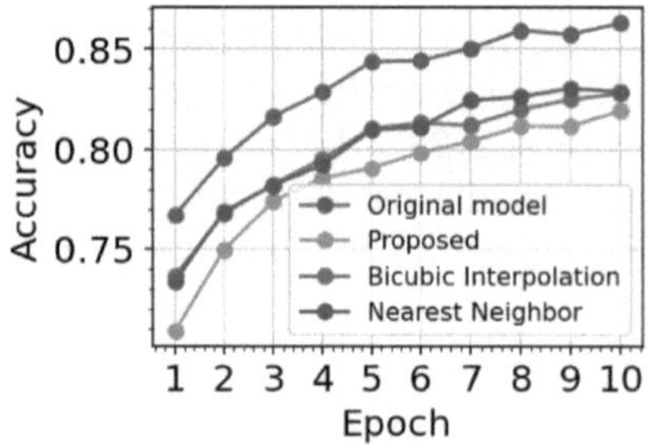

(a) Subset of 50% data used. Images resized to 28 × 14 pixels.

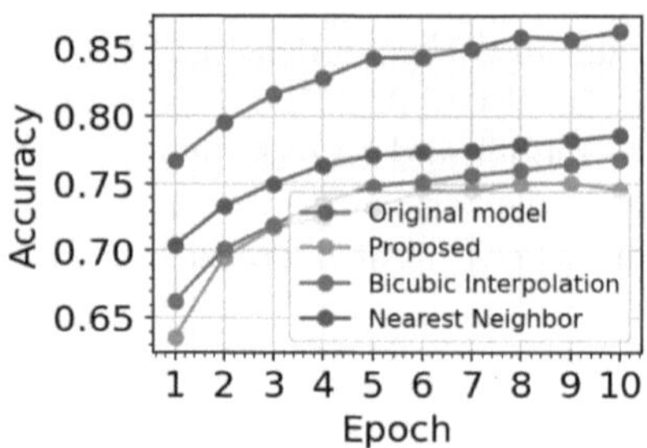

(b) Subset of 50% data used. Images resized to 14 × 14 pixels.

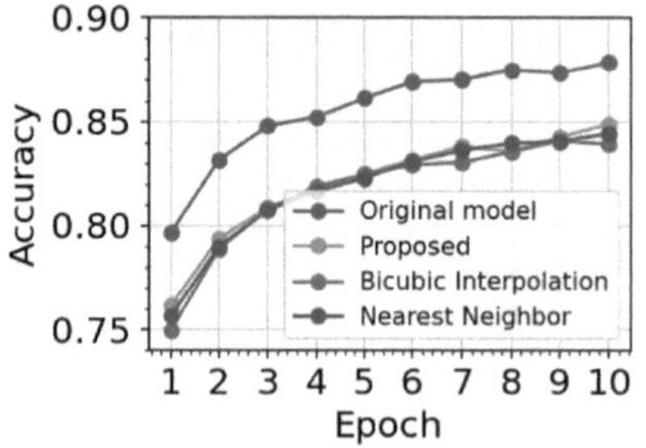

(c) Subset of 80% data used. Images resized to 28 × 14 pixels.

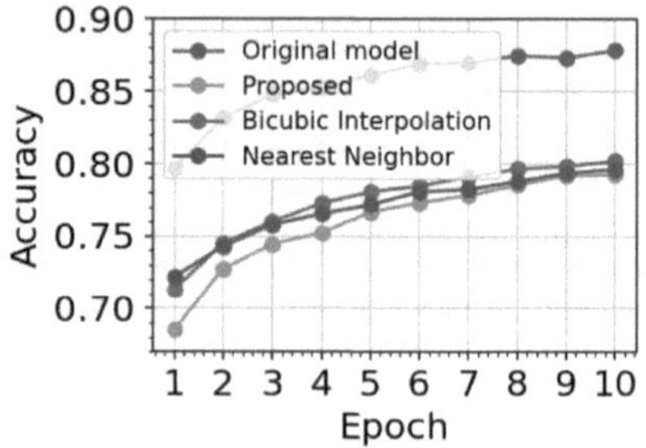

(d) Subset of 80% data used. Images resized to 14 × 14 pixels.

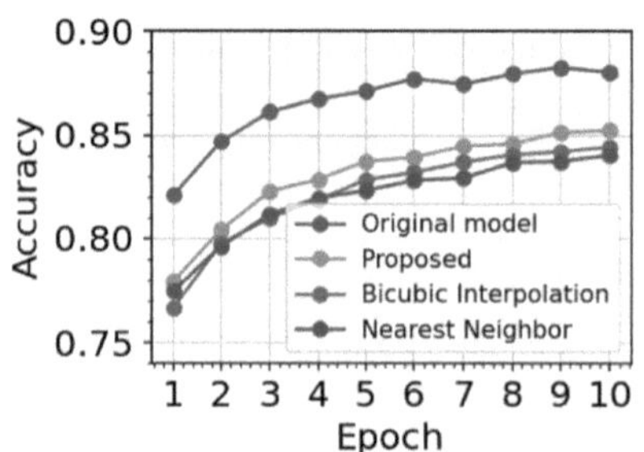

(e) Entire dataset used. Images resized to 28 × 14 pixels.

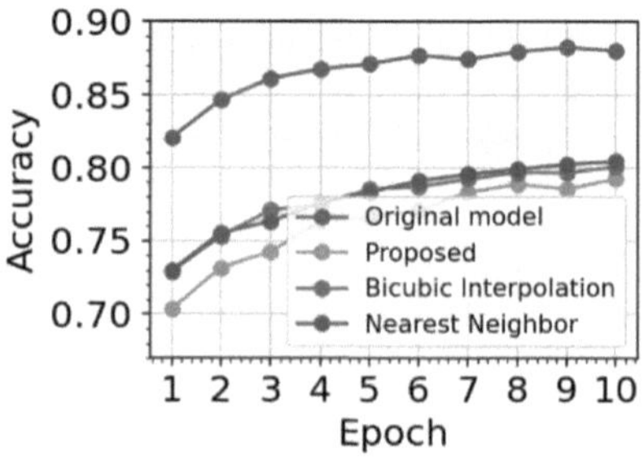

(f) Entire dataset used. Images resized to 14 × 14 pixels.

Fig. 9. Accuracy values obtained on validation data for models trained on different subsets of MNIST Fashion samples resized to different sizes.

Figures 8 and 9 show the F1 score respectively accuracy values for the models trained and validated on subsets of different sampling ratios of MNIST Fashion dataset. These experiments followed the same resizing procedures as those used in the MNIST Digits dataset experiments. The explanations provided for the results obtained on MNIST Digits dataset experiments also apply here: models' performance declined when smaller images were used but improved when larger subsets of data are used for training. The models trained on the images resized using the proposed method provided results similar to those trained on images resized with baseline methods. Although there was not a clear winner among the models trained on resized images (w.r.t. the employed resizing method), the results demonstrated that our approach provides consistent transformations, enabling the models to provide strong outcomes. While the proposed approach is similar to baseline models in terms of the model's training time and performance, it offers the advantage of faster image resizing, as presented in Table 1.

5   Conclusions

This paper presented the usage of the Cantor pairing function for image resizing, aiming to obtain smaller images to train CNN models quickly while preserving the performance of the model trained on the original data as much as possible. The Cantor pairing function was used to map two pixels into one, and combined with a quadratic-like downscaling method, the presented approach provided a significantly fast and lossless image transformation.

Multiple experiments were conducted on MNIST Digits and Fashion datasets to evaluate the performance of the proposed approach, considering image transformation to different sizes and training the models on subsets with different sampling ratios. The experiments showed that the models trained on images resized with the proposed method achieved F1 score and accuracy values similar to the models trained on images resized with interpolation-based methods. Regarding the models' training time, it scaled linearly with the number of pixels. Compared to the model trained on the original data, the model trained on images resized using the proposed approach showed slight decrease in F1 score and accuracy values, but this is a worthwhile trade-off considering the improvements in model's training time.

Although the transformed images maintain semantic (visual) consistency of the original ones, the main benefit of the proposed method is that its execution time was at least six times faster than that of the baseline methods. Future works will consider using the introduced approach for clustering problems. The order of pixels affects image's semantic elements and thus impacts model performance in image classification problems. However, clustering methods use commutative distance metrics, thus the order of the data attributes (the arrangement used for the pairing procedure) should not impact the clustering performance.

Disclosure of Interests. The authors declare no competing interests that are relevant to the content of this article.

Appendix

We provide an example for mapping two pixel values into one and recovering the original values, using the method proposed within this paper. Let us consider the Cantor pairing function from Eq. 1:

$$f : \mathbb{N} \times \mathbb{N} \to \mathbb{N}, \quad f(a,b) = \frac{(a+b) \cdot (a+b+1)}{2} + b$$

Let $v : \mathbb{N} \to [0, \infty)$ be the function associated to the quadratic-like downscaling transformation presented in Eq. 2 and function $\phi : \mathbb{N} \to [0, \infty)$ from Eq. 3:

$$v(z) = \frac{-1 + \sqrt{1 + 8 \cdot z}}{4}; \quad \phi(z) = \frac{-1 + \sqrt{1 + 8 \cdot z}}{2}$$

On some intervals, the function v is invertible, thus the inverse v^{-1} can be determined:

$$y = \frac{-1 + \sqrt{1 + 8 \cdot z}}{4} \quad | \ \cdot 4$$
$$4 \cdot y = -1 + \sqrt{1 + 8 \cdot z} \quad | \ +1$$
$$4 \cdot y + 1 = \sqrt{1 + 8 \cdot z} \quad |\ ^2 \quad (\text{with } z \geq 0)$$
$$(4 \cdot y + 1)^2 = 1 + 8 \cdot z \quad | \ -1$$
$$(4 \cdot y + 1)^2 - 1 = 8 \cdot z \quad | : 8$$
$$\frac{(4 \cdot y + 1)^2 - 1}{8} = z$$
$$v^{-1}(z) = \frac{(4 \cdot z + 1)^2 - 1}{8}$$

For a pair of two values $a, b \in \mathbb{N}$, the order of the transformations is:

$$(a, b) \longrightarrow f(a,b) = z \longrightarrow v(z) = z'$$

As an example, for the pair $(4, 11)$ the following values are obtained: $f(4, 11) = 131$, then $v(131) = 7.84706 = z'$. To recover the original pair $(4, 11)$, the order of the transformations is

$$z' \longrightarrow v^{-1}(z') = z \longrightarrow f^{-1}(z) \longrightarrow (a, b)$$

The following values are obtained: $v^{-1}(7.84706) = 131$. Concerning the inverse of the Cantor pairing function, according to [19], in order to recover the values $(4, 11)$ starting from $z = 131$, the auxiliary terms w and t are introduced:

$$w = \lfloor \phi(z) \rfloor = \lfloor 2 \cdot v(z) \rfloor = 15$$

$$t = \frac{w \cdot (w+1)}{2} = 120$$

Then, $b = z - t = 131 - 120 = 11$ and $a = w - b = 4$. The pair $(4, 11)$ was recovered. In the same manner, the rest of the original pixel values can be recovered from the resized image in order to reconstruct the original one. Also, the pairing configuration (usually used as the identity permutation) must be retained to determine which pixels (referred to by their indexes) were combined.

References

1. Code implementation available on GitHub (2025). https://github.com/DacianGoina/cnn-pairing
2. MNIST dataset (2025). https://yann.lecun.com/exdb/mnist/
3. Acharya, T., Tsai, P.S.: Computational foundations of image interpolation algorithms. Ubiquity **2007**, 4–1 (2007)
4. Amanatiadis, A., Andreadis, I.: A survey on evaluation methods for image interpolation. Measurement Sci. Technology - MEAS SCI TECHNOL 20 (2009). https://doi.org/10.1088/0957-0233/20/10/104015
5. Asamwar, R., Bhurchandi, K., Gandhi, A.: Interpolation of images using discrete wavelet transform to simulate image resizing as in human vision. Int. J. Autom. Comput. **7**, 9–16 (2010). https://doi.org/10.1007/s11633-010-0009-7
6. Bishop, C.M.: Pattern Recognition and Machine Learning (Information Science and Statistics). Springer-Verlag, Berlin, Heidelberg (2006)
7. Cégielski, P., Richard, D.: Decidability of the theory of the natural integers with the cantor pairing function and the successor. Theoret. Comput. Sci. **257**(1–2), 51–77 (2001)
8. Danon, D., Arar, M., Cohen-Or, D., Shamir, A.: Image resizing by reconstruction from deep features. Comput. Vis. Media **7**, 1–14 (2021). https://doi.org/10.1007/s41095-021-0216-x
9. Gao, H., et al.: Object classification using CNN-based fusion of vision and LIDAR in autonomous vehicle environment. IEEE Trans. Industr. Inf. **14**(9), 4224–4231 (2018)
10. Han, J.W., Choi, K.S., Wang, T.S., Cheon, S.H., Ko, S.J.: Wavelet based seam carving for content-aware image resizing. In: 2009 16th IEEE International Conference on Image Processing (ICIP), pp. 345–348 (2009)
11. Hirahara, D., Takaya, E., Kobayashi, Y., Ueda, T.: Effect of the pixel interpolation method for downsampling medical images on deep learning accuracy. J. Comput. Commun. **9**, 150–156 (2021). https://doi.org/10.4236/jcc.2021.911010
12. Hossain, M.J., Barkatullah, M., Monir, M.F., Ahmed, T.: Repercussion of image compression on satellite image classification using deep learning models. In: 2023 IEEE 98th Vehicular Technology Conference (VTC2023-Fall), pp. 1–5 (2023). https://doi.org/10.1109/VTC2023-Fall60731.2023.10333668
13. Hsieh, T.H., Kiang, J.F.: Comparison of CNN Algorithms on Hyperspectral Image Classification in Agricultural lands. Sensors **20**(6) (2020). https://doi.org/10.3390/s20061734
14. Manzoor, U., Nefti, S., Ferdinando, M.: Investigating the effectiveness of wavelet approximations in resizing images for ultrasound image classification **40**(10), 1–9 (2016). https://doi.org/10.1007/s10916-016-0573-7
15. Murphy, K.P.: Machine Learning: A Probabilistic Perspective. The MIT Press (2012)
16. Nam, S.H., et al.: Content-aware image resizing detection using deep neural network. In: 2019 IEEE International Conference on Image Processing (ICIP), pp. 106–110 (2019)
17. Rosenberg, A., Strong, H.: Addressing arrays by shells. IBM Technical Disclosure Bulletin **14**(10), 3026–3028 (1972)
18. Sabottke, C., Spieler, B.: The effect of image resolution on deep learning in radiography. Radiol. Artif. Intell. **2** (2020). https://doi.org/10.1148/ryai.2019190015
19. Szudzik, M.P.: The Rosenberg-Strong Pairing Function (2019)

20. Talebi, H., Milanfar, P.: Learning to resize images for computer vision tasks. In: 2021 IEEE/CVF International Conference on Computer Vision (ICCV), pp. 487–496 (2021). https://doi.org/10.1109/ICCV48922.2021.00055
21. Xiao, H., Rasul, K., Vollgraf, R.: Fashion-mnist: a novel image dataset for benchmarking machine learning algorithms (2017)
22. Yadav, S., Jadhav, S.: Deep convolutional neural network based medical image classification for disease diagnosis. J. Big Data **6** (2019). https://doi.org/10.1186/s40537-019-0276-2

A 3D Compact Shape Descriptor Based on Largest Intersection and Projection Signature

Florian Delconte[1], Jui-Ting Lu[1]([✉]), Phuc Ngo[1],
Isabelle Debled-Rennesson[1], and Bertrand Kerautret[2]

[1] Université De Lorraine, CNRS, LORIA, 54000 Nancy, France
`jui-ting.lu@loria.fr`
[2] Université Lumière Lyon 2, CNRS, Ecole Centrale De Lyon, INSA Lyon, Universite
Claude Bernard Lyon 1, LIRIS, UMR5205, 69007 Lyon, France

Abstract. Motivated by the performance of the shape descriptor based on the Largest Intersection and Projection (LIP) signature, we propose a three-dimensional extension in link to concrete application. By projecting 3D shape onto their principal planes and analyzing the resulting profiles, we extract compact and interpretable geometric features. Descriptors obtained from these features are then used for 3D object classification. The proposed method is evaluated on both the ModelNet benchmark and a real-world dataset of tree bark defects. Results show that the descriptor effectively captures shape variations while enabling a significant reduction in feature dimensionality.

Keywords: Geometric characteristics · Shape descriptor · Dimensionality reduction · Classification

1 Introduction

Accurate and interpretable geometric description of 3D shapes is essential for effective shape analysis. Modern sensing technologies such as LiDAR and RGB-D cameras provide dense 3D point clouds, which encode detailed geometric and spatial structures of real-world environments. Leveraging this structural richness is crucial for tasks such as object classification, mapping, and scene understanding. While detailed geometric information can significantly improve performance in such tasks, it often comes at the cost of increased computational complexity and processing time, necessitating a trade-off between descriptive power and efficiency.

To address this challenge, several hand-crafted geometric descriptors have been proposed to characterize 3D shapes. Spin Images [9] capture local surface geometry by projecting neighboring points onto a 2D histogram around a surface point. Shape Distributions [13] summarize global shape characteristics by computing statistical distributions over geometric properties such as distances

P. Balázs et al. (Eds.): IWCIA 2025, LNCS 15985, pp. 19–34, 2026.
https://doi.org/10.1007/978-3-032-19347-6_2

or angles between surface points. The Signature of Histograms of OrienTations (SHOT) [16] combines local histograms of surface normals with a unique signature structure, offering a robust and distinctive local descriptor.

Beyond global descriptors, intrinsic geometric attributes such as orientation, orientability, and linearity offer valuable insights into the structural organization of 2D and 3D shapes. The detection of dominant orientation, in particular, has been extensively studied through diverse approaches, including moments [18], symmetry axes [17], principal axes [8], or boundary points [21]. Measures of orientability, i.e., how strongly a shape exhibits a dominant orientation, have also been explored [7,11,15], providing insight into the structural coherence of a form. These concepts are particularly valuable in domains where shape regularity and alignment influence classification outcomes.

To evaluate these methods, several large-scale datasets for 3D object recognition, such as ModelNet [19], ShapeNet [2] and ScanNet [3], are commonly used. While deep learning models like PointNet [14], RS-CNN [10] and SPNet [20] perform well, they often require large amounts of annotated data and do not explicitly capture geometric structure. In contrast, geometric descriptors focus on structural information and interpretability, providing a complementary approach that can be advantageous in tasks requiring geometric precision.

In this paper, we propose a novel method for 3D shape description based on a geometry-based feature, LIP signature (Largest Intersection and Projection) [11]. More precisely, by projecting 3D shapes onto their principal planes and analyzing the resulting projection images, we derive compact and interpretable geometric features. In particular, the obtained vector contains only 57 features, compared to 3,243 in the full LIP representation. Unlike purely learning-based methods, our approach emphasizes the importance of geometric descriptors that can be extracted directly from the 3D structure. This improves generalization, maintains interpretability, and offers an efficient alternative or complement to learning-based pipelines.

Implementation details and reproducible evaluation procedures are available in the following GitHub repository:

https://github.com/FlorianDelconte/LIP3D

The paper is organized as follows. Section 2 provides a recall of the LIP signature in the 2D case before introducing its extension to 3D shapes (Sect. 3). Section 4 provides details on the experimental setup and presents the results of the shape descriptor on various datasets, followed by a discussion about the limitations of the proposed method.

2   LIP: Largest Intersection and Projection Signature

To lay the groundwork for our 3D extension, we begin by revisiting the concept of LIP signature in the two-dimensional case that was originally introduced in [11].

The LIP signature uses the projections of the shape in order to captures its geometric properties. The main points of the LIP signature computation

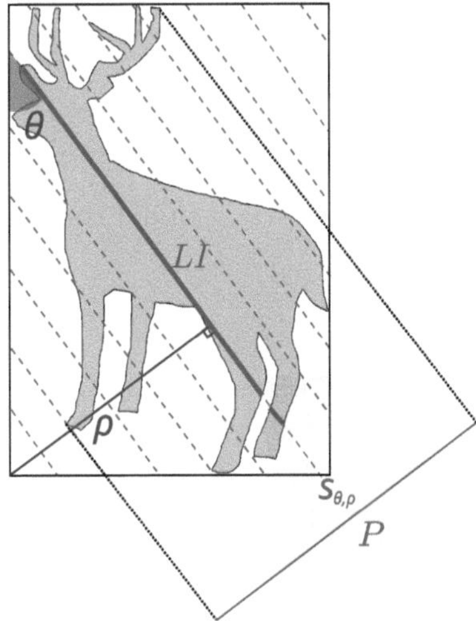

Fig. 1. The LIP signature computation: LIP_0^θ is the ratio between the largest intersection LI^θ and the projection P^θ of the shape for a fixed angle θ.

are illustrated on Fig. 1. Let $\mathcal{S}$ be a two dimensional shape, we denote by $s_{\theta,\rho}$ the intersection between the shape $\mathcal{S}$ and the line defined by an angle θ and a distance ρ from a fixed point. Let LI^θ denote the largest intersection $s_{\theta,\rho}$ of direction θ and P^θ denote the projection of $\mathcal{S}$ in the direction θ onto a line which is perpendicular to $s_{\theta,\rho}$. Then LIP_0^θ is defined as the ratio of LI^θ to P^θ. This LIP signature can expressed more formally using the following definition.

Definition 1. *Let M denote the number of discretization values in the interval $[0, \pi]$. Let $\Theta = \{\theta_i\}_{1 \leq i \leq M}$ be a set of directions. For a given $\theta = \theta_i$,*

$$LIP_0^\theta \equiv \frac{LI^\theta}{P^\theta}.$$

Then the LIP *signature is the following profile:*

$$LIP(\mathcal{S}, \Theta) = \left\{ LIP_0^\theta \right\}_{\theta \in \Theta}.$$

The projection-based nature of the LIP signature has a direct connection to the Radon transform [4] (see Fig. 2). The (discrete) Radon transform takes an image as input and generates a matrix as output. Each cell of the matrix has as its value the integral of the intersection of the line defined by ρ and α with the shape $\mathcal{D}$, as follows:

$$\int f(x)\delta(\rho - x \cdot \boldsymbol{n}(\alpha))\mathrm{d}x \tag{1}$$

where $\boldsymbol{n}(\theta) = (\cos\theta, \sin\theta)$, δ the dirac function and

$$f(x) = \begin{cases} 1 \text{ if } x \in \mathcal{D} \\ 0 \text{ otherwise.} \end{cases} \tag{2}$$

Furthermore, the LIP signature is invariant under rotation, translation, and scale changes.

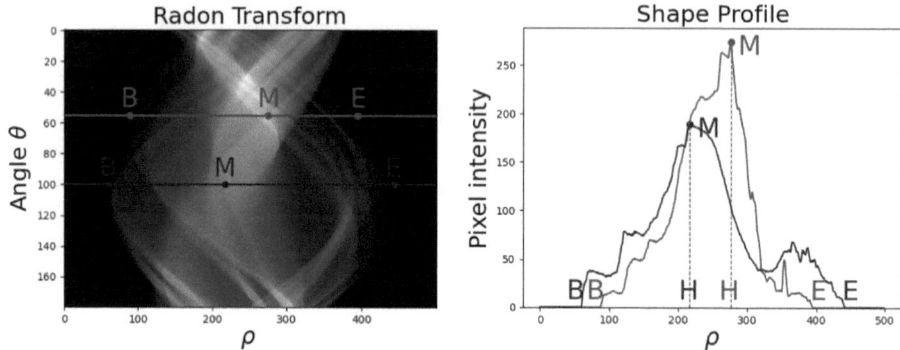

(a) Radon transform applied to the binary image of the deer used previously. The blue and green lines correspond to the profiles on the right, shown in their respective colors.

(b) The points B, H, E, and M used to calculate the characteristics LIP_0 to LIP_5.

Fig. 2. Correspondence between the Radon transform [4] and the LIP signature [11]. The two profiles correspond to slices of the Radon transform computed with two different projection angles θ.

By examining the profile along a fixed direction of the Radon transform (see Fig. 2), we identify four key points: B, E, M, and H. Point B (respectively E) marks the leftmost (respectively rightmost) extremity of the strictly positive part of the profile. Point M corresponds to the highest peak of the profile, while H is its projection on BE. These four points are guaranteed to exist since the profile is always positive, and we assume the shape is finite and enclosed within a bounding box. With these notations, we recall that LIP_0^θ is defined as the ratio $\frac{MH}{BE}$. The significance of LIP_0^θ lies in its elongation along a specific direction. Additionally, there is LIP_1^θ as the ratio $\frac{BH}{BE}$, which reflects the balance or symmetry on the two sides of the profile around its peak.

Statistical descriptors also provide insights on the geometry of the form. Let m_1 and m_2 denote the means, and σ_1 and σ_2 the standard deviations, of the left and right sides of the segment MH, respectively. To ensure invariance by transformations, these values are divided by BE. These characteristics correspond to LIP_2^θ through LIP_5^θ, respectively. See Table 1 for a summary of these characteristics.

Table 1. Summary of the LIP signature.

Characteristic	Location w.r.t. MH	Definition
LIP_0^θ	Global	$\frac{MH}{BE}$
LIP_1^θ	Left	$\frac{BH}{BE}$
LIP_2^θ	Left	$\frac{m_1}{BE}$
LIP_3^θ	Left	$\frac{\sigma_1}{BE}$
LIP_4^θ	Right	$\frac{m_2}{BE}$
LIP_5^θ	Right	$\frac{\sigma_2}{BE}$

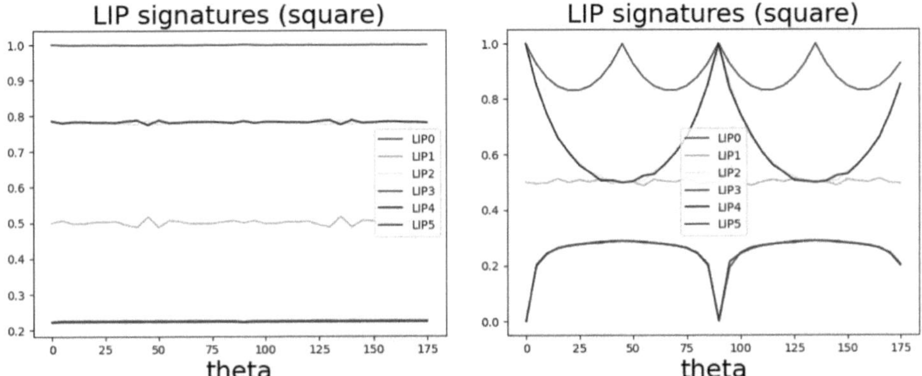

Fig. 3. Examples of LIP signature profiles for a disk (left) and a square (right). Due to the symmetry of the shapes, the LIP_2^θ and LIP_4^θ profiles are superimposed, as are the LIP_3^θ and LIP_5^θ profiles.

For a given LIP characteristic, a profile can be constructed by varying the angle θ on the horizontal axis and giving on the vertical axis its associated LIP value. Figure 3 illustrates such type of profiles obtained on two simple shapes: a disk and a square. The disk appears identical from any viewing direction, resulting in an almost constant characteristic profile. Some minor variations are present due to the discrete nature of the computation. In contrast, the LIP_0^θ profile for the square reveals the presence of its four corners.

In practice, the authors of [11] employ the LIP signature to describe 2D shapes. The resulting LIP features are classified using a Nearest Neighbor Classifier based on appropriate similarity measures. The following section presents the proposed extension for 3D shape processing.

3   Extension to 3D

The LIP signature has useful properties in 2D, such as robustness to noise and invariance to rotation and scale. To extend it to 3D, we propose a method based on the analysis of profile images. This extension involves (1) computing three principal directions on a 3D shape (set of voxels), and (2) generating the profile images from 2D projections of the shape along these directions.

3.1 Principal Planes

To construct representative views of a 3D object, we rely on three principal directions, denoted as d_1, d_2, and d_3. They are obtained by performing *Principal Component Analysis (PCA)* on the voxel representation. This involves translating the voxel coordinates to the barycenter of the shape, and then computing the covariance matrix. The decomposition of this matrix yields three orthogonal eigenvectors, sorted by decreasing eigenvalues, which capture the axes of greatest geometric variation within the shape.

For each direction d_i (with $i = 1, 2, 3$), a slicing plane is defined as the plane orthogonal to d_i. This construction allows us to generate three profile images that are intrinsically aligned with the object's geometry. Since the orientation is derived from PCA, the method is invariant to the initial pose of the object. Figure 4 shows an example of the three principal planes obtained from a 3D airplane model.

Fig. 4. Projection planes which are orthogonal respectively to the three principal directions d_1, d_2, and d_3, computed via PCA on the voxel set.

3.2 Profile Image

Once the three principal planes are defined, the next step is to compute the corresponding 2D *profile images*. Each profile image is computed by sweeping a plane through the object along a principal axis. The process, illustrated in Fig. 5, shows how planar traversal enables the construction of a profile image. This process is applied independently along each of the three directions, resulting in three distinct profile images. The overall procedure for constructing a profile image is described step by step in Algorithm 1.

Algorithm 1: Profile Image Generation

Input: Voxelized 3D image `volImage`, direction `normalDir`, image size `N`,
        maximum scan depth `maxScan`

Output: 2D binary profile image `profileImage`

`center` ← starting point computed by shifting the center of the voxel grid
domain along −`normalDir`;

`profileImage` ← 2D image over `aDomain2D`, initialized to zero;

`embedder` ← 2D-to-3D embedding using `center`, `normalDir`, and `N`;

`extractedImage` ← 2D view of `volImage` using `embedder`;

`firstFound` ← `false`;

`k` ← 0;

while $k <$ *maxScan or not firstFound* **do**
    `embedder`.shiftOriginPoint(`normalDir`);
    foreach $p \in aDomain2D$ **do**
        if $profileImage(p) == 0$ *and* $extractedImage(p) \neq 0$ **then**
            `profileImage(p)` ← 255;
            `firstFound` ← `true`;

    if $firstFound$ **then**
        $k \leftarrow k + 1$;

return $profileImage$;

The voxel grid and the profile images are designed to share the same resolution: the grid has dimensions $N \times N \times N$, and each profile image is of size $N \times N$. The parameter N controls the spatial resolution and is discussed in the following section. Figure 6 illustrates the generation of binary profile images from a 3D object, showing how each principal plane reflects a different geometric view of the shape. The LIP signature is then computed on each profile image in order to capture complementary geometric information of the 3D shape.

3.3   Compact Shape Descriptor

The generated profile images presented in the previous section allows to compute the shape descriptor. More precisely, our descriptor focuses on compactness by extracting only succinct information. For each profile image, we use the LIP signature introduced in Sect. 2, and compute the maximum, minimum, and median values of the six LIP profiles.

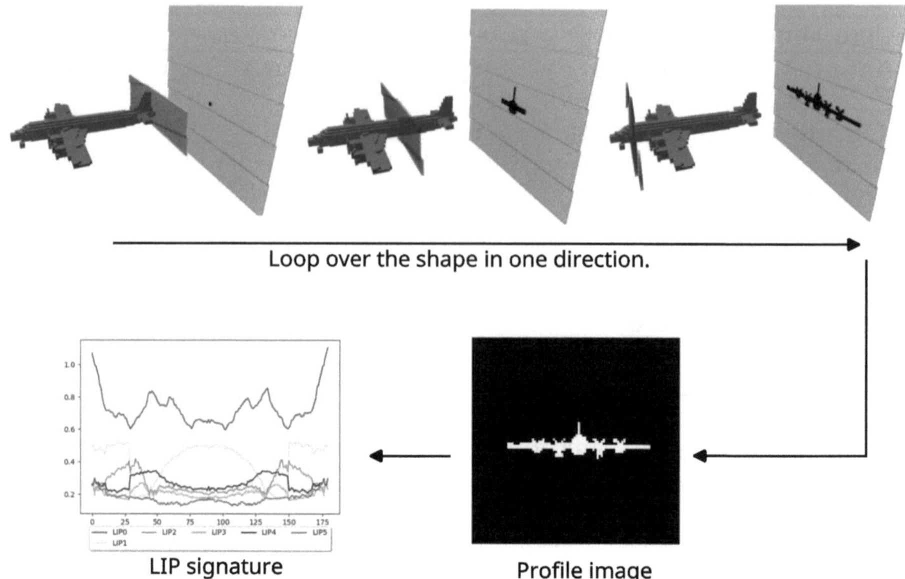

Fig. 5. Computation of profile image along a principal direction and the corresponding LIP signature. This process is repeated independently along each principal axis, producing three distinct profile images on which the LIP signature is computed.

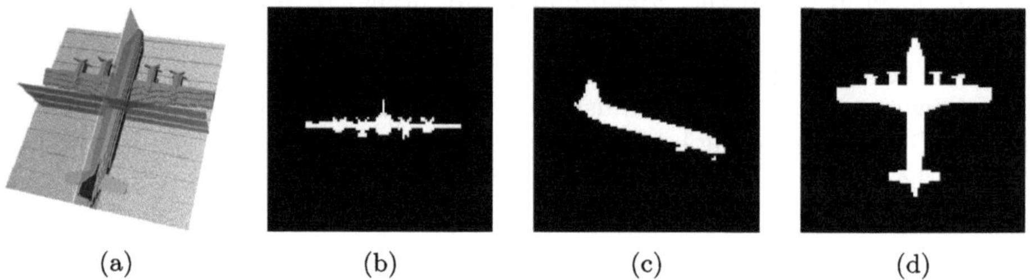

Fig. 6. Example of profile image generation from a voxelized 3D shape: (a) voxel grid; (b), (c), (d) profile images corresponding to the three principal planes.

Furthermore, we include a circularity measure to capture global shape properties. We now denote the 3D shape as $\mathcal{S}$ and $\mathcal{PS} = \mathcal{PS}_{d_i}$ $(i = 1, 2, 3)$ one of its profile image. The additional circularity characteristic is defined as

$$Cir(\mathcal{PS}) = \frac{4\pi \times \mathrm{A}(\mathcal{PS})}{(\mathrm{P}(\mathcal{PS}))^2},\qquad(3)$$

where $\mathrm{A}(\cdot)$ denotes the area and $\mathrm{P}(\cdot)$ stands for perimeter.

Formally, let Θ be the set of angles, resulting from the discretization of $[0, \pi]$. Our descriptor for the given profile image is then defined as:

$$\{Cir(\mathcal{PS})\} \cup \bigcup_{n \in \{0,\dots,5\}} \left\{ \min_{\theta \in \Theta}(LIP_n^\theta), \max_{\theta \in \Theta}(LIP_n^\theta), \operatorname{med}_{\theta \in \Theta}(LIP_n^\theta) \right\}. \tag{4}$$

In total, we collect 18 LIP-based features and 1 additional circularity characteristic per profile image. Thus, for a given shape, our descriptor consists of only 57 features. In contrast, using the full LIP profiles would lead to a significantly larger descriptor, with $(180 \times 6 + 1) \times 3 = 3,243$ features. Hence, our 3D shape descriptor offers a compact yet informative representation.

4   Experimental Setup and Results

The proposed 3D shape descriptor is evaluated in terms of compactness, discriminative power, and applicability. A first test is performed on a subset of ModelNet [19] with clearly distinct categories. Then the method is applied to real-world LiDAR data from the ANR WoodSeer project[1], focusing on external wood defects. Finally, we assess its limitations on the full ModelNet dataset, which includes visually similar shapes.

4.1   Dataset and Pre-processing

ModelNet [19] is a widely used benchmark for 3D object classification. It comprises 12,311 CAD models across 40 categories such as chairs, tables, sofas, and bathtubs. The dataset is split into 9,843 training models and 2,468 test models. For our experiments, we selected a simplified subset consisting of 3,981 CAD models distributed across 10 categories that are clearly distinguishable from one another. This simplified version reduces class ambiguity and allows for a more focused evaluation of the descriptive power of our compact descriptor. Furthermore, we also consider real-world data acquired in forest environments during the ANR WoodSeer project. This dataset includes 728 external wood defects (illustrated on Fig. 9), scanned using terrestrial LiDAR. Five categories of surface defects are represented: branches, scars, burls, small defects, and bark. In both datasets, the input data consist of 3D meshes which require a pre-processing step with voxelization.

To extract profile images, each mesh is first normalized by uniformly scaling it so that the diagonal of its bounding box is the same across all samples in a dataset, and then discretized into a regular voxel grid. This grid representation facilitates slicing and provides a uniform discretization of the shape, as described in Sect. 3. Voxelization converts the continuous surface into a binary volumetric representation, where each voxel corresponds to a unit cube intersecting at least one triangular face of the mesh. The finer the discretization, the more accurately the voxel representation approximates the original geometry. Figure 7 illustrates the effect of grid resolution on voxel quality.

---

[1] https://anr.fr/Projet-ANR-19-CE10-0011.

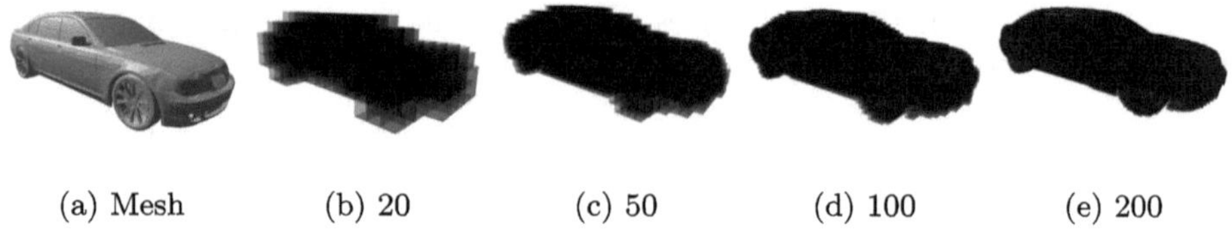

(a) Mesh   (b) 20   (c) 50   (d) 100   (e) 200

Fig. 7. Illustration of mesh voxelization depending on grid resolution: higher grid sizes yield finer discrete representations.

4.2   Experimental Setup

For both datasets, classification is performed using a Random Forest model [1] trained on the compact descriptors (see Sect. 3.3). To ensure a robust evaluation, each dataset is split into a training and a test set. Hyperparameter tuning is conducted by using 5-fold cross-validation on the training set. The explored hyperparameters include the number of trees, maximum tree depth, minimum samples required to split a node or remain in a leaf, and the number of features considered at each split. The best model is selected based on validation accuracy.

4.3   Evaluation on a Selected Subset of ModelNet

In this experiment, both the voxel grid resolution and the profile image size were set to 100, ensuring a consistent representation across all objects. Table 2 reports the classification results obtained on the selected ModelNet subset using the compact LIP-based descriptor. The Random Forest classifier achieves a over-all accuracy of 96%, with strong performance across most categories. Notably, classes such as *airplane, guitar,* and *toilet* achieve F1-scores of 1.00, 0.99, and 0.98 respectively. This demonstrates that the method effectively captures distinctive geometric features for well-structured shapes. Even classes like *cone, glass_box,* and *monitor* maintain F1-scores above 0.93, confirming the robustness of the proposed descriptor.

The confusion matrix in Fig. 8 provides additional insight into the model's behavior. The matrix shows true labels along the vertical axis and predicted labels along the horizontal axis. The values inside each cell correspond to the number of samples, while the color intensity reflects the proportion of correct predictions per class. Strong diagonal responses indicate successful classification, especially for clearly defined categories like *airplane* and *toilet*. Most confusion occurs between geometrically similar classes, such as *table* and *bed*, which can produce similar silhouettes in their profile images. These ambiguities highlight the inherent limitation of projection-based descriptors. Further analysis will be given in Sect. 4.5.

Table 2. Per-class performance comparison between LIP and LIP+Circularity features. Best F1-scores per class are highlighted in bold. The weighted average gives more importance to under-represented categories.

Class	LIP			LIP+Cir		
	Precision	Recall	F1-score	Precision	Recall	F1-score
airplane	0.99	1.00	**1.00**	0.99	1.00	**1.00**
bed	0.86	0.93	0.89	0.91	0.93	**0.92**
car	0.99	0.97	**0.98**	0.99	0.97	**0.98**
cone	1.00	0.95	**0.97**	1.00	0.95	**0.97**
door	0.90	0.95	**0.93**	0.90	0.95	**0.93**
glass_box	0.99	0.87	0.93	0.98	0.95	**0.96**
guitar	1.00	0.99	**0.99**	1.00	0.99	**0.99**
monitor	0.93	0.96	0.95	0.97	0.96	**0.96**
table	0.93	0.93	**0.93**	0.93	0.93	**0.93**
toilet	0.96	0.99	**0.98**	0.95	0.99	0.97
Accuracy	0.95			**0.96**		
Weighted avg	0.96	0.95	0.95	0.96	0.96	**0.96**

4.4 Application to External Wood Defects

The proposed method was also tested on a real-world dataset of external wood defects, acquired in forest conditions using terrestrial LiDAR. In forestry, surface defects such as scars, branches, or burls are key indicators of internal wood quality and commercial value (see Fig. 9). Automating their detection [5,6] prior to harvesting is thus highly relevant. Unlike clean synthetic meshes, these defects exhibit irregular and noisy geometries, for which our profile-based approach is well suited. For each sample, 2D profiles were generated from two principal planes, with voxel resolution and image size set to 200 to ensure stable PCA computation. As defects lie on the trunk surface, one direction is constrained by the tree's axis, and two orthogonal planes suffice to capture the relevant geometry. Compact LIP signatures were extracted from these profiles for classification. Figure 10 shows examples of five defect types with their 3D mesh and corresponding profile.

To better understand the contribution of each type feature in the compact LIP-based descriptor, we conducted an ablation study in the specific context of wood defect classification. The results revealed that while many features appear to be partially redundant, three of them consistently stand out as the most impactful: LIP_0, circularity, and orientation merit [11]. These features alone account for most of the discriminative power for analyzing surface defects in forest LiDAR data. However, this observation does not generalize to arbitrary 3D objects.

Using only the min, max, and median values of the LIP_0 signature from both principal profiles, together with circularity and orientation merit, yields

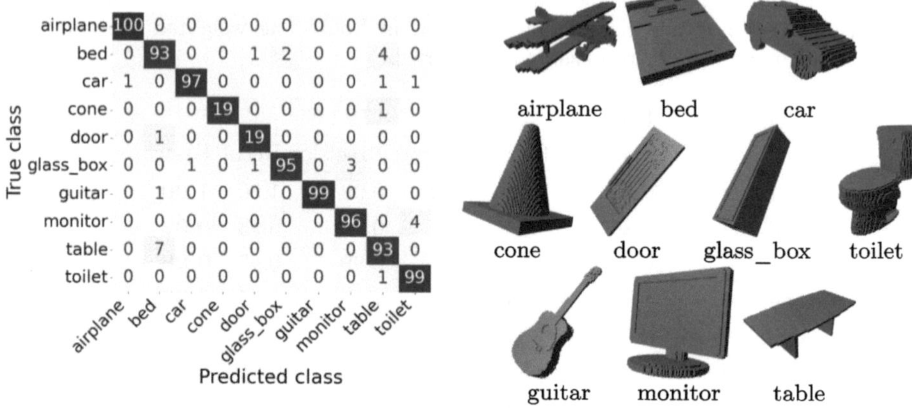

Fig. 8. Confusion matrix of classification (left) and voxel representations of the corresponding 3D objects (right) on the selected subset of ModelNet.

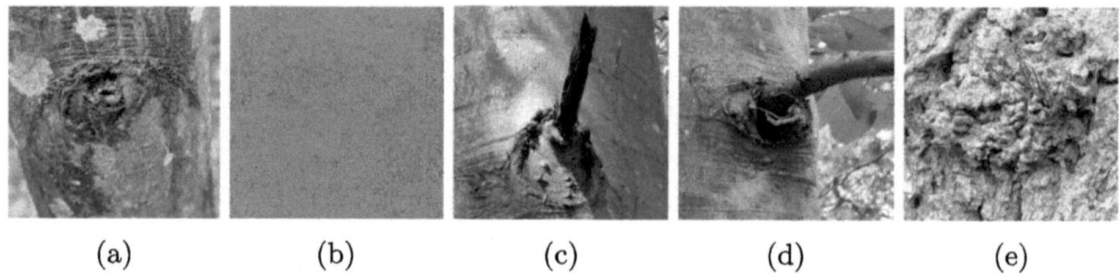

(a)   (b)   (c)   (d)   (e)

Fig. 9. Typical wood surface defects. Subfigures (a) and (b) show branch scars, (c) and (d) depict small branch outgrowths, and (e) shows a burl.

a 10-dimensional descriptor. Adding tree species, an important discriminative feature in this context, brings the total to 11 features. Moreover, we included the three most relevant features from the descriptor proposed V-T. Nguyen *et al.* in [12], resulting in a final feature vector of size 14.

The proposed method enables a reduction in feature dimensionality. While the original V-T. Nguyen descriptor [12] includes 17 components, the combined V-T. Nguyen + LIP descriptor requires only 14 features, 10 of which come from the compact LIP-based descriptor. As we can see in Table 3, despite this reduction, classification performance is maintained or even improved, which highlights our compact LIP-based descriptor is able to do discriminative task on real-world data. We observe an improved average performance, particularly for the *small defect* category, which was the primary target of our approach. The model also performs competitively on other classes such as *bark*, *branch*, and *burl*. However, a slight decrease in performance is noted for the *scar* class. This decline is mitigated by the higher number of scar samples in the dataset, which helps stabilize the weighted average.

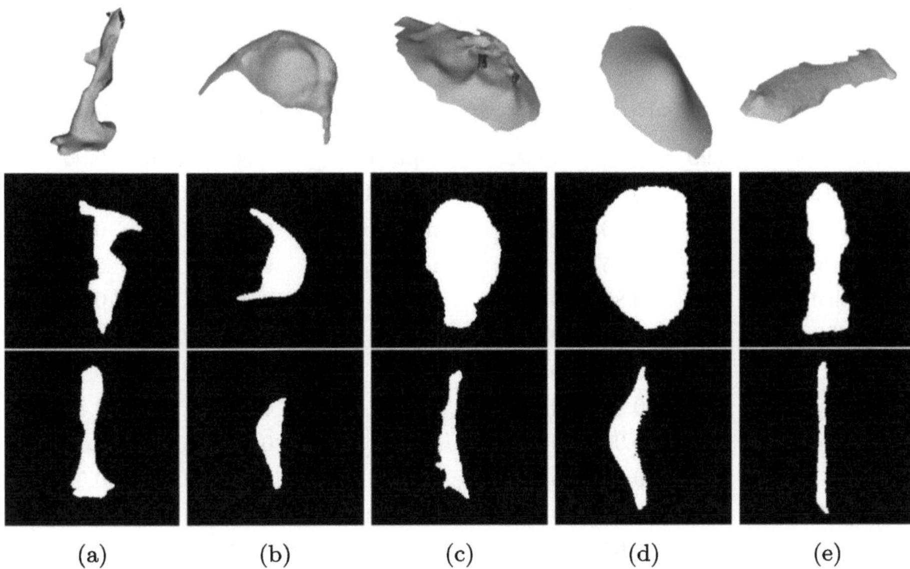

(a)   (b)   (c)   (d)   (e)

Fig. 10. Profile images for five wood defects: (a) branch, (b) scar, (c) burl, (d) small defect, (e) bark. Rows show the mesh, first, and second principal projections.

Table 3. Classification accuracy on the wood defect dataset (class sizes in parentheses). The second row combines V-T. Nguyen and LIP features. Bold indicates best per-class scores.

	Bark (11)	Branch (23)	Scar (48)	Burl (28)	Small Defect (23)	Weighted Avg.
V-T [12] (17)	0.76	0.77	**0.83**	0.64	0.84	0.78
V-T (3) + LIP (11)	**0.81**	**0.85**	0.82	**0.66**	**0.89**	**0.80**

4.5   Limitation et Discussion

While the proposed method achieves competitive performance on both synthetic and real-world datasets, several limitations must be acknowledged. Table 4 reports the classification performance on the full ModelNet40 dataset. The drop in accuracy from 96% (on the simplified ModelNet10 subset) to 73% on Model-Net40 can be explained by similar categories. Since our descriptor relies on 2D projections along principal planes, categories that share similar profile projections are difficult to distinguish.

Several sources of confusion can be explained by similarities in projected shape:

- **Flat or box-like objects:** *bathtub, sofa, bench, bed,* and *table* often produce wide, rectangular profiles that are geometrically similar.
- **Four-legged furniture:** *chair, stool, desk,* and *bed* share a common structural layout that leads to ambiguity in profile space.

Table 4. Classification on ModelNet40 [19] using the compact LIP-based descriptor. The weighted average gives more importance to under-represented categories.

Class	Precision	Recall	F1-score	Class	Precision	Recall	F1-score
airplane	0.99	1.00	1.00	laptop	0.90	0.90	0.90
bathtub	0.90	0.36	0.51	mantel	0.75	0.85	0.80
bed	0.70	0.86	0.77	monitor	0.87	0.90	0.88
bench	0.41	0.75	0.53	night_stand	0.62	0.56	0.59
bookshelf	0.57	0.80	0.66	person	0.72	0.68	0.70
bottle	0.85	0.86	0.86	piano	0.76	0.58	0.66
bowl	0.75	0.90	0.82	plant	0.78	0.70	0.74
car	0.96	0.92	0.94	radio	0.67	0.20	0.31
chair	0.62	0.90	0.73	range_hood	0.92	0.59	0.72
cone	0.89	0.80	0.84	sink	0.80	0.40	0.53
cup	0.64	0.45	0.53	sofa	0.48	0.94	0.64
curtain	0.62	0.65	0.63	stairs	0.62	0.25	0.36
desk	0.56	0.28	0.37	stool	0.62	0.50	0.56
door	0.89	0.80	0.84	table	0.71	0.72	0.72
dresser	0.79	0.67	0.73	tent	0.33	0.25	0.29
flower_pot	0.00	0.00	0.00	toilet	0.80	0.96	0.87
glass_box	0.81	0.85	0.83	tv_stand	0.73	0.53	0.61
guitar	0.94	0.99	0.97	vase	0.64	0.62	0.63
keyboard	0.90	0.90	0.90	wardrobe	0.80	0.20	0.32
lamp	0.68	0.65	0.67	xbox	0.61	0.70	0.65

Summary	Precision	Recall	F1-score
Accuracy		0.73	
Weighted avg	0.75	0.73	0.72

- **Tall vertical forms:** *bookshelf, wardrobe, curtain,* and *door* yield elongated rectangular projections that are difficult to distinguish without finer structural and geometrical details.
- **Container-shaped classes:** *cup, bowl,* and *vase* can appear nearly identical from certain viewpoints.
- **Overlapping semantics:** *flower_pot* (F1-score: 0.00) is likely confused with *plant* (0.74), as their shapes are not clearly separable in voxelized or profile form.
- **Electronics and small devices:** *radio* is often misclassified, likely due to its compact rectangular shape being visually similar to *monitor* or *tv_stand*.
- **Musical instruments:** *guitar* is well identified due to its distinctive profile, whereas *piano* is confused with *bench*-like objects.

These observations underline the limitations of relying on the used of the three profile images. While effective for capturing global geometry, this approach may overlook finer discriminative details needed to separate structurally similar classes. Future work could address this by integrating further views or depth information.

5   Conclusion

We proposed a compact and interpretable descriptor for 3D shape analysis by extending the LIP signature. The method reduces the complexity of 3D shapes by projecting them onto three principal planes and summarizing their profiles with low-dimensional LIP features. This results in a compact vector representation that preserves essential geometric information while significantly reducing dimensionality. Our experiments show that the proposed descriptor is effective for 3D object classification, particularly when objects have clearly distinct shapes. However, its performance declines when object categories share similar forms and, as a result, produce similar projections. Full implementation is also available online, enabling reproducibility and integration for other applications.

In future works, our descriptor can be improved through two potential approaches. First, increasing the number of projection direction may allow for a more comprehensive characterization of geometric variability, thereby improving robustness. However, this comes with a higher computational cost. Second, a more general formulation of the LIP signature could be investigated: instead of relying on a 1D intensity histogram, one could compute a 2D intensity map over selected principal directions. In this formulation, LIP_0 becomes the ratio of the maximum intensity to the area of the projected surface. Although this approach may be computationally more demanding, it offers a geometrically richer representation of spatial shape variation.

References

1. Breiman, L.: Random forests. Mach. Learn. **45**, 5–32 (2001)
2. Chang, A.X., et al.: ShapeNet: An Information-Rich 3D Model Repository. Technical report arXiv:1512.03012 [cs.GR], Stanford University – Princeton University – Toyota Technological Institute at Chicago (2015)
3. Dai, A., et al.: ScanNet: Richly-annotated 3d reconstructions of indoor scenes. In: Proceedings of the Computer Vision and Pattern Recognition (CVPR). IEEE (2017)
4. Deans, S.R.: The Radon transform and some of its applications. Courier Corporation (2007)
5. Delconte, F., et al.: Tree defect segmentation using geometric features and cnn. In: International Workshop on Reproducible Research in Pattern Recognition, pp. 80–100 (2021)
6. Delconte, F.: CNN-based method for segmenting tree bark surface singularites. Image Process. Line **12**, 1–26 (2022)

7. Ha, V., Moura, J.: Efficient 2D shape orientation. In: Proceedings 2003 International Conference on Image Processing, vol. 1, pp. I-225 (2003)
8. Lin, J.C.: The family of universal axes. Pattern Recogn. **29**(3), 477–485 (1996)
9. Johnson, A.E., Hebert, M.: Using spin images for efficient object recognition in cluttered 3d scenes. IEEE Trans. Pattern Anal. Mach. Intell. **21**(5), 433–449 (1999)
10. Liu, Y., Fan, B., Xiang, S., Pan, C.: Relation-shape convolutional neural network for point cloud analysis. In: IEEE Conference on Computer Vision and Pattern Recognition (CVPR), pp. 8895–8904 (2019)
11. Nguyen, T.P., Nguyen, X.S.: Shape measurement using lip-signature. Comput. Vis. Image Underst. **171**, 83–94 (2018)
12. Nguyen, V.T., Constant, T., Kerautret, B., Debled-Rennesson, I., Colin, F.: A machine-learning approach for classifying defects on tree trunks using terrestrial lidar. Comput. Electron. Agric. **171**, 105332 (2020)
13. Osada, R., Funkhouser, T., Chazelle, B., Dobkin, D.: Shape distributions. ACM Trans. Graph. **21**(4), 807–832 (2002)
14. Qi, C.R., Su, H., Mo, K., Guibas, L.J.: PointNet: Deep learning on point sets for 3D classification and segmentation. arXiv preprint arXiv:1612.00593 (2016)
15. Rosin, P.L.: Measuring the orientability of shapes. In: Kropatsch, W.G., Kampel, M., Hanbury, A. (eds.) Computer Analysis of Images and Patterns, pp. 620–627. Springer, Berlin Heidelberg, Berlin, Heidelberg (2007)
16. Salti, S., Tombari, F., Di Stefano, L.: Shot: unique signatures of histograms for surface and texture description. Comput. Vis. Image Underst. **125**, 251–264 (2014)
17. Tsai, W.H., Chou, S.L.: Detection of generalized principal axes is rotationally symmetric shapes. Pattern Recogn. **24**(2), 95–104 (1991)
18. Žunić, J., Hirota, K., Dukić, D., Aktaş, M.A.: On a 3D analogue of the first hu moment invariant and a family of shape Ellipsoidness measures. Mach. Vision Appl. **27**(1), 129–144 (2016)
19. Wu, Z., et al.: 3D shapenets: a deep representation for volumetric shapes. In: Proceedings of the IEEE conference on computer vision and pattern recognition, pp. 1912–1920 (2015)
20. Yavartanoo, M., Kim, E.Y., Lee, K.M.: SPNet: Deep 3D object classification and retrieval using stereographic projection. In: Computer Vision - ACCV 2018, pp. 691–706 (2019)
21. Žunić, J., Stojmenovic, M.: Boundary based shape orientation. Pattern Recogn. **41**, 1768–1781 (2008)

Boustrophedon Pushdown Automata for Two-dimensional Picture Languages

Henning Fernau[1(✉)], R. Jennifer Rose[2], Robinson Thamburaj[2],
and D. Gnanaraj Thomas[2]

[1] Abteilung Informatikwissenschaften, Fachbereich 4,
Universität Trier, 54286 Trier, Germany
fernau@uni-trier.de
[2] Madras Christian College, Chennai - 600059, India
robinson@mcc.edu.in

Abstract. In the area of picture languages, many mechanisms have been studied to describe language classes, often inspired by what has been done before for string languages. Following this tradition, we are now introducing a pushdown automaton (PDA) extension of two previously defined variants of finite automata (FA) working on arrays, namely boustrophedon finite automata (BFA) and returning finite automata (RFA), leading to BPDA and RPDA, respectively. While BFA and RFA characterize the same class of array languages, BPDA and RPDA turn out to describe incomparable classes of array languages. We also prove that BPDA and RPDA can be viewed as a proper generalization of context-free matrix languages, which is interesting, as it is known that (apart from a transposition) BFA and RFA characterize the regular matrix languages.

Keywords: Array languages · Boustrophedon and returning automata · Context-free matrix grammars

1 Introduction

Picture languages or (two-dimensional) array languages are a generalization of formal string languages to two dimensions. They play an important role in the area of pattern recognition and image processing. Several syntactic models for the recognition and generation of rectangular-shaped pictures have been introduced and studied. A simple method of scanning pictures row by row which changes direction on reaching a boundary symbol, called Boustrophedon Finite Automaton (BFA) was introduced by Fernau *et al.* [2]. A BFA computes rectangular-shaped digitized pictures by moving its head one by one visiting each position exactly once. An alternative model called Returning Finite Automaton (RFA) scans pictures row by row always from left to right. It was shown that these two models describe the same class of picture languages. In this paper, we introduce Boustrophedon Pushdown Automata (or BPDA) and Returning Pushdown Automata (or RPDA), incorporating a pushdown memory feature to

P. Balázs et al. (Eds.): IWCIA 2025, LNCS 15985, pp.35–50, 2026.
https://doi.org/10.1007/978-3-032-19347-6_3

recognize rectangular pictures. Both BPDA and RPDA are a natural extension of the BFA and RFA. The pushdown feature in our model helps to recognize certain class of picture languages that are not recognizable by BFA. We prove that BPDA and RPDA describe two different classes of picture languages, unlike BFA and RFA describing the same class of picture languages. Secondly, we show that the class of picture languages recognized by RPDA properly includes the class of picture languages generated by context-free matrix grammars (CFMG) of Siromoney *et al.* [10]. This is interesting as CFMG can be viewed as another way to generalize regularity in array languages towards context-freeness. Moreover, apart from a transposition, BFA and RFA describe the same class of array languages as regular matrix grammars (RMG) do.

2   Preliminaries

We assume some basic knowledge of formal (string) languages on the side of the reader, as it is contained, for instance, in classical textbooks like [7]. We denote the empty string by ε and the set of all strings over the alphabet Σ as Σ^*. The reversal of a string w is written as w^R. It is well-known that finite automata characterize the regular languages and that pushdown automata (PDA) characterize the context-free languages. As an intermediate class, Greibach discussed *blind counter automata* (BCA) in [6]. BCA can be best described as a finite automaton with an integer (representing the value of the counter) associated to each configuration. When executing a transition, this counter may be either incremented or decremented, or it keeps its value. At the beginning of the computation, the counter has value zero, and in order to accept a word, also at the very end, the counter should be zero (and additionally, a final state should have been reached). Observe that the automaton cannot test this counter for being zero (or containing any other value) in the course of the computation. Still, this automaton model is strong enough to accept the well-known non-regular languages $\{a^n b^n \mid n \geq 1\}$ or even

$$\{w \in \{a, b\}^* \mid w \text{ contains as many } a\text{'s as } b\text{'s}\}.$$

The second example illustrates the fact that blind counter automata do use positive as well as negative numbers. Likewise, one can attach a blind counter (or even several blind counters) to a pushdown automaton; the corresponding device is able to accept, for example, the well-known non-context-free language $\{a^n b^n c^n \mid n \geq 1\}$. An *array* (or *picture*) over the alphabet Σ is a rectangular object $A = [a_{i,j}]_{m,n} = \begin{matrix} a_{1,1} & a_{1,2} & \cdots & a_{1,n} \\ a_{2,1} & a_{2,2} & \cdots & a_{2,n} \\ \vdots & \vdots & \ddots & \vdots \\ a_{m,1} & a_{m,2} & \cdots & a_{m,n} \end{matrix}$ containing entries $a_{i,j} \in \Sigma$ with m rows and n columns. The set of all arrays with m rows and n columns, containing entries from Σ, is denoted by Σ_m^n. If $n \geq 1$ is arbitrary, we can also write Σ_m^+ and if also $m \geq 1$ is arbitrary, we write Σ_+^+. In a sense, Σ_+^+ collects all pictures over the alphabet Σ. We call $L \subseteq \Sigma_+^+$ an *array language*.

Many operations can be defined for arrays and array languages. In this paper, we focus on one simple unary operation, which is *transposition*, denoted by T, i.e., $\mathrm{T}(A) = \begin{matrix} a_{1,1} & a_{2,1} & \cdots & a_{m,1} \\ a_{1,2} & a_{2,2} & \cdots & a_{m,2} \\ \vdots & \vdots & \ddots & \vdots \\ a_{1,n} & a_{2,n} & \cdots & a_{m,n} \end{matrix}$. We can lift this operation to sets of arrays (i.e., array languages) and even to array language families. One can define automata that recognize pictures, see [4,5,2] in historical order. For instance, we will define below finite automata and pushdown automata that scan pictures row by row. This way, one can also define array languages. Comparing these is the main theme of this paper.

We occasionally also employ two binary (partial) operations. Let $A := [a_{i,j}]_{m,n}$ and $B := [b_{i,j}]_{m',n'}$ be two pictures over Σ. The *column concatenation* of A and B, denoted by $A \; B$, is undefined if $m \neq m'$ and is otherwise obtained by writing A to the left of B, yielding the picture

$$\begin{matrix} a_{1,1} & a_{1,2} & \cdots & a_{1,n} & b_{1,1} & b_{1,2} & \cdots & b_{1,n'} \\ a_{2,1} & a_{2,2} & \cdots & a_{2,n} & b_{2,1} & b_{2,2} & \cdots & b_{2,n'} \\ \vdots & \vdots & \ddots & \vdots & \vdots & \vdots & \ddots & \vdots \\ a_{m,1} & a_{m,2} & \cdots & a_{m,n} & b_{m',1} & b_{m',2} & \cdots & b_{m',n'} \end{matrix}$$

The *row concatenation* of A and B, denoted by $A \ominus B$, is undefined if $n \neq n'$ and is otherwise obtained by writing A above of B. For a picture A and $k, k' \in \mathbb{N}$, by A^k we denote the k-fold column-concatenation of A, by A_k we denote the k-fold row-concatenation of A, and $A^k_{k'} = (A^k)_{k'}$. This can be lifted to array languages and then this in particular explains the notation $\Sigma^k_{k'}$ that was previously introduced.

3 Array Pushdown Automata

In this section, we introduce two variants of array pushdown automata, returning and boustrophedon pushdown automata, based on the following general formal syntax definition. Then, we explain how they work, introducing their semantics formally and with some examples.

Definition 1. *An* array pushdown automaton, *or APDA for short, is a 7-tuple* $M = (Q, \Sigma, \Gamma, \delta, q_0, \lhd, \#)$ *where*

- *Q is a finite set of states;*
- *Σ is the input alphabet with $\# \notin \Sigma$ ($\#$ will serve as the boundary symbol);*
- *Γ is the pushdown storage alphabet;*
- *$\delta : Q \times (\Sigma \cup \{\#, \varepsilon\}) \times \Gamma \to 2^{Q \times \Gamma^*}$ is the transition function;*
- *$q_0 \in Q$ is the initial state;*
- *$\lhd \in \Gamma$ is the start symbol of the pushdown.*

To $P \in \Sigma^n_m$, we can associate $P^\# \in \Sigma^{n+2}_m$ by setting $P^\# = \#_m \oplus P \oplus \#_m$. The automaton begins to read the picture P from the top left corner in the initial state q_0, with the start symbol $\lhd$ as the pushdown, and processes the picture row by row. During the computation, it is more convenient to think of working on $P^\#$, as this allows the automaton to read the boundary symbol $\#$. A computation by an APDA is accepted if the automaton has an empty pushdown store and it has read the whole input, including the final symbol $\#$. We describe a *configuration* by a quadruple (q, P, μ, γ), where $q \in Q$ is the current state, P describes the array that has yet to be worked on, $1 \le \mu \le m + 1$ describes the current row (where $\mu = m + 1$ is only possible after having read the input completely), and $\gamma \in \Gamma^*$ is the current pushdown content. In order to describe P, we use a *blank symbol* $\square$, so that $P \in (\Sigma \cup \{\square\})_+^+$. But we can be more precise here, and we have (for the first time) differentiated between returning and boustrophedon pushdown automata.

- The first $\mu - 1$ rows only contain $\square$.
- If the automaton is an RPDA or if $\mu - 1$ is even, then the μth row of $P^\#$ is in $\{\#\} \oplus \{\square\}^j \oplus \Sigma^{n-j} \oplus \{\#\}$ for some $0 \le j \le n$. Then, the automaton is currently reading the $j + 1^{\text{st}}$ symbol in that row of P. If the automaton is a BPDA and if $\mu - 1$ is odd, then the μth row of $P^\#$ is in $\{\#\} \oplus \Sigma^{n-j} \oplus \{\square\}^j \oplus \{\#\}$ for some $0 \le j \le n$. Then, the automaton is currently reading the $n - j^{\text{th}}$ symbol in that row P.
- Rows $\mu + 1, \ldots, m$ contain only symbols from Σ.

We denote the currently read symbol as $crs(P^\#)$ and its position within $P^\#$ as $(\mu, crc(P^\#))$, where crc is the currently read column. We can think of row μ as the *current row*. The currently read input symbol is always the first non-blank symbol in the movement of the automaton over the picture. Hence, in the beginning, $P \in \Sigma_+^+$, so that (in the previous list of conditions) $\mu = 1$. This means that the *initial configuration* is $(q_0, P, 1, \triangleleft)$. Similarly, the set of *final accepting configurations* of M equals

$$\text{FIN}_M = \{(q, \square_m^n, m + 1, \varepsilon) \mid q \in Q \wedge m, n \ge 1\}.$$

Notice that there is no currently read symbol defined for final accepting configurations.

Next, we formally define a *transition step* $\vdash_M$ between two configurations (q, P, μ, γ) and (q', P', μ', γ'), implicitly differentiating between RPDA and BPDA. Namely, $(q, P, \mu, \gamma) \vdash_M (q', P', \mu', \gamma')$ if and only if one of the two conditions described in the following is satisfied. As in the string case, silent transitions do not read any letter from the input, which explains the condition $P' = P$. When actually reading a letter, $P' \ne P$, but the exact shape of P' depends on the type of APDA (RPDA versus BPDA).

silent transition $\mu' = \mu$ and $P' = P$ and there exists $(q', \alpha) \in \delta(q, \varepsilon, Y)$, where Y is a prefix of the pushdown store γ, i.e., $\gamma = Y\gamma''$, and $\gamma' = \alpha\gamma''$.

non-silent transition There exists $(q', \alpha) \in \delta(q, crs(P^\#), Y)$, where Y is a prefix of γ, such that $\gamma = Y\gamma''$, and $\gamma' = \alpha\gamma''$ and
    - either $crs(P^\#) \in \Sigma$, $\mu' = \mu$ and P' equals P but for the position $(\mu, crc(P^\#))$ where a symbol from Σ (in P) is replaced by $\square$ in P',
    - or $crs(P^\#) = \#$, $\mu' = \mu + 1$ and $P' = P$.

Notice that a computation of an APDA will stop immediately when the pushdown is empty, because the transition function δ requires to read a pushdown symbol. Silent transitions do not consume any input and one may need these to empty the pushdown. This allows us to define

$$L(M) = \{P \in \Sigma_+^+ \mid (q_0, P, 1, \triangleleft) \vdash_M^* (q, \square_m^n, m, \varepsilon) \mid q \in Q \wedge m, n \geq 1\}.$$

Observe how this language (i.e., the semantics of M) differs when M is seen as a BPDA or when M is viewed as an RPDA.

Remark 1. Boustrophedon and returning finite automata (BFA / RFA) as introduced in [2] can be seen as special cases of BPDA and RPDA by considering $\Gamma = \{\triangleleft\}$, forbidding silent transitions, and assuming that the pushdown store is only changed by eventually popping the symbol $\triangleleft$, which is the only symbol that is ever on the pushdown store. In this way, the final state acceptance defined for BFA and RFA can be easily simulated, as this popping will be only allowed in final states (when reading #).

We will write $\mathcal{L}(\text{RFA})$, $\mathcal{L}(\text{BFA})$, $\mathcal{L}(\text{BPDA})$ and $\mathcal{L}(\text{RPDA})$ to denote the class of array languages accepted by these devices.

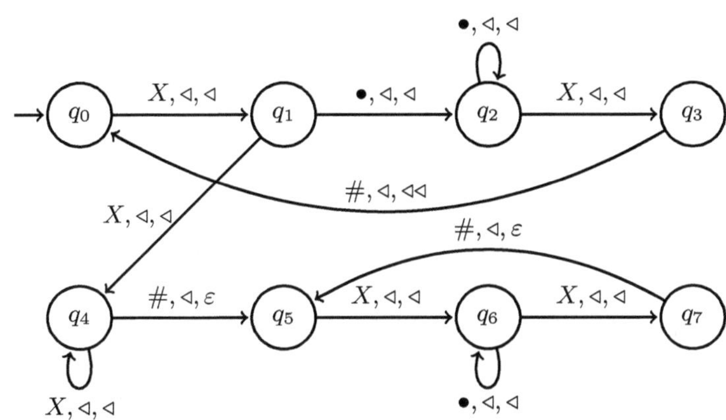

Fig. 1: Transition diagram of M_H that recognizes the language L_H.

Example 1. The set L_H of tokens H of different sizes and proportions (maintaining symmetry of the figure) can be formally defined as

$$L_H = \left\{ (X \oplus (\bullet)^n \oplus X)_m \ominus X^{n+2} \ominus (X \oplus (\bullet)^n \oplus X)_m \mid n, m \geq 1 \right\} \cup X^+.$$

Any row of any picture from L_H can be read from left to right or from right to left, without changing the acceptance of the picture. Therefore, we can design an APDA that can work both as a BPDA and as an RPDA and accepts L_H in both cases. An example for such an APDA is provided in Figure 1. In this figure, a triple like $\#, \triangleleft, \triangleleft\triangleleft$ attached to an arrow between two states means that when transitioning from one state, e.g., q_3, to the other one, say, q_0, the automaton reads $\#$ in the input (array), finds $\triangleleft$ on the pushdown and replaces this symbol by $\triangleleft\triangleleft$. A sample computation can be seen below in Figure 2. There, we interpret M_H as a BPDA.

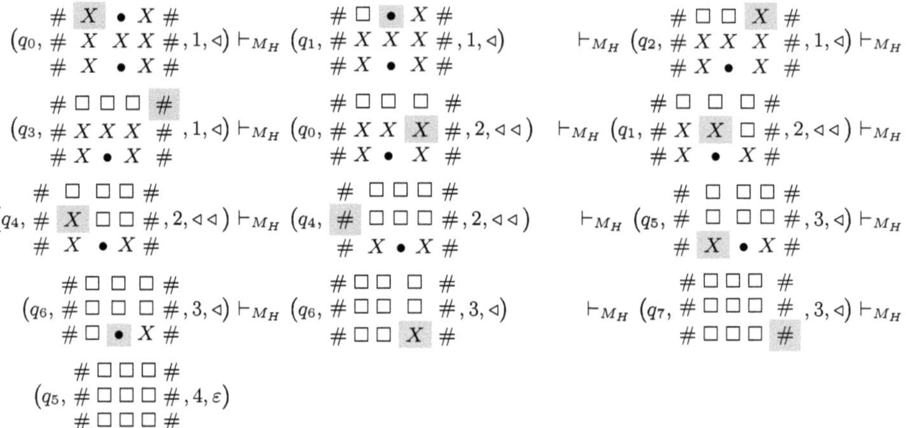

Fig. 2: A sample computation, interpreting M_H as a BPDA.

Theorem 1. $\mathcal{L}(\text{BFA}) = \mathcal{L}(\text{RFA}) \subsetneq \mathcal{L}(\text{BPDA}) \cap \mathcal{L}(\text{RPDA})$.

Proof. The equality $\mathcal{L}(\text{BFA}) = \mathcal{L}(\text{RFA})$ was shown in [2]. There, one can also find a horizontal pumping lemma (Lemma 40) for this array language class. This proves (by zero-pumping rows) that L_H is not contained in this class. Remark 1 shows the claimed inclusion $\mathcal{L}(\text{RFA}) \subseteq \mathcal{L}(\text{BPDA}) \cap \mathcal{L}(\text{RPDA})$. By Example 1, the strictness of the inclusion follows. $\square$

Clearly, the strictness of the inclusion stated in the previous theorem already follows from what is known from the string case, as strings can be easily considered as arrays with a single row. Also, it is easy to check that the acceptance of such arrays by RPDA and BPDA coincides with the classical acceptance mechanism of pushdown automata working on strings, here with empty pushdown acceptance. Nonetheless, Example 1 is interesting because of the obvious connections to character recognition, a classical topic in image analysis. Another interesting example is the following one.

Example 2. Consider the language $L_{(010)}$ of squares of odd side length with 1 in the center and 0 elsewhere. That is,

$$
L_{(010)} = \left\{
\begin{array}{l}
 \quad \quad 0\,0\,0\,0\,0\,0\,0 \\
 \quad 0\,0\,0\,0\,0 \quad 0\,0\,0\,0\,0\,0\,0 \\
0\,0\,0 \quad 0\,0\,0\,0\,0 \quad 0\,0\,0\,0\,0\,0\,0 \\
0\,1\,0, \quad 0\,0\,1\,0\,0, \quad 0\,0\,0\,1\,0\,0\,0, \quad \dots \\
0\,0\,0 \quad 0\,0\,0\,0\,0 \quad 0\,0\,0\,0\,0\,0\,0 \\
 \quad 0\,0\,0\,0\,0 \quad 0\,0\,0\,0\,0\,0\,0 \\
 \quad \quad 0\,0\,0\,0\,0\,0\,0
\end{array}
\right\}
$$

$L_{(010)}$ is accepted by the APDA $M_{(010)} = (Q, \Sigma, \Gamma, \delta, q_0, \triangleleft, \#)$ where $Q = \{q_0, q_1\}$, $\Sigma = \{0, 1\}$, $\Gamma = \{\triangleleft, 0, 1\}$, $q_0 \in Q$ is the start state, $\triangleleft \in \Gamma$ is the start symbol of the pushdown and the δ-transitions are given as follows:

$$\delta(q_0, 0, \triangleleft) = \{(q_0, 0\triangleleft)\}$$
$$\delta(q_0, 0, 0) = \{(q_0, 00)\}$$
$$\delta(q_0, \#, 0) = \{(q_0, 0)\}$$
$$\delta(q_0, 1, 0) = \{(q_1, 0)\}$$

$$\delta(q_1, 0, 0) = \{(q_1, \varepsilon)\}$$
$$\delta(q_1, \#, 0) = \{(q_1, 0)\}$$
$$\delta(q_1, \#, \triangleleft) = \{(q_1, \varepsilon)\}$$

It is irrelevant whether we interpret $M_{(010)}$ as a BPDA or as an RPDA. Again by Lemma 40 of [2], $L_{(010)} \notin \mathcal{L}(\text{RFA})$.

Remark 2. Based on the classical power set construction, it can be seen that deterministic BFA and RFA are as powerful as their nondeterministic counterparts, also see [2]. However, as observed above, single-row arrays correspond to strings, and therefore, the classical result that deterministic push-down automata are strictly weaker than nondeterministic ones (see [7]) also implies the same results for deterministic BPDA or RPDA. In this context, observe that Example 1 and Example 2 are indeed deterministic APDA. This is also interesting, as it is known from the string case that deterministic pushdown automata accepting with empty pushdown are strictly weaker than those accepting with final states (a concept not formally defined in this paper), while both acceptance conditions are equivalent in the nondeterministic string case.

4   Boustrophedon versus Returning APDA

In this section, we compare the two row-reading modes described by the words *boustrophedon* and *returning* against each other. We will show that, in contrast to the situation for finite automata, these two concepts yield incomparable array language classes in the case of pushdown automata. To prove this result, we need a number of auxiliary results that allow us to link APDA to the theory of pushdown string automata that are equipped with additional blind counters.

Let $M = (Q, \Sigma, \Gamma, \delta, q_0, \triangleleft, \#)$ be an RPDA. Assume that $L(M) \subseteq (\Sigma_2)^*$, i.e., M accepts only arrays with two rows. Let $\Sigma_\# = \Sigma \cup \{\#\}$. Let $M_{\text{flat}} = (Q, \Sigma_\#, \Gamma, \delta, q_0, \triangleleft)$ be the *flattened PDA* associated to the RPDA M. For an array language $L \subseteq (\Sigma_2)^*$, define $L_{\text{flat}} = \left\{ w_1 \# w_2 \in \Sigma_\#^* \mid \begin{bmatrix} w_1 \\ w_2 \end{bmatrix} \in L \right\}$ as the *flattened version* of L. This operation has been employed before to study array languages, for instance, in Sec. 6 of [3].

Lemma 1. *For each RPDA M with input alphabet Σ and $L(M) \subseteq (\Sigma_2)^*$,*

$$L(M)_{\textit{flat}} \subseteq L(M_{\textit{flat}})$$

and

$$L(M)_{\textit{flat}} = L(M_{\textit{flat}}) \cap \{w_1 \# w_2 \mid w_1, w_2 \in \Sigma^* \wedge |w_1| = |w_2|\}.$$

Proof. Consider a language L of arrays with only two rows accepted by an RPDA $M = (Q, \Sigma, \Gamma, \delta, q_0, \triangleleft, \#)$ such that $L = L(M)$. The flattened PDA associated to M is $M_{\text{flat}} = (Q, \Sigma_\#, \Gamma, \delta, q_0, \triangleleft)$. By definition, we have $L(M)_{\text{flat}} = \left\{ w_1 \# w_2 \in \Sigma_\#^* \mid \begin{bmatrix} w_1 \\ w_2 \end{bmatrix} \in L(M) \right\}$. Let $w \in L(M)_{\text{flat}}$. Then $w = w_1 \# w_2 \in \Sigma_\#^*$ such that $\begin{bmatrix} w_1 \\ w_2 \end{bmatrix} \in L(M)$. This implies that $|w_1| = |w_2|$ where $w_1, w_2 \in \Sigma^*$. Thus $L(M)_{\text{flat}} \subseteq L(M_{\text{flat}}) \cap \{w_1 \# w_2 \mid w_1, w_2 \in \Sigma^* \wedge |w_1| = |w_2|\}$. Now, let $w \in L(M_{\text{flat}}) \cap \{w_1 \# w_2 \mid w_1, w_2 \in \Sigma^* \wedge |w_1| = |w_2|\}$. As $w = w_1 \# w_2 \in L(M_{\text{flat}})$ and $|w_1| = |w_2|$, $\begin{bmatrix} w_1 \\ w_2 \end{bmatrix} \in L(M)$. Thus, $L(M_{\text{flat}}) \cap \{w_1 \# w_2 \mid w_1, w_2 \in \Sigma^* \wedge |w_1| = |w_2|\} \subseteq L(M)_{\text{flat}}$. $\qquad\square$

Moreover, we can show the following result:

Lemma 2. *For each RPDA M with input alphabet Σ and $L(M) \subseteq (\Sigma_2)^*$, the flattened language $L(M)_{\textit{flat}}$ can be accepted by a pushdown automaton that is equipped with an additional blind counter that makes only one turn.*

Proof. Let $L(M) = \left\{ \begin{bmatrix} w_1 \\ w_2 \end{bmatrix} \mid w_1, w_2 \in \{a, b\}^* \right\}$ be the language accepted by an RPDA $M = (Q, \Sigma, \Gamma, \delta, q_0, \triangleleft, \#)$. We construct a PDA $M' = (Q, \Sigma_\#, \Gamma, \delta', q_0, \triangleleft)$ equipped with a blind counter to accept the flattened language

$$L(M)_{\text{flat}} = \left\{ w_1 \# w_2 \in \Sigma_\#^* \mid \begin{bmatrix} w_1 \\ w_2 \end{bmatrix} \in L(M) \right\},$$

where δ' is a transition function from $Q \times (\Sigma_\# \cup \{\varepsilon\}) \times \Gamma$ into finite subsets of $Q \times \Gamma^* \times \{0, 1, -1\}$. M' processes $w_1 \# w_2$ similar to the machine M according to the δ'-transitions, in addition to which it has a counter which is initially 0. M' simply adds 1 to the counter at each step when it reads the contents of w_1 and when M' reads a $\#$, it adds 0 to the counter and when M' reads the contents of w_2, it subtracts 1 from the counter at each step of its computation. When the machine completes reading the entire input, emptying the stack, the counter should return to 0, which ensures that the length of w_1 equals the length of w_2 as required. $\qquad\square$

We now develop a similar concept for BPDA. Let $M = (Q, \Sigma, \Gamma, \delta, q_0, \triangleleft, \#)$ be a BPDA. Assume that $L(M) \subseteq (\Sigma_2)^*$, i.e., M accepts only arrays with two rows. Let $\Sigma_\# = \Sigma \cup \{\#\}$. Let $M_{\text{flat}} = (Q, \Sigma_\#, \Gamma, \delta, q_0, \triangleleft)$ be the *flattened PDA* associated to the BPDA M. For an array language $L \subseteq (\Sigma_2)^*$, define

$$L_{\text{flat}^B} = \left\{ w_1 \# w_2^R \in \Sigma_\#^* \mid \begin{bmatrix} w_1 \\ w_2 \end{bmatrix} \in L \right\}$$

as the *B-flattened version* of L. Observe that for any word $w = w_1 \# w_2 \in \Sigma^* \{\#\} \Sigma^*$, we can define the *right-hand reversal* $\mathrm{rhr}(w) = w_1 \# w_2^R$ if $\# \notin \Sigma$. We can lift this operation to languages.

Now, consider the following kind of inverse operation to flattening: If $L \subseteq \Sigma^+ \{\#\} \Sigma^+$ and $\# \notin \Sigma$, then $\mathrm{layer}(L)$ is the 2-row array language defined by

$$\left\{ \begin{bmatrix} w_1 \\ w_2 \end{bmatrix} \mid w_1 \# w_2 \in L \wedge |w_1| = |w_2| \right\}.$$

Observe that $P = \mathrm{layer}(P_{\text{flat}})$ for any 2-row picture language P. Moreover, we can define $\mathrm{rhr}(P) = \mathrm{layer}(\mathrm{rhr}(P_{\text{flat}})) = \mathrm{layer}(P_{\text{flat}^B})$, which means that the operator rhr would now reverse the lower row of any of the 2-row arrays that belong to P. This allows us to observe:

Lemma 3. *For any $L \subseteq (\Sigma_2)^*$, $L_{\text{flat}^B} = \mathrm{rhr}(L_{\text{flat}})$.*

Lemma 3 immediately entails:

Proposition 1. *For any array language $L \subseteq \Sigma_2^*$, $\mathrm{rhr}(L) = \mathrm{layer}(L_{\text{flat}^B}) \in \mathcal{L}(\mathrm{BPDA})$ if and only if $L = \mathrm{layer}(\mathrm{rhr}(L_{\text{flat}})) \in \mathcal{L}(\mathrm{RPDA})$.*

This proposition allows us to directly translate Lemma 1 and Lemma 2 into the world of BPDA. This yields:

Lemma 4. *For each BPDA M with input alphabet Σ and $L(M) \subseteq (\Sigma_2)^*$,*

$$L(M)_{\text{flat}^B} \subseteq L(M_{\text{flat}})$$

and

$$L(M)_{\text{flat}^B} = L(M_{\text{flat}}) \cap \{w_1 \# w_2^R \mid w_1, w_2 \in \Sigma^* \wedge |w_1| = |w_2|\}.$$

Lemma 5. *For each BPDA M with input alphabet Σ and $L(M) \subseteq (\Sigma_2)^*$, the flattened language $L(M)_{\text{flat}^B}$ can be accepted by a pushdown automaton that is equipped with an additional blind counter that makes only one turn (on the counter).*

Let us briefly explain this automaton type. A pushdown automaton that is equipped with an additional blind counter, being in a certain state, can read the input symbol and the topmost symbol on the pushdown store, and it can then move into a successor state, replace what has been read on the pushdown store by a string of pushdown symbols, and also increment or decrement its counter (or keeps its value). An input is accepted if both the pushdown is empty and the counter stores zero (as in the beginning). Such an automaton makes only one turn on its counter if during its run, it first increments the counter (or keep its value) in a first phase, but at some point, it switches into a second phase where it may only decrement the counter (or keep its value).

Crucial to our main result of this section is the following (unfortunately yet unpublished) theorem, see [11]. The proof of this result is based on the well-quasi ordering pumping technique that was introduced in [1].

Theorem 2 (Zetzsche). *The marked copy language $\{w\#w \mid w \in \{a,b\}^+\}$ cannot be accepted by any pushdown automaton equipped with an arbitrary (but fixed) number of additional blind counters.*

Theorem 3. $\mathcal{L}(\text{RPDA})$ *and* $\mathcal{L}(\text{BPDA})$ *are incomparable but not disjoint.*

Proof. The non-emptiness of intersection of $\mathcal{L}(\text{RPDA})$ and $\mathcal{L}(\text{BPDA})$ was made clear in Theorem 1. For the incomparabilities, we can use the following arguments.

1. Consider $L = \left\{ \begin{bmatrix} w \\ w \end{bmatrix} \mid w \in \{a,b\}^+ \right\}$. We can construct a BPDA M_L that will push all symbols that it reads from $\{a,b\}$ on the pushdown when scanning the first row containing w. Upon reading the border marker $\#$, it will change state and then continue reading the second row (where we expect to see w again) from right to left. Upon scanning this second row, M_L will compare the top symbol of the pushdown store with the currently read symbol and will continue working only if both symbols coincide. Clearly, $L(M_L) = L$. Hence, $L \in \mathcal{L}(\text{BPDA})$.

   Now, assume for the sake of contradiction that $L \in \mathcal{L}(\text{RPDA})$. As $L \in (\{a,b\}_2)^+$, L_{flat} can be accepted by a pushdown automaton that is equipped with an additional blind counter by Lemma 2. Since $L_{\text{flat}} = \{w\#w \mid w \in \{a,b\}^+\}$, this contradicts with Theorem 2. Hence, $\mathcal{L}(\text{BPDA})\backslash\mathcal{L}(\text{RPDA}) \neq \emptyset$.

2. By Proposition 1, $L' = \left\{ \begin{bmatrix} w \\ w^R \end{bmatrix} \mid w \in \{a,b\}^+ \right\} = \text{rhr}(L) \in \mathcal{L}(\text{RPDA})$. By Proposition 1, $L' \in \mathcal{L}(\text{RPDA})$ if and only if $\text{rhr}(L') = L \in \mathcal{L}(\text{BPDA})$. Hence, by the first item, $L' \in \mathcal{L}(\text{RPDA}) \setminus \mathcal{L}(\text{BPDA})$. $\qquad\square$

5   Context-free Matrix Languages

We recall the necessary definitions of matrix languages in the context of array languages, as introduced in [10]. We will define this class of array languages in terms of automata, but as the class of context-free matrix grammars also characterizes it, we will abbreviate the corresponding class as $\mathcal{L}(\text{CFMG})$.

Definition 2. *A* pushdown matrix automaton, *or PDMA for short, can be specified as an 10-tuple*

$$M = (Q, \Sigma, T, \Gamma, \delta, \delta', Q_{init}, F', \triangleleft, \#)$$

where the finite state set Q can be partitioned into $Q = \overline{Q} \cup \bigcup_{i=1}^{k} Q_i$. Each Q_i contains an initial state q_i and a final state f_i (which could be identical). Let $Q_{init} = \{q_i \mid i = 1, 2, \ldots, k\}$ and $F' = \{f_i \mid i = 1, 2, \ldots, k\}$ be the set of transition states. $\overline{Q}$ has an initial state q_0. $T = \{t_1, t_2, \ldots, t_k\}$ is the finite set of first storage symbols and $t_i \in T$ corresponds to Q_i. Γ is the finite set of second storage symbols. $\triangleleft \in \Gamma$ is the bottom symbol of the second (pushdown) storage. $\# \notin \Sigma$ is the end marker. For $i = 1, 2, \ldots, k$, δ_i is a mapping from

$Q_i \times \Sigma$ *into finite subsets of* $Q_i \times \{\varepsilon\}$ *and from* $\{f_i\} \times \{\#\}$ *into finite subsets of* $(S \cup \{q_0\}) \times \{t_i\}$. *For brevity, we sometimes suppress the index of* δ_i, *leading to the definition of a partial function* $\delta : \bigcup_{i=1}^{k} Q_i \times \Sigma \to Q \times T^*$. δ' *is a mapping from* $\overline{Q} \times (T \cup \{\varepsilon\}) \times \Gamma$ *into finite subsets of* $\overline{Q} \times \Gamma^*$.

The PDMA M works on an array P over the alphabet Σ, i.e., $P \in \Sigma_m^n$, in two phases:

1. M scans the picture P (or, more precisely, $P_\# = P \ominus \#^n$) column-by-column, starting with reading the first column top-down, then the second column top-down, etc. More precisely, at the very beginning, M guesses the finite automaton specified by δ_{i_1} that will be in charge by choosing the initial state $q_{i_1} \in Q_{\text{init}}$. Then, it reads the first column. After reading this column, M will have reached a state, say, p_{i_1}. (If we want to formally specify the notion of a configuration, we can indicate the reading of an input symbol by replacing it by $\square$ as formalized above for APDA.) Then, M notices the lower border by reading the border symbol $\#$. If $p_{i_1} = f_{i_1}$, M will replace $\#$ by t_{i_1} and then guess a new start state $q_{i_2} \in Q_{\text{init}}$ and hence a finite automaton that is used to work on the second column, etc.

2. After the whole array has been processed this way, we formally see a configuration whose picture can be described by an array $\square_m^n \ominus T^n$ from $\square_+^+ \ominus T^+$. Moreover, the automaton should have chosen $q_0 \in Q_{\text{init}}$ as the next start state, as now the PDA $M_{PDA} = (\overline{Q}, T, \Gamma, \delta', q_0, \triangleleft)$ starts its work on the only row that is not filled with blanks.[3]

Now, an array language belongs to $\mathcal{L}(\text{CFMG})$ if and only if there is some PDMA accepting it. Notice that only in the second working phase a PDMA really makes use of its pushdown capabilities. If we replace the underlying pushdown automaton M_{PDA} by a finite automaton (with final-state acceptance), the corresponding devices are known as finite-state matrix automata (FSMA), and they characterize the class $\mathcal{L}(\text{RMG})$ of regular matrix array languages, as introduced in [10]. Instead of giving a complete formal definition, we explain the concept by a small example.

Example 3. Consider the FSMA $M = (Q, \Sigma, T, \delta, \delta', Q_{\text{init}}, F, F', \#)$, where F is the set of final states of the underlying DFA $M_{DFA} = (\overline{Q}, T, \delta', q_0, F)$. We have $T = \{t_1\}$ and $L(M_{DFA}) = T^+$, so that $\overline{Q} = \{q_0, q_0^f\}$, $F = \{q_0^f\}$, $\delta'(q_0, t_1) = q_0^f$ and $\delta'(q_0^f, t_1) = q_0^f$ suffices. Apart from this, we only have one other DFA involved, namely $M_1 = (Q_1, \Sigma, \delta_1, q_1, F')$, with $Q_1 = \{q_1, q_1^a, q_1^b, q_1^f\}$, $F' = \{q_1^f\}$ and $\delta_1(q_1, a) = q_1^a$, $\delta_1(q_1, b) = q_1^b$, $\delta_1(q_1^a, a) = \delta_1(q_1^b, b) = q_1^f$. Obviously, $L(M_1) = \{aa, bb\}$. Together, these two trivial automata describe

$$L(M) = (\{a \ominus a\} \cup \{b \ominus b\})^+ = \left\{ \begin{bmatrix} w \\ w \end{bmatrix} \mid w \in \{a, b\}^+ \right\},$$

---

[3] In the original paper [10], PDMA contained the specification of a final state set for this PDA, but we use again acceptance with empty pushdown here. It is well-known that for nondeterministic PDA working on strings, acceptance by final states and acceptance by empty pushdown are equivalent.

showing a relation to an array language that was a crucial example in proving Theorem 3.

In [8], more general mechanisms have been introduced. For the simplicity of presentation, we are not considering them in this paper.

Notice that in the following theorem, we need the transposition operator T as defined in the beginning. This is in line with the characterization

$$\mathrm{T}(\mathcal{L}(\mathrm{RMG})) = \mathcal{L}(\mathrm{RFA}) = \mathcal{L}(\mathrm{BFA})$$

known from the finite-automaton case, see [2].

Theorem 4. $\mathrm{T}(\mathcal{L}(\mathrm{CFMG})) \subsetneq \mathcal{L}(\mathrm{RPDA}) \cap \mathcal{L}(\mathrm{BPDA})$.

Proof. First, consider the language $L_{(010)} = \mathrm{T}(L_{(010)})$. By Example 2, $L_{(010)} \in \mathcal{L}(\mathrm{RPDA}) \cap \mathcal{L}(\mathrm{BPDA})$, but $L_{(010)} \notin \mathcal{L}(\mathrm{CFMG})$ due to the Iteration Theorems given in [9].

Next, we show the inclusion $\mathrm{T}(\mathcal{L}(\mathrm{CFMG})) \subsetneq \mathcal{L}(\mathrm{RPDA})$. This claim is obviously equivalent to proving $\mathcal{L}(\mathrm{CFMG}) \subsetneq \mathrm{T}(\mathcal{L}(\mathrm{RPDA}))$. Let $L \in \mathcal{L}(\mathrm{CFMG})$. Let $M = (Q, \Sigma, T, \Gamma, \delta, \delta', S, F', \lhd, \#)$ be a PDMA accepting L. As in Definition 2, let Q be partitioned into $Q = \overline{Q} \cup \bigcup_{i=1}^{k} Q_i$. To prove the claim, we construct an RPDA $\hat{M} = (\hat{Q}, \Sigma, \hat{\Gamma}, \hat{\delta}, \widehat{q_0}, \lhd, \#)$ to accept $\mathrm{T}(L)$. The basic idea is that $\hat{M}$ simulates both phases of M in parallel. Therefore, let $\hat{Q} = \overline{Q} \times \left(\bigcup_{i=1}^{k} Q_i \right) \cup \{\widehat{q_0}\}$.

We also 'recycle' some symbols from M in $\hat{M}$, like the border symbol $\#$ and the start symbol $\lhd$ of the pushdown; more generally, $\hat{\Gamma} = \Gamma$. Whenever M reads symbols from Σ (i.e., it works in the first phase), $\hat{M}$ works completely alike. Hence, for $q \in \bigcup_{i=1}^{k} Q_i$ and $a \in \Sigma$, and for all $\bar{q} \in \overline{Q}$ and $Y \in \Gamma$,

$$\hat{\delta}((\bar{q}, q), a, Y) = \{((\bar{q}, p), Y) \mid p \in \delta(q, a)\}.$$

Here, recall that $q \in Q_i$ for some unique i, with $1 \leq i \leq k$, so that $\delta(q, a) = \delta_i(q, a)$ in that case. In order to determine this index i, we set $\varphi_M(q) = i$, hence defining the function $\varphi_M : \bigcup_{i=1}^{k} Q_i \to \{1, 2, \ldots, k\}$. As an initialization, we use silent transitions

$$\hat{\delta}(\widehat{q_0}, \varepsilon, \lhd) = \{((q_0, q_i), \lhd) \mid 1 \leq i \leq k\}$$

that perform the first guess (where to start) that M is doing at the beginning.

When M reaches the bottom end of the ith column, $\hat{M}$ has to reach the right end of the ith row, as $\hat{M}$ is aiming at the recognition of the transposition of $L(M)$. Then, both automata read the border symbol $\#$. Hence, in the first phase, M prints the symbol t_i corresponding to the current transition function δ_i and also guess a start symbol q_j corresponding to the next chosen transition function δ_j. In the second phase, M will read t_i, update the state of the underlying pushdown automaton, and also change the pushdown storage according to δ. As promised, this is done in parallel in the simulating automaton $\hat{M}$. To this end, define $\hat{\delta}((\bar{q}, q), \#, Y)$ as

$$\{((\bar{p}, q_j), \gamma) \mid 1 \leq j \leq k, q = f_{\varphi_M(q)}, q_j \in \delta_{\varphi_M(q)}(q, \#), (\bar{p}, \gamma) \in (\delta'(\bar{q}, t_{\varphi_M(q)}, Y)\}.$$

Notice that the condition $q = f_{\varphi_M(q)}$ checks if M has actually reached a final state in the first phase. As the underlying pushdown automaton accepts with empty pushdown as does the RPDA $\hat{M}$, it is clear by a simple induction that an array A is accepted by M if and only if $\mathrm{T}(A)$ is accepted by $\hat{M}$.

Last, in order to prove the inclusion $\mathrm{T}(\mathcal{L}(\mathrm{CFMG})) \subsetneq \mathcal{L}(\mathrm{BPDA})$, we would have to formally distinguish whether or not M is working on an odd- or even-numbered column in the first phase. This can be kept track of by considering

$$\{0,1\} \times \overline{Q} \times \left(\bigcup_{i=1}^{k} Q_i \right) \cup \{\widehat{q_0}\}$$

as the state set of the simulating BPDA $\hat{M}$. From the start state $\widehat{q_0}$, $\hat{M}$ would now silently transit into some $(1, q_0, q_i)$. When simulating odd-numbered columns, $\hat{M}$ behaves as described before for RPDA. Then, upon reading a border marker, the first component of the state will switch from 1 to 0. Likewise, instead of selecting a start state, it will now select a final state in the third component as $\{f_j\}$. (The simulation of the underlying PDA will be the same.) When simulating even-numbered columns, $\hat{M}$ will have to simulate the movements of M (working in the first phase) backwards on the corresponding even-numbered row. Now, when reaching a border marker again, $\hat{M}$ has to verify if the current set of states contains the initial state and only then, it proceeds with the simulation of the work of M on odd-numbered rows again; also, in parallel, $\hat{M}$ will simulate the underlying pushdown automaton M_{PDA}. The technical details of this construction are more sophisticated, but the description given so far should suffice to reconstruct these. Also, the difference between odd- and even-numbered rows was similarly used in the simulation proving $\mathcal{L}(\mathrm{RFA}) \subseteq \mathcal{L}(\mathrm{BFA})$ in [2, Theorem 20]; actually, the current simulation is somewhat easier, as we have only one final state to consider.

These three points together allow us to prove the theorem. $\square$

The reader might have wondered why we had to apply the transposition operation in the previous theorem. This is due to the following fact:

Proposition 2. *The two array language families* $\mathcal{L}(\mathrm{RPDA}) \cup \mathcal{L}(\mathrm{BPDA})$ *and* $\mathcal{L}(\mathrm{CFMG})$ *are incomparable.*

Proof. As we have seen before, Example 2 gives an example of a language that is even in $\mathcal{L}(\mathrm{RPDA}) \cap \mathcal{L}(\mathrm{BPDA})$ but not in $\mathcal{L}(\mathrm{CFMG})$. Consider the language $L = (\{a \ominus a \ominus a\} \cup \{b \ominus a \ominus b\})^+$. Similar as in Example 3, $L \in \mathcal{L}(\mathrm{RMG}) \subset \mathcal{L}(\mathrm{CFMG})$ can be seen. Also, as the second row is always only filled with a, one understands that $L \in \mathcal{L}(\mathrm{RPDA})$ if and only if $L \in \mathcal{L}(\mathrm{BPDA})$, because an APDA M can be interpreted either as a BPDA or an RPDA and will either accept L in both interpretations or not.

Now, consider the string language $L' = \{w \# a^{|w|} \# w \mid w \in \{a,b\}^+\}$ that can be associated to L. Moreover, consider the family $\mathcal{L}$ of string languages accepted by pushdown automata with an arbitrary number of blind counters.

If $L \in \mathcal{L}(\mathrm{RPDA}) \cup \mathcal{L}(\mathrm{BPDA})$, then $L' \in \mathcal{L}$ can be seen similarly to Lemma 2 and Lemma 5 as explained in the following. We can use two blind counters, incrementing both when reading the prefix $u \in \{a, b\}^+$ of some word $u\#v\#w$ (hypothetically from L'), decrementing the first one when reading the middle part $v \in \{a\}^+$ and decrementing the second one when reading the suffix w. Moreover, the pushdown is working as for the APDA. As in Lemma 2, the counters are used to check the length conditions only. Also, the class $\mathcal{L}$ is easily seen to be closed under rational transductions (i.e., it forms a trio, see [7]). Consider the transducer τ that works in the following way:

1. First, when seeing symbols from $\{a, b\}$, it will just output them as read, staying in the initial state q_0.
2. Upon reading the first $\#$, it will output $\#$ but then it will switch to the state q_1.
3. When being in state q_1, it will suppress any output when reading a's in the following; upon reading b's in this phase, it will move into a non-final state, say, q_2, and stop.
4. Upon reading a second $\#$, it will not output anything, but it will switch to the final state q_f and continue: when seeing symbols from $\{a, b\}$, it will just output them as read, staying in the state q_f.
5. If the transducer reads another $\#$ while being in state q_f, it will switch into the non-final state q_2 and stop.

It is not hard to understand that $\tau(L') = \{w\#w \mid w \in \{a, b\}^+\}$. Hence, if $L' \in \mathcal{L}$, then also $\tau(L')$ would belong to $\mathcal{L}$, contradicting Theorem 2. $\square$

The preceding proof also shows the following result.

Corollary 1. *The two families $\mathcal{L}(\mathrm{RPDA}) \cap \mathcal{L}(\mathrm{BPDA})$ and $\mathcal{L}(\mathrm{CFMG})$ are incomparable.*

6   Conclusions

We have introduced two interpretations of array pushdown automata (APDA), namely, boustrophedon and returning APDA, i.e., BPDA and RPDA. We have explored their inter-relations as well as their relations to other models defining array language classes, especially those related to finite or pushdown automata, see Figure 3. There are many more things to be investigated and these are also on our agenda; we only highlight two of these future research topics below.

- For many applications, it is important that certain properties are decidable and if so, that they are decidable efficiently. These aspects have been discussed for finite array automata in the literature, see (e.g.) [2] in the case of BFA and RFA but still need to be studied for BPDA and RPDA. Even for PDMA, these questions are largely unexplored. In this context, it might be also interesting to restrict the attention to deterministic variants as they are particularly important for practical applications. Notice that we did not make much use of nondeterminism in our examples.

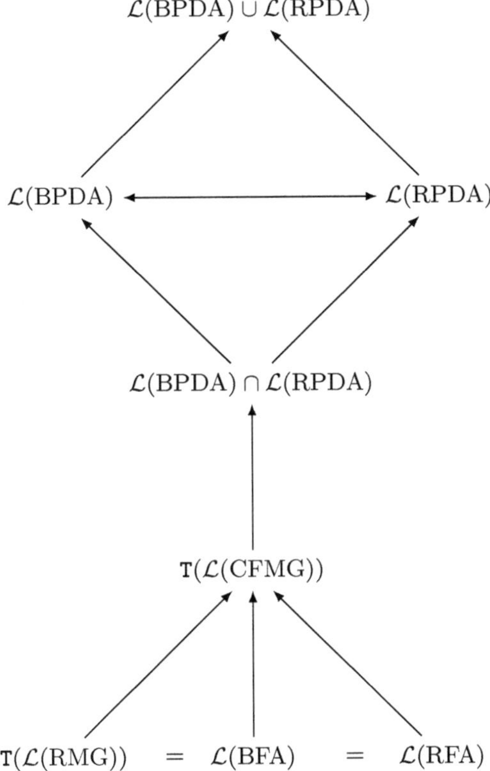

Fig. 3: Hierarchy among classes of array languages recognized by two-dimensional automaton models mentioned in this paper. An arrow $\mathcal{L}(X) \to \mathcal{L}(Y)$ indicates the relation $\mathcal{L}(X) \subsetneq \mathcal{L}(Y)$; this propagates by transitivity; the double arrow $\mathcal{L}(X) \leftrightarrow \mathcal{L}(Y)$ indicates $\mathcal{L}(X)$ and $\mathcal{L}(Y)$ are incomparable but not disjoint.

– We also want to discuss operations on array languages and closure properties of the new array language families. This also brings along the question of nice criteria of non-membership for our array language classes in order to prove non-closure properties. So far, we mostly rely on Theorem 2 whose proof uses quite heavy proof machinery. More elementary techniques are needed, as the pumping lemmas known for BFA, see [2].

Acknowledgements. We thank the anonymous reviewers for many useful comments that helped improve our submitted manuscript.

References

1. Anand, A., Schmitz, S., Schütze, L., Zetzsche, G.: Verifying unboundedness via amalgamation. In: Sobocinski, P., Lago, U.D., Esparza, J. (eds.) Proceedings of the 39th Annual ACM/IEEE Symposium on Logic in Computer Science, LICS. pp. 4:1–4:15. ACM (2024)
2. Fernau, H., Paramasivan, M., Schmid, M.L., Thomas, D.G.: Simple picture processing based on finite automata and regular grammars. Journal of Computer and System Sciences **95**, 232–258 (2018)
3. Fernau, H., Paramasivan, M., Thomas, D.G.: Picture scanning automata and group actions on pictures. Romanian Journal of Information Science and Technology **21**(3), 238–248 (2018)
4. Giammarresi, D., Restivo, A.: Two-dimensional finite state recognizability. Fundamenta Informaticae **25**, 399–422 (1996)
5. Giammarresi, D., Restivo, A.: Two-dimensional languages. In: Rozenberg, G., Salomaa, A. (eds.) Handbook of Formal Languages, Volume III, pp. 215–267. Berlin: Springer (1997)
6. Greibach, S.: Remarks on blind and partially blind one-way multicounter machines. Theoretical Computer Science **7**, 311–324 (1978)
7. Hopcroft, J.E., Ullman, J.D.: Introduction to Automata Theory, Languages, and Computation. Reading (MA): Addison-Wesley (1979)
8. Krithivasan, K., Siromoney, R.: Array automata and operations on array languages. International Journal of Computer Mathematics 4(A), 3–40 (1974)
9. Krithivasan, K., Siromoney, R.: Characterizations of regular and context-free matrices. International Journal of Computer Mathematics 4(A), 229–245 (1974)
10. Siromoney, G., Siromoney, R., Krithivasan, K.: Abstract families of matrices and picture languages. Computer Graphics and Image Processing **1**, 284–307 (1972)
11. Zetzsche, G.: Blind counters do not help a pushdown automaton to accept the copy language (Apr 2025), personal communication

The Injectivity of Affine Mappings Restricted to Convex Sets

Zoltán Domokos$^{(\boxtimes)}$

Department of Image Processing and Computer Graphics, University of Szeged,
Árpád tér 2., 6720 Szeged, Hungary
`domokos@inf.u-szeged.hu`

Abstract. Hereby, the problem of deciding the injectivity of an affine mapping restricted to sets taken from a finite family of convex sets is studied. We propose sufficient and necessary conditions on injectivity. As the method presented highly depends on how the intersections of the sets under study can be detected, we propose a relaxation for sets that can be bounded by open d-balls of arbitrary dimensionality.

Keywords: Inverse problems · Affine mappings · Injectivity · Projection angle selection

1 Introduction

In Computed Tomography, the field of projection angle selection intends to address the problem of finding a low number of projection angles in a way that a given dataset can still be reconstructed perfectly or with relatively small error. Some interesting applications of this approach include discrete [11] and binary tomography [10]. Uniqueness conditions for finite lattice grids were studied in e.g. [2,4].

However, to the best knowledge of the author, given images of arbitrary graylevels as elements of an N-dimensional Euclidean space, reducing this search space to a finite union $S \subset \mathbb{R}^N$ of convex sets, and fusing such prior knowledge with angle selection would be a novel approach. Additionally, the discretization scheme of the Radon-transform is assumed to follow a ray-driven approach (see [6]). In order to take a step towards the aforementioned model, the problem of deciding the injectivity of a given transform, restricted to some subsets of the whole search space needs to be covered, which is the main contribution of our work.

An important related result is the Gale-Nikaido Theorem, a global univalence theorem for orientation preserving local diffeomorphisms, where the domain of the mapping is a rectangular region [5]. Unfortunately, this theorem cannot be generalized to domains that have a more complex shape [1]. In this work, the key difference from former results, which are based on the Gale-Nikaido Theorem, is the fact that we restrict our discussion to affine mappings, enabling us to decide

P. Balázs et al. (Eds.): IWCIA 2025, LNCS 15985, pp. 51–64, 2026.
https://doi.org/10.1007/978-3-032-19347-6_4

injectivity, even if sets exhibiting a significantly wider variety of shapes can be chosen as the restricted domain. More specifically, we consider indexed sets of bounded convex sets given in affine subspaces of a finite-dimensional Euclidean space. Taking a non-injective affine mapping, and an indexed set of finitely many bounded convex sets, the question whether the mapping restricted to the union of some elements taken from the indexed set is injective will be discussed in details.

2 Assessment of Key Contributions

We consider finite-dimensional Euclidean affine spaces $\left(E, \vec{E}, + \right)$, where the vector space $\vec{E}$ is an affine space over itself. Therefore, the additive group action $+$ is assumed to be the vector addition in $\vec{E}$, and we will not make a distinction between E and $\vec{E}$. Additionally, elements x in E are treated either as a point or a vector interchangeably.

Keeping the conventions outlined above in mind, affine mappings $D : E \to E$ can be viewed as a composition $T_c \circ L$, where $T_c : E \to E$ is a translation by a vector $c \in E$ and $L : E \to E$ is a linear transform centered at the origin. As mentioned before, we will study the injectivity of D, restricted to elements of an indexed set of finitely many bounded convex sets $\mathcal{S} := \{S_i : i \in I\}$. In this context the affine hulls $aff\,(S_i)$ are referred to frequently, which necessitates introducing the shorthand notation A_i to denote the affine hull of a given set S_i. Moreover, the linear subspace associated to A_i is denoted by V_i.

Our main problem, will be addressed in Sect. 4, and it can be assessed formally as follows:

Problem 1. In an N-dimensional Euclidean space E, let S be a subset which equals the union of all bounded convex sets S_i taken from the indexed set $\mathcal{S} := \{S_i : i \in I\}$, where I is of finite cardinality. Given an affine mapping $D : E \to E$ and an element $y \in D\,(S)$, the preimage $D\,|_S^{-1}\,(y)$ has to be found, given that the mapping $D\,|_{A_i} : A_i \to E$ is injective for all $S_i \in \mathcal{S}$.

While solving Problem 1, it is beneficial to acquire knowledge about the injectivity of restrictions to sets $S_i \cup S_j$ and build a graph that captures these pieces of information. Consequently, we intend to find sufficient conditions on the injectivity of the restricted mapping $D\,|_{S_i \cup S_j}$, for a given triplet D, S_i, S_j, where S_i, S_j are contained in $\mathcal{S}$. The proposed conditions are declared in Theorem 1.

Difficulties might arise when one attempts to analyze injectivity in line with Theorem 1 directly, as one has to decide whether there is an intersection between bounded convex sets of arbitrary shape. Due to this reason, less tight conditions will be proposed, assuming that a pair of relatively open d_i, d_j-balls B_i, B_j exists, such that $S_i \subseteq B_i$, $S_j \subseteq B_j$. Additionally, the affine hull of B_i equals the affine hull of S_i. Similarly, for B_j and A_j, $aff\,(B_j) = A_j$ holds. Furthermore, a pair of relatively open d_i', d_j'-balls B_i', B_j' is also given, which satisfies $D\,(S_i) \subseteq B_i'$, $D\,(S_j) \subseteq B_j'$ and $aff\,(B_i') = D\,(A_i)$, $aff\,(B_j') = D\,(A_j)$. In Theorem 2, we propose sufficient conditions on the injectivity of $D\mid_{S_i \cup S_j}$, provided the sets B_i, B_j, B_i', B_j', and the mapping D.

In Sect. 4, we present some results of our empirical analysis on the applicability of the proposed theoretical framework. Furthermore, it will be shown how our model can be applied to solve the following problem.

Problem 2. Given an indexed set of projection operators $\mathcal{D}$ and an indexed set of bounded convex sets $\mathcal{S}$, find the shortest sequence $\mathcal{D}'$ of elements in $\mathcal{D}$, which guarantees that $D\mid_S$ is injective, where $D = \sum_{i=0}^{i=|\mathcal{D}'|} D_i'$.

Problem 2 can be viewed as an extension of the projection angle selection problem, incorporating prior information to reduce the search space to a well-defined subset. Even if the model we propose is only applicable for small problem instances, it might be a valuable starting point towards real-world applications in future work.

3   Conditions on Injectivity

It can be derived from a more general, still very simple formula, mentioned in several fundamental works – e.g. in [9] – that the injectivity of $D\mid_{S_i \cup S_j}$ implies the following equation:

$$D\left(S_i\right) \cap D\left(S_j\right) = D\left(S_i \cap S_j\right) \ . \tag{1}$$

Equation(1) on its own is necessary, not sufficient. However, assuming the injectivity of $D\mid_{S_i}$ and $D\mid_{S_j}$ leads to a set of sufficient conditions. We prove this in the current section, and derive a corollary that is essential for solving Problem 1 algorithmically. The relaxed conditions on injectivity, which rely on another straightforward corollary, are also stated in this section.

3.1   Injectivity over a Finite Union of Sets

The following theorem is the cornerstone of our work. We only prove that the conditions declared here are sufficient, as the other way round the proof is trivial.

Theorem 1. *Given a pair of sets S_i, S_j and an affine mapping D, the restricted mapping $D\mid_{S_i \cup S_j}$ is injective if and only if $D\mid_{S_i}$ and $D\mid_{S_j}$ are injective and S_i, S_j satisfies Eq. (1).*

Proof. Supposing that $D\mid_{S_i}$, $D\mid_{S_j}$ are injective and Eq. (1) holds for S_i, S_j, while $D\mid_{S_i \cup S_j}$ is not injective. We denote the image of $x \in S_i$ and $x' \in S_j$ by y. Notice that y must be an element of $D\left(S_i\right) \cap D\left(S_j\right)$. As Eq. (1) is assumed to hold, $y \in D\left(S_i \cap S_j\right)$. Therefore, the intersection of $D\mid_{S_i \cup S_j}^{-1}(y)$ and $S_i \cap S_j$ cannot be empty. This fact and the injectivity of $D\mid_{S_j}$ implies $x' \in S_i \cap S_j$. Then, $x' \in S_i$ is implied. As $D\mid_{S_i}$ is injective, $x = x'$ must be true, which contradicts our premise that the mapping D restricted to $S_i \cup S_j$ is not injective. $\square$

In Theorem 1, we only rely on knowledge about the injectivity of the mapping restricted separately to the sets S_i, S_j. However, it is not always possible to prove the injectivity of these two restricted mappings in finite steps. Therefore, the following obvious consequence of Theorem 1 might be more suitable in many settings.

Corollary 1. *The mapping $D\mid_{S_i \cup S_j}$ is injective if $D\mid_{A_i}, D\mid_{A_j}$ are injective and Eq. (1) holds for S_i, S_j.*

Another straightforward result that follows from Theorem 1 will be beneficial, when we define the injectivity conditions on sets contained in d-balls later on. Clearly, when the equation below holds:

$$D(A_i) \cap D(A_j) = D(A_i \cap A_j) \ , \tag{2}$$

and $D\mid_{A_i}, D\mid_{A_j}$ are injective, $D\mid_{A_i \cup A_j}$ must also be injective, which implies the injectivity of $D\mid_{S_i \cup S_j}$. This fact is formalized in Corollary 2, which we could have derived based on the definition of affine subspaces, associated vector spaces and standard linear algebra. Still, it is worth noticing how convenient it is to obtain the following corollary from Theorem 1 instead.

Corollary 2. *Let $S_i, S_j \in \mathcal{S}$ be a pair of sets such that $D\mid_{A_i}, D\mid_{A_j}$ are injective. If the affine hulls satisfy Eq. (2), then $D\mid_{S_i \cup S_j}$ is injective.*

Turning our attention towards less tight conditions on injectivity, let S_i, S_j be bounded and convex sets. It is always possible to find B_i, B_j relatively open d_i, d_j-balls, such that $S_i \subseteq B_i$, $S_j \subseteq B_j$, and $A_i = aff(B_i)$, $A_j = aff(B_j)$ holds. Similarly, the images $D(S_i), D(S_j)$ are subsets of the d_i', d_j'-balls B_i', B_j', satisfying $D(A_i) = aff(B_i')$, $D(A_j) = aff(B_j')$. Note that the balls including the images are generally not assumed to be the images of B_i, B_j, even if it might happen in some special cases. As a first step towards the relaxed injectivity conditions, we address the simplest case, when $S_i \cap S_j$ equals the empty set.

Lemma 1. *Let $S_i, S_j \in \mathcal{S}$ be bounded convex sets and B_i, B_j be relatively open d_i, d_j-balls, contained by the affine hulls of S_i and S_j respectively. Furthermore, $S_i \subseteq B_i$, $S_j \subseteq B_j$ must hold. Therefore, $aff(B_i)$ equals A_i, $aff(B_j)$ equals A_j. The images $D(S_i), D(S_j)$ are bounded by B_i', B_j', such that $D(A_i) = aff(B_i')$, $D(A_j) = aff(B_j')$ holds. If $B_i \cap B_j = \emptyset$, $B_i' \cap B_j' = \emptyset$ and $D\mid_{A_i}, D\mid_{A_j}$ are injective, then $D\mid_{S_i \cup S_j}$ is injective.*

The lemma stated above, combined with Corollary 2 enables us to declare a set of sufficient conditions on injectivity for sets bounded by d-balls. The conditions declared in Theorem 2 are relatively loose. However, an algorithm that it is guaranteed to terminate with a result in finite steps can be constructed considering these conditions, which is a meritable advantage of this approach.

Theorem 2. *Let $S_i, S_j \in \mathcal{S}$ be bounded convex sets and B_i, B_j be relatively open d_i, d_j-balls, contained by the affine hulls of S_i and S_j respectively. Furthermore, $S_i \subseteq B_i$, $S_j \subseteq B_j$ must hold. Therefore, $aff(B_i)$ equals A_i, $aff(B_j)$ equals A_j. The images $D(S_i), D(S_j)$ are bounded by B_i', B_j', such that $D(A_i) = aff(B_i')$, $D(A_j) = aff(B_j')$ holds. Given that $D|_{A_i}$ and $D|_{A_j}$ are injective, if the conjunction of $B_i \cap B_j = \emptyset$ and $B_i' \cap B_j' = \emptyset$ is true or Eq. (2) holds for A_i, A_j, then $D|_{S_i \cup S_j}$ must be injective.*

Proof. The mappings $D|_{A_i}$ and $D|_{A_j}$ are assumed to be injective. Consequently, the first condition is an application of Lemma 1, while the second condition is an application of Corollary 2. Both are sufficient conditions on the injectivity of $D|_{S_i \cup S_j}$. $\qquad\square$

The sets S_i are assumed to be known as a set of constraints given in closed form. E.g., assuming that all the S_i are polytopes, they can be defined by the set of vertices of the polytopes (vertex representation). As in our work an enclosing ball is not necessarily the smallest, a rather simple approach to finding it would be the calculation of the polytope centroid, which determines the center of the ball, followed by finding the vertex of highest distance from the centroid. The highest distance shall equal the radius of the ball. The process has a computation complexity that is linear with respect to the number of vertices.

Assuming that A_i is not known, while the S_i polytope's vertices are given, it is easy to retrace the problem of finding A_i to computing a basis of the associated linear subspace V_i, which can be solved by e.g., Gaussian elimination. Therefore, when the center of the ball is determined as the centroid of S_i and the radius as the highest distance from the centroid, while A_i is computed by Gaussian elimination, the complexity of finding the ball B_i is dominated by the complexity of Gaussian elimination.

Decision about the emptiness of pairwise non-degenerate ball-ball intersections is trivial in Euclidean spaces, while a generalization that considers balls contained by lower dimensional subspaces is a rather simple problem, which involves computing the intersection of hyperplanes. It is also easy to decide whether Eq. (2) holds, for example by obtaining a basis of the vector spaces associated to $D(A_i) \cap D(A_j)$ and $A_i \cap A_j$. Both problems can be retraced to calculating the basis of intersections of linear subspaces, for this purpose, well-known algorithms - e.g. the Zassenhaus-algorithm [7] - can be utilized.

4   Reconstruction Algorithm

In this section, we present a reconstruction algorithm that aims to restore the preimage of a given element. Moreover, as the discussion can be reasonably simplified if we visualize the information about injectivity conditions being satisfied or violated by specific pairs of sets, we propose a graph based representation.

4.1   Obtaining the Conjunction of the Preimage and a Single Set

When $D\mid_{A_i}$ is injective for each set $S_i \in \mathcal{S}$, the preimage of any $y \in D(S)$ is of finite cardinality. In order to reconstruct the preimage of a given element y, we need to take the family of sets $\mathcal{S}$ and the mapping D. We assume that all the $D\mid_{A_i}$ mappings are injective, otherwise the preimage would obviously not be of finite cardinality. In this section, the reconstruction algorithm is described and its correctness is proved.

Algorithm 1. Single Set Reconstruction

Input : S_i, D, y
Output : X_i
$A_i \leftarrow aff(S_i)$
$\alpha_i = \textbf{Sample}(A_i)$
$w_i \leftarrow T^{-1}_{D(\alpha_i)}(y)$
$V_i \leftarrow T^{-1}_{\alpha_i}(A_i)$
if $w_i \in L(V_i)$ **then**
$\quad X_i \leftarrow T_{\alpha_i}\left(L\mid_{V_i}^{-1}(w_i)\right)$
$\quad$ **if** $X_i \subseteq S_i$ **then**
$\quad\quad$ **return** X_i
$\quad$ **end if**
end if
return $\emptyset$

Lemma 2. *Given that $D\mid_{A_i}$ is injective, Algorithm 1 returns $D\mid_{S_i}^{-1}(y)$.*

Proof. Let us assume that $y \notin D(S_i)$. In this case, $y \notin D\mid_{S_i}(S_i)$ is obviously true, which implies $D\mid_{S_i}^{-1}(y) = \emptyset$. It is easy to see that $y \notin D\mid_{A_i}(S_i)$ is also implied, which leads to $w_i \notin L\mid_{V_i}\left(T^{-1}_{\alpha_i}(S_i)\right)$. From the latter and the fact that $D\mid_{A_i}$ and $L\mid_{V_i}$ are injective, it follows that $T_{\alpha_i}\left(L\mid_{V_i}^{-1}(w_i)\right)$ cannot be included in S_i. Consequently, the algorithm returns the empty set when $y \notin D(S_i)$.

Now, we assume $y \in D(S_i)$ and prove that the algorithm returns $D\mid_{S_i}^{-1}(y)$. In this case, $w_i \in L\left(T^{-1}_{\alpha_i}(S_i)\right)$, where $T^{-1}_{\alpha_i}(S_i) \subseteq V_i$. As a consequence, the first condition is satisfied. Moreover, it is also true that $w_i \in L\mid_{V_i}\left(T^{-1}_{\alpha_i}(S_i)\right)$. As $L\mid_{V_i}$ is injective, $T_{\alpha_i}\left(L\mid_{V_i}^{-1}(w_i)\right)$ has a unique element and must be contained in S_i, which allows passing the second condition and returning X_i as a result. $\square$

Reconstruction for a single set, as of Algorithm 1, is performed as follows. First, the point y is translated by the vector that points to $D(\alpha_i)$. If the result w_i is an element of $L(V_i)$, then $L\mid_{V_i}(z_i) = w_i$, written as a system of linear equations, must be consistent. Furthermore, $D\mid_{A_i}$ is assumed to be injective, from which the injectivity of $L\mid_{V_i}$ follows. Therefore, the system of linear equations has a unique solution. After solving this equation, z_i has to be translated back to A_i. The result is accepted if it is in S_i. By iterating through the elements of $\mathcal{S}$, and merging all partial results, we get the preimage of $D(S)$. However, this is not the most efficient way of solving our problem.

4.2 Reconstruction of the Preimage

When we have an indexed set of d_i-balls bounding the elements of $\mathcal{S}$ and another indexed set of d_i'-balls bounding the images, it is possible to construct a graph to capture the structural traits of injectivity for a given $\mathcal{S}$ and D. This graph $G(\mathcal{S}, \mathcal{E})$, with the elements of $\mathcal{S}$ as nodes, has an edge in $\mathcal{E}(G)$ if and only if the conditions on the injectivity of $D\mid_{S_i \cup S_j}$, as declared in Theorem 2 are not satisfied.

Definition 1. *Let $\mathcal{S} := \{S_i : i \in I\}$ be an indexed set of bounded and convex sets, $\mathcal{B} := \{B_i : i \in I\}$ be an indexed set of relatively open d_i-balls and $\mathcal{B}' := \{B_i' : i \in I\}$ an indexed set of relatively open d_i'-balls, such that $S_i \subseteq B_i$, $D(S_i) \subseteq B_i'$, $aff(B_i) = A_i$ and $aff(B_i') = D(A_i)$ for any $i \in I$. Then, $G = (\mathcal{S}, \mathcal{E})$ is an undirected graph, where an edge exists between two nodes S_i, S_j if and only if the sufficient conditions on injectivity, as declared in Theorem 2, are not satisfied for B_i, B_j, B_i', B_j' with respect to the affine mapping D.*

Exploiting the additional information about injectivity, encoded in the graph defined above, can be practical in order to avoid iterating through all the sets S_i as we restore the preimage. For this purpose, we rely on the following lemma.

Lemma 3. *Consider a mapping D, a family of convex sets $\mathcal{S}$, where all $D\mid_{A_i}$ are injective. The preimage of any $D(S_i)$ with respect to $D\mid_S$ is always contained in S'. Here, S' is the union of all the sets S_j satisfying $(S_i, S_j) \in \mathcal{E}(G)$.*

Algorithm 2. Reconstruction

  Input : D, G, y
  Output : X
  for S_i in $\mathcal{S}$ **do**
      $X \leftarrow$ **Single Set Reconstruction** (S_i, D, y)
      if $X \neq \emptyset$ **then**
         for $(S_i, S_j) \in \mathcal{E}(G)$ **do**
            $X \leftarrow X \cup$ **Single Set Reconstruction** (S_j, D, y)
         end for
         break
      end if
  end for
  return X

Instead of iterating all the S_i in $\mathcal{S}$, Algorithm 2 seeks the first S_i for which $S_i \cap D\,|_S^{-1}(y)$ is not empty. After that, it traverses all the sets S_j that are adjacent to S_i in the graph $G(\mathcal{S}, \mathcal{E})$. Right after these final steps, the algorithm terminates.

Theorem 3. *Taking an affine mapping D, an indexed set of bounded convex sets $\mathcal{S}$, if all $D\,|_{A_i}$ are injective, then Algorithm 2 always returns $D\,|_S^{-1}(y)$ for any element of $D(S)$.*

Proof. As of Lemma 2, for a given S_i, the set X_i returned by Algorithm 1 equals $D\,|_{S_i}^{-1}(y)$. Furthermore, it must be included in S_i, and the premise that $D\,|_{A_i}$ is injective implies $|D\,|_{S_i}^{-1}(y)| = 1$. As of Lemma 3, another element x_j of the preimage can exist only in some S_j if (S_i, S_j) is in $\mathcal{E}(G)$. Therefore, it is sufficient to find the first set S_i for which Algorithm 1 returns a nonempty result, then traversing its neighbors provides the complete preimage. $\square$

Note that the graph is proposed to be constructed based on Theorem 2. Still, similar constructions that consider Corollary 1 or even Theorem 1 are also possible and might turn out to be valuable tools to characterize the injectivity of D, when restricted to subsets of $\mathcal{S}$.

5   Numerical Experiments

In order to illustrate how our findings could be beneficial in answering some interesting questions, a series of numerical experiments have been carried out. First, we performed a series of exploratory studies that demonstrate how the relative distance and dimensionality of convex sets in a given collection influence the connectedness of $G(\mathcal{S}, \mathcal{E})$. Furthermore, a rather simple and easy to follow example will be provided, which sheds light to a new way of treating the selection of projection angles in CT, assuming that there is prior information needed to be taken into account while selecting the angles.

5.1   Gaussian Random Projections over Collections of Simplices

We studied an arrangement $\mathcal{S}$ of simplices in an N-dimensional Euclidean space, which was mapped by Gaussian Random Projections [3], for varying number of simplices K and dimensions of the affine hull of the image $dim(D(S))$, where the latter is denoted by M. Each simplex S_i was assigned the dimensionality M. Then, all simplices were individually transformed by a random orthogonal transform, taken from the $O(N)$ Haar distribution [8]. Finally, a unique translation was applied, where the translation vector has components drawn from a uniform random distribution, scaled by a constant factor T.

In Fig. 1, the results of the experiment described above can be seen. For some selected values of T, numerical values are also presented in Table 1. Notice that the average proportion of isolated nodes, initially increases sharply as the translation scale factor T is increasing. However, for a given T, each curve has

Table 1. The cardinality of the isolated nodes in $G(\mathcal{S}, \mathcal{E})$, divided by the number of all nodes is averaged in each case, as Gaussian Random Projections are applied to simplices in an N-dimensional space, where $N = 7$. The corresponding variance can also be seen. Numerical values are provided for two different number of simplices K and three projection dimensions M as a function of the translation factor T.

T	$M = 1$				$M = 3$				$M = 5$			
	K = 5		K = 10		K = 5		K = 10		K = 5		K = 10	
	avg	var	avg	var	avg	var	avg	var	avg	var	avg	var
1.0	0.0685	0.0148	0.0685	0.0148	0.0015	0.0003	0.0015	0.0003	0.0005	0.0001	0.0005	0.0001
2.0	0.3165	0.0749	0.3165	0.0749	0.1005	0.0232	0.1005	0.0232	0.0700	0.0207	0.0700	0.0207
3.0	0.4240	0.0778	0.4240	0.0778	0.3355	0.0677	0.3355	0.0677	0.3180	0.0651	0.3180	0.0651
4.0	0.5315	0.0832	0.5315	0.0832	0.5619	0.0934	0.5619	0.0934	0.5545	0.0886	0.5545	0.0886
5.0	0.6329	0.0781	0.6329	0.0781	0.6945	0.0762	0.6945	0.0762	0.7735	0.0676	0.7735	0.0676
6.0	0.6505	0.0805	0.6505	0.0805	0.8159	0.0555	0.8159	0.0555	0.9015	0.0356	0.9015	0.0356
7.0	0.7049	0.0728	0.7049	0.0728	0.8939	0.0384	0.8939	0.0384	0.9175	0.0291	0.9175	0.0291
8.0	0.7245	0.0682	0.7245	0.0682	0.9149	0.0296	0.9149	0.0295	0.9635	0.0142	0.9635	0.0141
9.0	0.7555	0.0713	0.7555	0.0713	0.9275	0.0292	0.9275	0.0292	0.9655	0.0129	0.9655	0.0129

an inflection. The location of the inflection varies by M, the dimension of the subspace in which the sets S_i are projected. For a smaller M, inflection occurs earlier, while convergence is significantly slower. We also investigated how the number of simplices K affects the shape of the curves. As expected, higher K values generally result in slower convergence.

5.2 Projection Angle Selection and Search Space Reduction

If we take a look at Fig. 2, a subset of 3×3 images can be seen, where the support size is at most two. It is known initially that the search space can be reduced to images with supports seen in Fig. 2. Moreover, it is also available as prior information that for a given support, pixel intensities are only allowed to take values in a given range. These pieces of information are represented as an indexed set of hypercubes $\mathcal{S}$. We assume that another indexed set, containing projections of different angles is also provided. Generally, each S_i cannot be mapped injectively by any of the linear mappings in $\mathcal{D} := \{D_i : i \in \{0, 1, 2\}\}$, where D_0, D_1, D_2 are the projections by angles 0, $\pi/4$ and $\pi/2$. In the current setting, projection rays are modeled by equidistant parallel lines. There are three rays for a given rotation angle. In case of D_0, these are parallel to the vertical axis and are crossing the centers of pixels that lie in the corresponding columns. The rotation center is defined as the top-left corner of the central pixel, while rotation is performed clockwise.

As we can see in Fig. 3, when prior information about the allowed support is available, while intensity ranges - in our case defined as $[0.01, 1.0]$ - are identical,

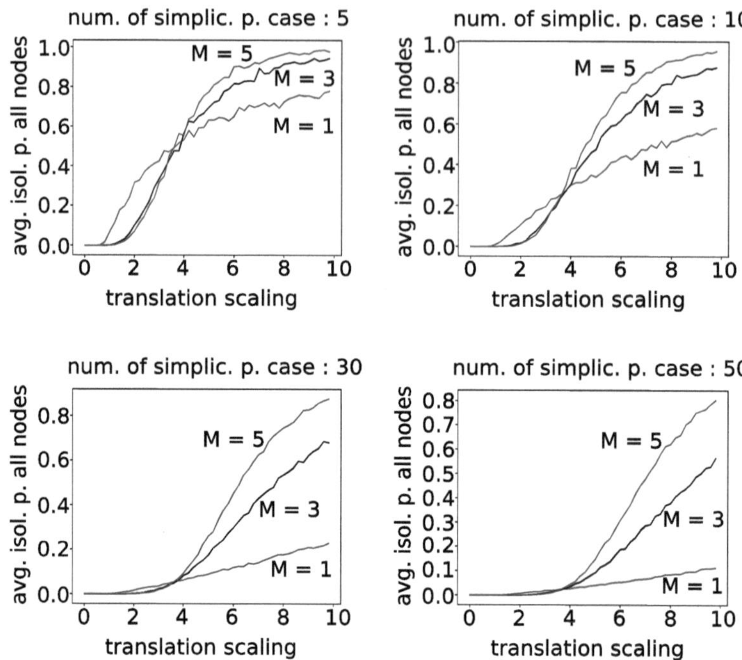

Fig. 1. The cardinality of the isolated nodes in $G(\mathcal{S}, \mathcal{E})$, divided by the number of all nodes is averaged in each case, as Gaussian Random Projections are applied to simplices in an N-dimensional space, where $N = 7$. In each case, 400 independent arrangements were averaged. By increasing the number of simplices or the projection dimension, inflection and convergence is delayed.

Algorithm 3. Projection Ordering

Input : $\mathcal{D}, \mathcal{S}, \Sigma$
$s_{min}, \sigma_{min} \leftarrow |\mathcal{D}|, \sigma_0$
for $\sigma \in \Sigma$ **do**
    $s, D_{sum} \leftarrow 0, \mathbf{0}$
    for $i \in I_{\mathcal{D}}$ **do**
        $D_{sum} \leftarrow D_{sum} + D_{\sigma(i)}$
        $G \leftarrow \mathbf{Construct}\,(D_{sum}, \mathcal{S})$
        $\mathcal{S}_{iso} \leftarrow \mathbf{Isolated}\,(G)$
        if $|\mathcal{S}_{iso}| = |\mathcal{S}|$ **then**
            $s \leftarrow i$
            break
        end if
    end for
    if $s < s_{min}$ **then**
        $s_{min}, \sigma_{min} \leftarrow s, \sigma$
    end if
end for
return s_{min}, σ_{min}

Fig. 2. Example images from all hypercubes S_i, indexed by $I := \{0, ..., 7\}$, from left to right, in row-major order. Notice that the isolated, four-connected and eight-connected groups have different intensity ranges and supports are restricted to only a few combinations of two pixels. Four-connected pixels tend to be brighter, intensity values are within $[0.7, 0.8]$. Eight-connected pixels are of medium intensity $[0.5, 0.6]$, while isolated pixels are the darkest $[0.3, 0.4]$.

a single projection is not sufficient to guarantee that the conditions on injectivity are satisfied. However, any two-element combination of angles that contains $\pi/4$ would lead to a projection D that is guaranteed to satisfy injectivity over the subdomain S. In case additional information about intensity ranges – see Fig. 2 – is introduced, a single projection angle $\pi/4$ is sufficient, as seen in Fig. 4. Here, we discuss an exhaustive search strategy that finds the shortest sequence of projections, when prior information is given as a collection of convex sets.

Algorithm 3 gets an indexed set of bounded convex sets $\mathcal{S}$, representing our initial knowledge that enables search space reduction, and another indexed set of allowed projections $\mathcal{D}$. Additionally, the set Σ of all possible permutations over the set of indices in $\mathcal{D}$ is also provided as input. We assume that the mapping $D = \sum_{i=0}^{i=|\mathcal{D}|} D_i$ and $\mathcal{S}$ satisfy sufficient conditions on the injectivity of $D\mid_S$. The outer loop iterates all the possible permutations σ, which we will use in the inner loop to iterate the elements of $\mathcal{D}$ in the order given by the current permutation. In each step of the inner loop, the current projection $D_{\sigma(i)}$ is added to D_{sum}. After that, we construct a graph, taking D_{sum} and the set of nodes $\mathcal{S}$. It is easy to see that whenever all the nodes in this graph are isolated, the set S is mapped injectively by D_{sum}, when D_{sum} is restricted to S. Therefore, the number of projections – denoted by s – needed to be applied to guarantee exact recovery equals the number of steps needed to be taken in the inner loop to construct a graph with all isolated nodes. The best score s_{min} and the corresponding permutation σ_{min} is returned. Given σ_{min}, s_{min} and $\mathcal{D}$, the projection with a minimal number of angles can be simply constructed.

In the following theorem, the correctness of Algorithm 3 is stated. Note that we refer to Theorem 2, while it would also be possible to replace these conditions by the ones stated in Corollary 1. It is easy to see that replacing Theorem 2 by Corollary 1 simultaneously in the theorem and the proof keeps them being consistent.

Theorem 4. *Let $\mathcal{S}$ be an indexed set of bounded convex sets $S_i \subseteq E$ and $\mathcal{D}$ be an indexed set of projections $D_i : E \to E$. Additionally, let $\sum_{i=0}^{i=|\mathcal{D}|} D_i$ and $\mathcal{S}$ satisfy the conditions declared in Theorem 2. Algorithm 3 returns an integer s and a permutation σ, where s is minimal and $\sum_{i=0}^{i=s} D_{\sigma(i)}$ and $\mathcal{S}$ satisfies the conditions declared in Theorem 2.*

Proof. First, it will be shown that the result returned is always feasible. The algorithm constructs G, given $\mathcal{S}$ and the sum of some elements in $\mathcal{D}$, based on Theorem 2. Therefore, any combination of D_{sum} and $\mathcal{S}$ leads to a graph with all isolated nodes if and only if D_{sum} restricted to $\mathcal{S}$ is guaranteed to be injective as of Theorem 2. As $\sum_{i=0}^{i=|\mathcal{D}|} D_i$ and $\mathcal{S}$ is assumed to satisfy the conditions given in Theorem 2, G must consist of all isolated nodes in at most $|\mathcal{D}|$ steps for each permutation.

Now, we prove that the number of projections needed is minimal. Let us assume that there exists a permutation σ such that $D, \mathcal{S}$, where $D = \sum_{i=0}^{i=s} D_{\sigma(i)}$, satisfies the conditions given in Theorem 2, while s is smaller than s_{min}. As no permutations are skipped, $D, \mathcal{S}$ must lead to a graph G with all isolated nodes. In this case, the result returned must be smaller or equal to s_{min}, which is a contradiction. $\qquad\square$

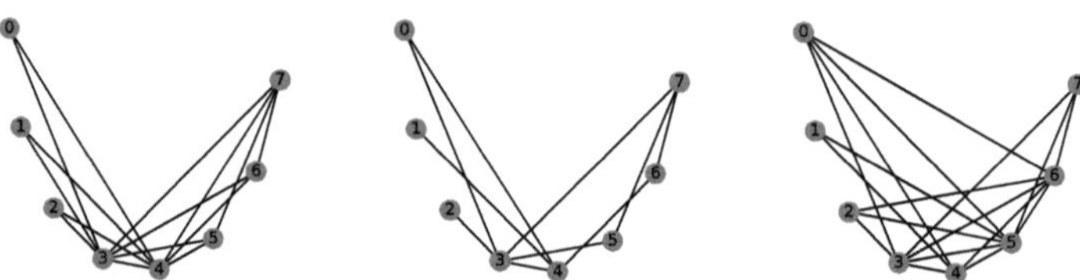

Fig. 3. Graphs constructed considering the conditions on injectivity as of Theorem 2, with prior information about the allowed support, given that intensity ranges are identical, defined as $[0.01, 1]$. A single projection angle 0 (left), $\pi/4$ (middle) or $\pi/2$ (right) is insufficient to guarantee that the conditions on injectivity are satisfied, as the corresponding graphs have edges.

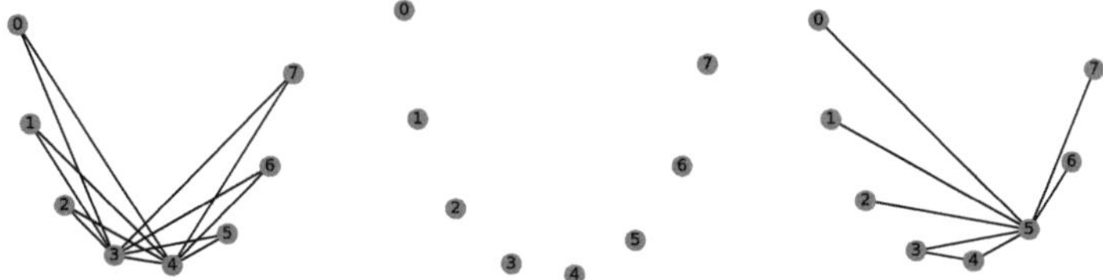

Fig. 4. Graphs constructed considering the conditions on injectivity as of Theorem 2, with prior information about the allowed support and additional, more specific information about disjunct intensity ranges – see Fig. 2. Here, a single projection angle $\pi/4$ is sufficient.

Based on our current results, it is only feasible to handle small objects, due to the fact that the algorithm is an exhaustive search over all possible permutations of $\mathcal{D}$ and some other restrictive premises made in the proposed theorems. Still, it is a merit of the algorithms we presented that they are exact, and can be considered as a valuable starting point to analyze possible relaxations of the discussed subproblems, which in the long term might lead to practical applications.

6 Summary

An algorithm was proposed to reconstruct the preimage of an element in the image of an affine mapping, and the conditions under which restoration of the preimage is possible were presented. We also provided a possible relaxation to the original problem, as the latter involves computing the intersection of convex sets, which might be practically infeasible in some cases.

Empirical studies were carried out to explore the potential of our method and identify future research directions. We found that the average proportion of isolated nodes grows non-linearly as translation magnitude is increasing, has an inflection at a well-defined location, then convergence rate depends on the dimensionality of the affine hull of the image and the number of the sets consisting the union to which the analyzed mapping is restricted. We observed a strong dependency between the projection dimension and the location of the inflection.

As an application-related example, it was shown how our method could be used to take into account prior information in projection angle selection, while seeking the inversion of the discrete Radon-transform. The research direction that would open up by the application of our method in this field could be a sequence of steps taken towards building an unsupervised machine learning pipeline that identifies the collection of bounding sets, for which an approximately smallest set of projection angles could be extracted.

Acknowledgments. This research was supported by Project no. TKP2021-NVA-09. Project no. TKP2021-NVA-09 has been implemented with the support provided by the Ministry of Innovation and Technology of Hungary from the National Research, Development and Innovation Fund, financed under the TKP2021-NVA funding scheme.

References

1. Aleksandrov, V.: On the fundamental gale-nikaido-inada theorem on the injectivity of mappings. Sibirsk. Mat. Zh. **35**, 715–718 (1994)
2. Ascolese, M., Dulio, P., Pagani, S.M.: Uniqueness and reconstruction of finite lattice sets from their line sums. Discret. Appl. Math. **356**, 293–306 (2024)
3. Bingham, E., Mannila, H.: Random projection in dimensionality reduction: applications to image and text data. In: Proceedings Of The Seventh ACM SIGKDD International Conference On Knowledge Discovery And Data Mining, pp. 245–250 (2001)
4. Brunetti, S., Dulio, P., Peri, C.: Uniqueness results for grey scale digital images. Fund. Inform. **172**(2), 221–238 (2020)
5. Gale, D., Nikaido, H.: The jacobian matrix and global univalence of mappings. Math. Ann. **159**(2), 81–93 (1965)
6. Jacobs, F., Sundermann, E., De Sutter, B., Christiaens, M., Lemahieu, I.: A fast algorithm to calculate the exact radiological path through a pixel or voxel space. J. Comput. Inf. Technol. **6**(1), 89–94 (1998)
7. Luks, E.M., Rákóczi, F., Wright, C.R.: Some algorithms for nilpotent permutation groups. J. Symb. Comput. **23**(4), 335–354 (1997)
8. Mezzadri, F.: How to generate random matrices from the classical compact groups. Not. Am. Math. Soc. **54**(5), 592–604 (2007)
9. Munkres, J.: Topology james munkres second edition. Pearson Eductaion Limited (2014)
10. Nagy, A., Kuba, A.: Reconstruction of binary matrices from fan-beam projections. Acta Cybernet. **17**(2), 359–383 (2005)
11. Varga, L., Balázs, P., Nagy, A.: Projection Selection Algorithms for Discrete Tomography. In: Blanc-Talon, J., Bone, D., Philips, W., Popescu, D., Scheunders, P. (eds.) ACIVS 2010. LNCS, vol. 6474, pp. 390–401. Springer, Heidelberg (2010). https://doi.org/10.1007/978-3-642-17688-3_37

Fully Parallel 3D Curve-Thinning on $(18, 12)$ Pictures of the FCC Grid

Noel Nagy and Kálmán Palágyi$^{(\boxtimes)}$

Department of Image Processing and Computer Graphics, University of Szeged,
Szeged, Hungary
{nagynoel,palagyi}@inf.u-szeged.hu

Abstract. Curve-thinning is an iterative reduction to extract center-lines from binary/segmented objects. Parallel thinning algorithms are composed of parallel reductions, and the fully parallel approach uses the same parallel reduction in each thinning phase. This paper presents the first fully parallel 3D curve-thinning algorithm acting on the unconventional face-centered cubic (FCC) grid. Our algorithm combines a sufficient condition for topology-preserving parallel reductions acting on $(18, 12)$ pictures of the FCC grid with curve-endpoint preservation, hence its topological correctness is guaranteed for all possible pictures. In order to reduce the count of unwanted side branches in the produced center-lines, a novel iteration-level endpoint re-checking process is proposed.

Keywords: Digital topology · Skeletonization · Curve-thinning ·
Face-centered cubic (FCC) grid

1 Introduction

Skeletonization plays a key role in a broad range of problems in image processing and computer vision [19]. It means extraction of *skeletal features* from (segmented) binary objects. The most important and most frequently used skeletal feature is the *centerline* that is a line-like 1D representation of both 2D and 3D objects [6].

Curve-thinning is an iterative process for producing centerlines: the outermost layer of an object is deleted, and the entire process is repeated until stability is reached (i.e., no further points are deleted) [13,14]. *Reduction* operators never change a *white* point in a *binary picture* to a *black* one, and a *parallel reduction* changes (or *deletes*) certain black points to white simultaneously [10]. A curve-thinning algorithm is said to be *fully parallel* if it applies the same parallel reduction at every iteration [10].

3D digital images are usually sampled on the regular *cubic grid* that is dual to $\mathbb{Z}^3$. In this article, however, our focus is on an unconventional 3D sampling scheme, the *face-centered cubic (FCC) grid* is taken into consideration [14]. The point $(x, y, z) \in \mathbb{Z}^3$ is an element of the FCC grid if

$$(x + y + z) \bmod 2 = 0.$$

P. Balázs et al. (Eds.): IWCIA 2025, LNCS 15985, pp. 65–82, 2026.
https://doi.org/10.1007/978-3-032-19347-6_5

Gau and Kong pointed out some advantageous properties of the FCC grid over the cubic grid [9], and Edelsbrunner et al. showed that the FCC grid provides the densest sphere packing [7]. Due to its favorable geometrical and topological properties, the importance of the FCC grid shows an upward tendency [1–5,8, 15,16,18,20,21].

This paper presents the first fully parallel 3D curve-thinning algorithm acting on the FCC grid. Our algorithm fulfills two important requirements: topology preservation and maximality.

The rest of this paper is organized as follows: Sect. 2 gives an outline of the concepts and existing results that we will need. Then in Sect. 3 we propose our curve-thinning algorithm. In Sect. 4, the two crucial properties of our algorithm are discussed. Section 5 is dedicated to the computationally efficient implementation of the proposed algorithm. In Sect. 6, results on some test pictures produced by our algorithm are presented. Finally, we round off the paper with some concluding remarks.

2 Basic Notions and Results

Here, we recall the key concepts of digital topology that apply specifically to the FCC grid [9,13,14], and review the existing results that the proposed thinning algorithm relies on [12].

Let us denote $\mathbb{F} \subset \mathbb{Z}^3$ the set of points in the FCC grid. An (18,12) picture on $\mathbb{F}$ is a quadruple $(\mathbb{F}, 18, 12, B)$, where $B \subseteq \mathbb{F}$ denotes the set of *black points*; each point in $\mathbb{F} \setminus B$ is a *white point*; *18-adjacency* relation is assigned to B; and *12-adjacency* belongs to $\mathbb{F} \setminus B$, see Fig. 1.

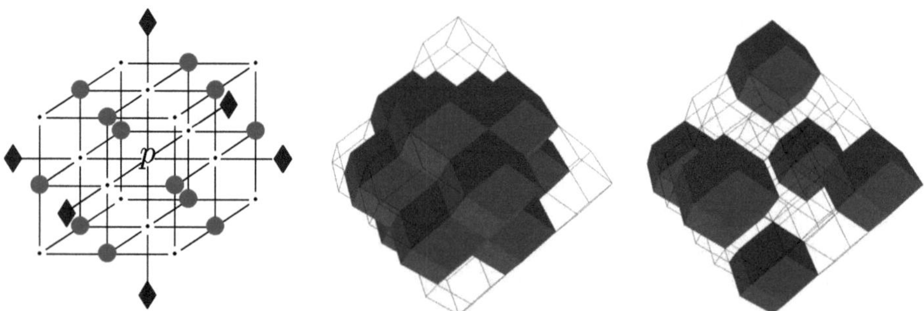

Fig. 1. The twelve points marked '$\bullet$' are 12-adjacent to the central point p, and form the set $N_{12}(p)$; the six further points marked '$\blacklozenge$' and the twelve points in $N_{12}(p)$ form the set $N_{18}(p)$, and they are 18-adjacent to p (left). (Note that points marked '$\cdot$' are not in $\mathbb{F}$.) The voxel representations associated with $N_{12}(p)$ (middle) and $N_{18}(p) \setminus N_{12}(p)$ (right) in which each voxel is a rhombic dodecahedron (i.e., a convex polyhedron with 12 congruent rhombic faces).

It can be readily seen that both of the adjacency relations are symmetric, thus their reflexive-transitive closure induces two equivalence relations on B

and $\mathbb{F} \setminus B$, and the generated equivalence classes are, respectively, called *black 18-components* and *white 12-components*. For brevity, we will refer to a black 18-component as an *object* in the rest of this paper. For practical reasons, we assume that B contains finitely many points, therefore there is a unique infinite white 12-component which is said to be the *background*, and a finite white 12-component is called a *cavity*.

A black point $p \in B$ is a *border point* in $(18, 12)$ pictures if there is at least one white point that is 12-adjacent to p (see Figs. 2 *a-b* and 3 *a-b*), it is an *interior point* if it is not a border point (see Fig. 3 *c*), it is said to be *isolated* if all the eighteen points in $N_{18}(p)$ are white (see Fig. 3 *a*), and it is an *endpoint* if $N_{18}(p)$ contains exactly one black point (see Fig. 2 *a*).

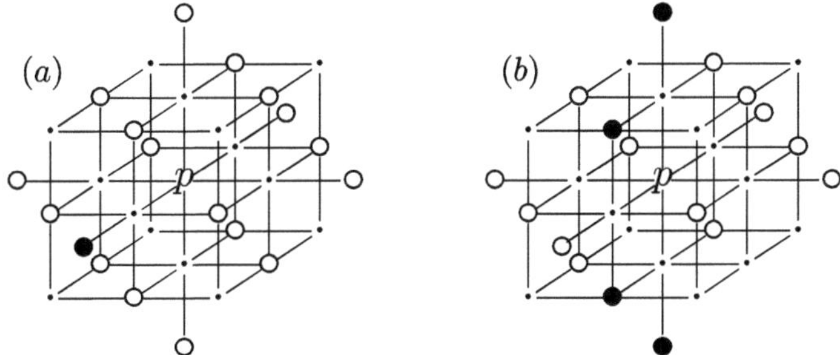

Fig. 2. Examples of simple points. It is easy to check that the central black point p is simple in both cases, since both conditions of Theorem 1 hold. Note that p is an endpoint in configuration (a).

Topology preservation is a crucial concern in thinning algorithms. It is necessary to guarantee that all parallel reductions of the thinning process preserve the topology for all possible pictures. A parallel reduction acting on 2D pictures is *topology-preserving* if it never splits an object into two or more, never deletes completely any object, never merges a cavity with the background or another cavity, and never creates a cavity [13]. Since 3D objects may contain *holes* or *tunnels* (like solid tori), a topology-preserving reduction in 3D neither creates nor eliminates a hole, and it does not merge any holes with each other [13].

Conditions for topology preservation were obtained with the help of the concept a *simple point* [13,14]. A single black point is simple if its deletion preserves the topology. In [9], Gau and Kong characterized simple points in $(18, 12)$ pictures on the FCC grid:

Theorem 1. (Proposition 4.3 of [9]) *Black point $p \in B$ in picture $(\mathbb{F}, 18, 12, B)$ is simple if and only if the both of the following conditions hold:*

1. *$N_{18}(p) \cap B$ forms exactly one 18-component.*
2. *$N_{12}(p) \setminus B$ forms exactly one 12-component.*

Figures 2 and 3, respectively, give some illustrative examples of simple and non-simple points. Notations: points marked '•' and 'o', respectively, refer to black and white points, and points denoted by '·' are not in $\mathbb{F}$.

We state three easy consequences of Theorem 1:

Lemma 1. *The simpleness for a point p depends only on the eighteen points in $N_{18}(p)$.*

Lemma 2. *Only non-isolated border points may be simple.*

Lemma 3. *Endpoints are simple points.*

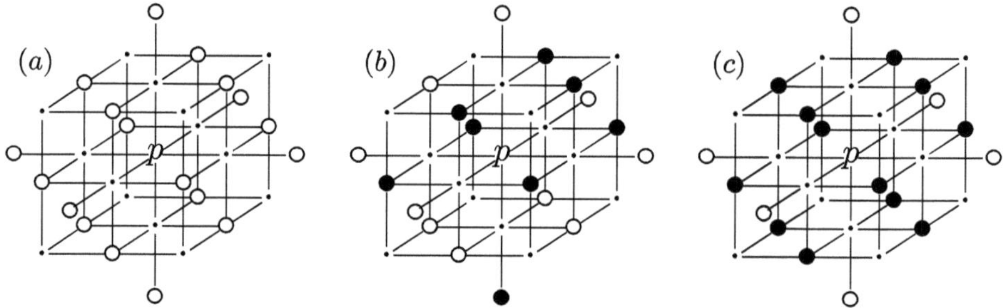

Fig. 3. Examples of non-simple black points. Condition 1 of Theorem 1 is violated in (a), in which the central point p is an isolated point; both conditions of Theorem 1 are violated in (b); and condition 2 of Theorem 1 is violated in (c), in which p is an interior point.

Since a parallel reduction can delete not just a single simple point, we need to guarantee topology preservation when a number of black points are deleted simultaneously. In [12], Karai et al. gave an *asymmetric point-based sufficient condition* for topology-preserving parallel reductions for $(18, 12)$ pictures on the FCC grid. That condition is called *asymmetric* because it takes into account a strict ordering relation, and it is said to be *point-based* since it examines the deletability of individual points. Point-based conditions can serve a methodology for designing topology-preserving parallel reductions and parallel thinning algorithms as well. The proposed curve-thinning algorithm (see Sect. 3) is based on the the above mentioned condition. That asymmetric point-based sufficient condition (see Theorem 2) uses the lexicographical order '$\prec$' between two distinct points $p = (p_x, p_y, p_z)$ and $q = (q_x, q_y, q_z)$ in $\mathbb{Z}^3$:

$$p \prec q \Leftrightarrow (p_z < q_z) \vee (p_z = q_z \wedge p_y < q_y) \vee$$
$$(p_z = q_z \wedge p_y = q_y \wedge p_x < q_x)$$

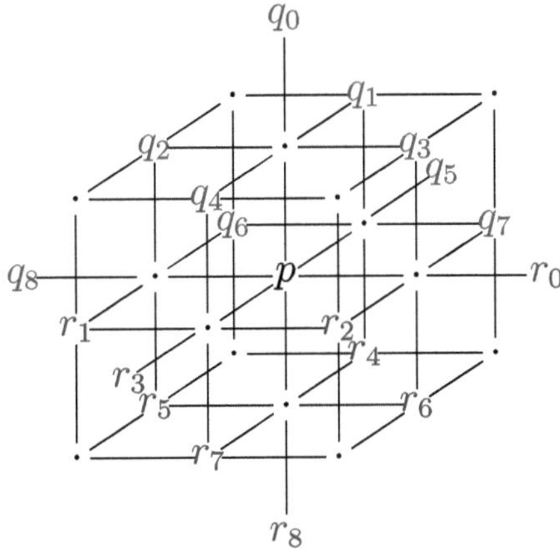

Fig. 4. The employed lexicographical order '$\prec$' between the central point $p \in \mathbb{F}$ and the elements in $N_{18}(p)$: $q_i \prec p$ and $p \prec r_i$ ($i \in \{0, 1, \ldots, 8\}$).

Figure 4 is to illustrate the employed relation '$\prec$' for the FCC grid. Before presenting the applied condition (see Theorem 2), we must also define the concept of the smallest element of a finite set of points in $\mathbb{Z}^3$: point p is the *smallest element* of the set of n points $\{p, s_1, \ldots, s_{n-1}\}$ if $p \prec s_j$ ($n = 2, 3, \ldots$; $j = 1, \ldots, n-1$).

Theorem 2. (Theorem 5.4 of [12]) *A parallel reduction preserves the topology for an arbitrary picture* $(\mathbb{F}, 18, 12, B)$ *if each point* $p \in B$ *deleted by that reduction satisfies all of the following conditions:*

1. *Point p is simple for B.*
2. *Let q be a point in B such that $q \in N_{12}(p)$, q is simple for B. Then p is simple for $B \setminus \{q\}$, or $q \prec p$.*
3. *Let q and r be two points in B such that p, q, and r are mutually 12-adjacent, both points q and r are simple for B, q is simple for $B \setminus \{r\}$. (Note that in this case r is simple for $B \setminus \{q\}$.) Then p is simple for $B \setminus \{q, r\}$, or p is not the smallest element of $\{p, q, r\}$.*
4. *Point p is not the smallest element of an object consisting of four mutually 12-adjacent points, see Fig. 5.*
5. *Point p is not the smallest element consisting of an object of mutually 18-adjacent points in which the elements are not mutually 12-adjacent (i.e., in such an object, there are two points that are not 12-adjacent to each other), see Fig. 6.*

Fig. 5. The two possible sets of four mutually 12-adjacent points.

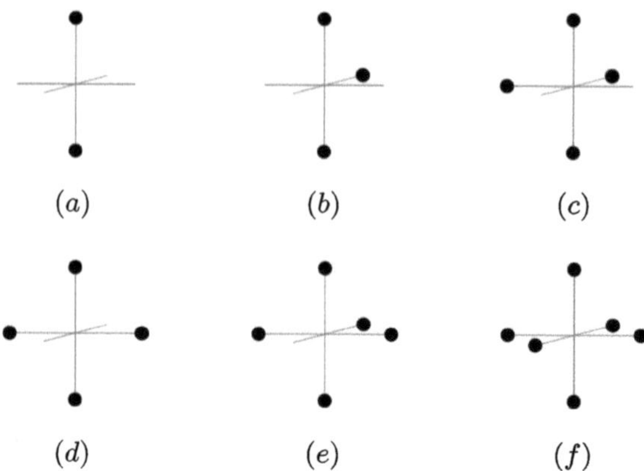

Fig. 6. Six of the 37 possible sets consisting of mutually 18-adjacent points in which there are two points that are not 12-adjacent to each other. The remaining 31 cases are reflected and rotated versions on these six sets. Note that such sets can consist of at most six points, and only contain two elements in case (a).

3   The Proposed Curve-Thinning Algorithm

In this section, we present our fully parallel 3D curve-thinning algorithm acting on $(18, 12)$ pictures of the $\mathbb{F}$ grid, see Algorithm 1.

Algorithm 1: FP(k) ($k \in \{1, 2, \ldots, 17\}$)

---

1 **Input**: picture $(\mathbb{F}, 18, 12, X)$
2 **Output**: picture $(\mathbb{F}, 18, 12, Y)$
3 $Y \leftarrow X$
4 **repeat**
5 　　// Phase 1 - condition-oriented parallel reduction
6 　　$D1 \leftarrow \{\, p \mid p$ is not an *endpoint* and it is *condition-deletable* for $Y \,\}$
7 　　$Z \leftarrow Y \setminus D1$
8 　　// Phase 2 - endpoint-related parallel reduction
9 　　$D2 \leftarrow \{\, p \mid p$ is *end-deletable* with parameter k for $(Y, Z)\}$
10 　　$Y \leftarrow Z \setminus D2$
11 **until** $D1 = \emptyset$;

The proposed algorithm **FP**(k) ($k \in \{1, 2, \ldots, 17\}$) uses two new concepts to be defined (see lines 6 and 9 in Algorithm 1):

Definition 1. *A black point is said to be* condition-deletable *for a picture if it satisfies all conditions of Theorem 2 (i.e., it fulfills the asymmetric point-based sufficient condition considered).*

Definition 2. *Let Y and $Z \subset Y$ be two sets of (black) points. Then point $p \in Z$ is said to be* end-deletable *for the pair of sets (Y, Z) if the following conditions all hold:*

1. *$N_{18}(p) \cap Z$ is a singleton set.*
2. *$N_{18}(p) \cap Y$ contains at least $k+1$ elements. (Recall that k is the only parameter of our algorithm.)*
3. *If $N_{18}(p) \cap Z = \{q\}$ and $N_{18}(q) \cap Z = \{p\}$, $q \prec p$.*

Line 3 in Algorithm 1 initializes the set of black points Y in the output picture that contains the produced centerline. The kernel of the **repeat** cycle (see lines 5–10) corresponds to one iteration step that is composed of two parallel reductions:

- The first reduction deletes all *condition-deletable points* that are not endpoints at a time from the actual picture $(\mathbb{F}, 18, 12, Y)$, see lines 6 and 7.
- The second reduction deletes simultaneously all *end-deletable points* from the picture $(\mathbb{F}, 18, 12, Z)$ (i.e., the picture resulted by the first phase of the actual iteration step), see lines 9 and 10. Notice that this second phase also depends on picture $(\mathbb{F}, 18, 12, Y)$ (i.e., the input picture of the actual iteration) and the parameter k (i.e., the minimal number of deleted points from $N_{18}(p)$).

The operation of the condition-oriented reduction (see lines 6 and 7) is easy to understand, but the deletion rule of the subsequent endpoint-related phase (see lines 6 and 7) needs explanation:

- Condition 1 of Definition 2 means that p is an endpoint in the picture resulted by the condition-oriented reduction.
- Condition 2 requires that at least k points in $N_{18}(p)$ were deleted in the first phase of the current iteration step.
- Condition 3 is crucial to ensure the topological correctness of the endpoint-related reduction. The smallest element of two-element objects (i.e. objects formed by two endpoints) cannot be *end-deletable*. Consequently, such objects cannot be deleted completely.

It can be readily seen that if $D1 = \emptyset$ (see line 6), then $D2 = \emptyset$ (see line 9). That is why we could write $D1 = \emptyset$ in line 11 (instead of $D1 \cup D2 = \emptyset$). The proposed algorithm terminates if no points are deleted in the current iteration (see line 11). Since all considered input pictures contain finitely many black points, algorithm $\mathbf{FP}(k)$ always terminates.

The biggest challenge in designing topology-preserving curve-thinning algorithms is dealing with the endpoints arose during the iterative process. If we always delete them, we will not produce centerlines, but rather solve the problem of *shrinking to a topological equivalent* [11]. If we preserve all the emerging endpoints, then (due to the constraint of topology preservation) new branches will necessarily grow from them on the centerlines. That is why, algorithm $\mathbf{FP}(k)$ uses a novel iteration-level endpoint re-checking process through the endpoint-related reduction. It is to reduce the count of parasitic side branches in the produced centerlines. The effect of the endpoint re-checking can be controlled by choosing the parameter k of our algorithm. It can be readily seen that $k = 1$ leads to the centerline with the fewest branches, while for $k = 17$, no endpoint is deleted by the endpoint reduction, as illustrated in Fig. 7.

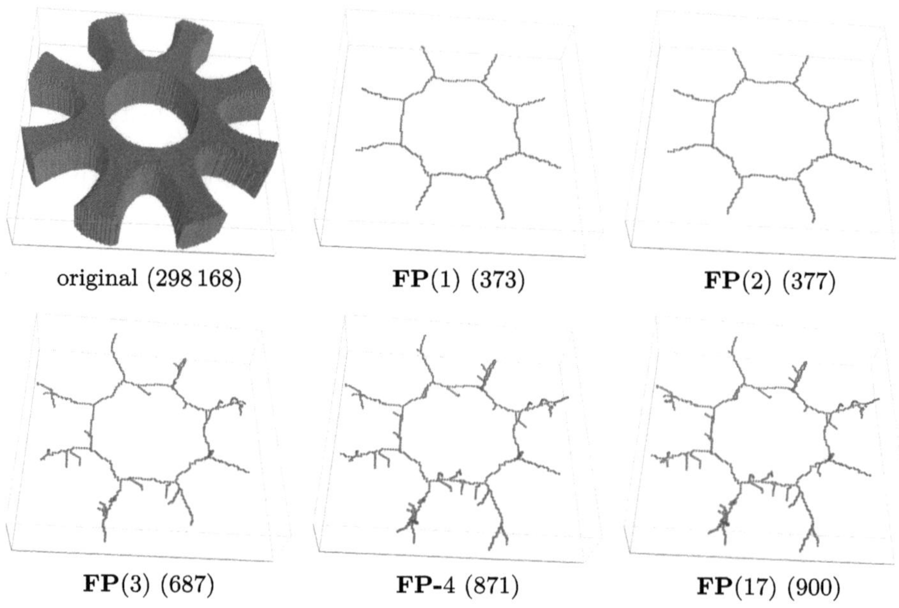

original (298 168)	**FP**(1) (373)	**FP**(2) (377)
FP(3) (687)	**FP**-4 (871)	**FP**(17) (900)

Fig. 7. A $45 \times 191 \times 191$ image of a gear (a) and its centerlines produced by the proposed algorithm **FP**(k) for $k = 1, 2, 3, 4, 17$ (b)-(f). Note that for $k = 17$, the endpoint re-checking process is absolutely ineffective. Numbers of parenthesis are the counts of black points.

4 Topology Preservation and Maximality

Topology preservation is a key issue in skeletonization approaches, including thinning. Thus, there is a fairly general agreement that a thinning algorithm is absolutely useless if it cannot preserve the topology for all possible pictures. Maximality is another important property of (2D and 3D) curve-thinning: an algorithm is said to be *maximal* if the centerlines it produces contain only non-simple points and (simple) endpoints (i.e., it is capable of producing one point thin centerlines for all possible pictures). We now show that the proposed algorithm meets both requirements.

Let us first state some easily understood properties of the set of points in conditions 3, 4, and 5 of Theorem 2, which we will need in the proofs of Theorems 3 and 4.

Proposition 1. *Consider a set of points consisting of three mutually 12-adjacent points (see condition 3 of Theorem 2). Then no element of that set is an endpoint.*

Proposition 2. *Consider the two possible objects consisting of four mutually 12-adjacent points (see Fig. 5 and condition 4 of Theorem 2). Then each point of these objects is simple and a non-endpoint.*

Proposition 3. *Consider the 37 possible objects of mutually 18-adjacent points in which the elements are not mutually 12-adjacent (see Fig. 6 and condition 5 of Theorem 2). Then the following are true:*

- *Each point of these objects is simple.*
- *All elements of objects consisting of more than two points are not endpoints (see Fig. 6b-f).*

We are now ready to state and prove our key theorems:

Theorem 3. *Algorithm* **FP**(k) *is topology-preserving.*

Proof. The following two cases need to be considered:

- The condition-oriented parallel reduction (see lines 6 and 7 in Algorithm 1) is topology-preserving.
- The endpoint-related parallel reduction (see lines 9 and 10 in Algorithm 1) is also topology-preserving.

Since the deletion rule (see Definition 1) of the condition-oriented parallel reduction is derived from Theorem 2, it is self-evident that the condition-oriented reduction preserves the topology. Therefore, we can turn to the examination of the second case.

Let us assume that a point p is *end-deletable* (see Definition 2). Then we need to verify whether all the conditions of Theorem 2 are satisfied.

- Since only endpoints may be *end-deletable* (see condition 1 of Definition 2) and endpoints are simple by Lemma 3, p is simple and condition 1 of Theorem 2 holds.
- Let q be the only black point being 12-adjacent to the endpoint p.
  - If q is also an endpoint and $q \prec p$, point q is not *end-deletable* by condition 3 of Definition 2. Therefore, we do not need to examine the simplicity of p after the deletion of q.
  - If q is also an endpoint and $p \prec q$, deletion of p violates condition 3 of Definition 2(i.e., p is not an *end-deletable* point), and we arrive at a contradiction.
  - If q is not an endpoint, q is not *end-deletable* by condition 1 of Definition 2. Therefore, we do not need to examine the simplicity of p after deleting q.

Thus condition 2 of Theorem 2 is satisfied.

- By Propositions 1 and 2, if p is an element of a set of three or four mutually 12-adjacent black points, p is not and endpoint. Then condition 1 of Definition 2 is violated, and we arrive at a contradiction. Thus conditions 3 and 4 of Theorem 2 hold.
- Let us assume that p is an element of an object $\mathcal{O}$ of mutually 18-adjacent points in which the elements are not mutually 12-adjacent.

- If $\mathcal{O} = \{p, q\}$ contains two points (see Fig. 6 a), both points p and q are endpoints.

  * If $p \prec q$, condition 3 of Definition 2 is violated (i.e., p is not an *end-deletable* point), and we arrive at a contradiction.
  * If $q \prec p$, q is not *end-deletable* by condition 3 of Definition 2. Therefore, we do not need to examine the simplicity of p after deleting q.

- If $\mathcal{O}$ contains more than two points (see Fig. 6 b-f), none of its elements – including p – is an endpoint by Proposition 3. Thus condition 1 of Definition 2 is violated (i.e., p is not an *end-deletable* point), and we arrive at a contradiction. Thus condition 5 of Theorem 2 is also satisfied.

Since all the conditions of Theorem 2 hold, the endpoint-related parallel reduction is also topology-preserving.

Since both kinds of reductions are topology-preserving, the proof is done. $\square$

After the topological correctness of the algorithm, let us address the question of maximality.

Theorem 4. *Algorithm* **FP**(k) *is maximal.*

Proof. We will prove this theorem by contradiction. Let us examine picture $(\mathbb{F}, 18, 12, C)$, which contains a centerline produced by our algorithm, and let $S \subseteq C$ be the set of all simple points that are not endpoints. Let us assume that $S \neq \emptyset$, and let p be the (uniquely existing) element of S such that for any $q \in S \setminus \{p\}$, $q \prec p$. Henceforth, we show that point p can be deleted by the condition-oriented parallel reduction (i.e., it is *condition-deletable* for B). Let us examine the conditions of Theorem 2 with respect to p:

- Since p is simple, condition 1 of Theorem 2 holds.
- Let $q \in C \cap N_{12}(p)$ be a simple point.
  - If $q \notin S$, it is an endpoint in $C \setminus S$. Since q cannot be deleted by the condition-oriented parallel reduction, we do not need to examine the simplicity of p after the deletion of q.
  - If $q \in S$, then $q \prec p$. Then condition 2 of Theorem 2 is satisfied.
- If $\{p, q, r\}$ is a set of three mutually 12-adjacent simple points, then no element of the set is an endpoint by Proposition 1. Then, on the one hand, $q \in S$ and $r \in S$, and on the other hand, $q \prec p$ and $r \prec p$. Thus condition 3 of Theorem 2 holds.

- If $\{p, q, r, s\}$ is an object of four mutually 12-adjacent points, $\{p, q, r, s\} \subseteq S$ (i.e., all elements of the object are simple and non-endpoints) by Proposition 2. Then, $q \prec p$, $r \prec p$, and $s \prec p$. Thus condition 4 of Theorem 2 is satisfied.
- Let us assume that p is an element of object $\mathcal{O}$ consisting of mutually 18-adjacent points in which the elements are not mutually 12-adjacent.
  - If $\mathcal{O} = \{p, q\}$ contains two points (see Fig. 6 a), both points p and q are endpoints. Since p is not and endpoint, we arrive at a contradiction.
  - If $\mathcal{O}$ contains more than two points (see Fig. 6 $b - f$), by Proposition 3, all elements of the object are in S (i.e., they are all simple and non-endpoints). Then, $q \prec p$ for any $q \in \mathcal{O} \setminus \{p\}$. Thus condition 5 of Theorem 2 holds.

Since all the conditions of Theorem 2 are satisfied, p can be deleted from C. Thus, C is not the centerline produced by our algorithm, and the proof by contradiction is completed. $\square$

5   Computationally Efficient Implementation

There is a general and computationally efficient implementation scheme for arbitrary thinning algorithms [17]:

- Since all thinning algorithms can only delete certain boundary points from the current picture, it is not necessary to evaluate interior points for possible deletion. Therefore, we can use a linked list for storing the border points in the current picture, eliminating the need to repeatedly traverse the array that stores the entire picture.
- Implementing parallel thinning algorithms requires a second list. It stores all the 'deletable' points of the current parallel reduction.
- In order to store both the input and output images of the thinning phase in a single array, we separate the evaluation and deletion phases: First, we move all the 'deletable' point from the first list to the second one. Then all points in the second list are deleted from the image array, and the first list is updated. All interior points that are adjacent (in our case: 12-adjacent) to a 'deletable' point become border points, and these brand new border points are to be inserted into the second list.

Both kinds of parallel reductions in algorithm $\mathbf{FP}(k)$ deletes only certain simple points that are border points by Lemma 2. Let us recall that the *condition-deletable* points are simple points (see condition 1 of Theorem 2, and all *end-deletable* points are also simple (see condition 1 of Definition 2 and Lemma 3). Since simpleness of a point p depends on $N_{18}(p)$ for $(18, 12)$ pictures of the FCC grid (see Lemma 1), the corresponding Boolean function can be evaluated for all possible 2^{18} configurations, and the results can be stored in the pre-calculated (unit time access) look-up-table $SIMPLE\_FCC1812$. We assigned a unique integer code called $config\_code \in [0, 1, \ldots, 2^{18} - 1]$ to each configuration in $N_{18}(p)$, and $SIMPLE\_FCC1812[config\_code] = 1$ if point p is simple, and

$SIMPLE\_FCC1812[config\_code] = 0$, otherwise. Based on Fig. 4, the code was defined with the following formula:

$$config\_code = \sum_{i=0}^{8} 2^i \cdot q_i + 2^9 \cdot \sum_{i=0}^{8} 2^i \cdot r_i \ .$$

At the beginning of each iteration step, for each element of the list containing border points, we calculate the configuration code of the given border point and the count of black points that is 18-adjacent to it, and we also store these two numbers in the list element. These stored numbers speeds up the verification of the conditions of Theorem 2 and the execution of the second reduction of the iteration steps.

6 Results

The proposed fully parallel curve-thinning algorithm $\mathbf{FP}(k)$ $(k \in \{1, 2, \ldots, 17\})$ was tested for various shapes. We selected ten illustrative examples shown in Figs. 8, 9, 10, 11, 12, 13, 14, 15, 16 and 17 and the centerline of each image was produced by choosing $k = 1$. The triplet (o, i, t) gives the count of black points in the picture, the required numbers of iteration steps, and the computation time (in msec) on a usual PC.

We can conclude that in all ten cases the produced centerlines are well-positioned, they contain all the essential branches and are free from unwanted side branches. The phenomenon that the centerlines of 'symmetric' shapes did not become 'perfectly symmetric' results from the fact that the deletion rule of the first phase of the iteration steps was derived from an asymmetric point-based sufficient condition for topology-preserving parallel reductions (see Theorem 2).

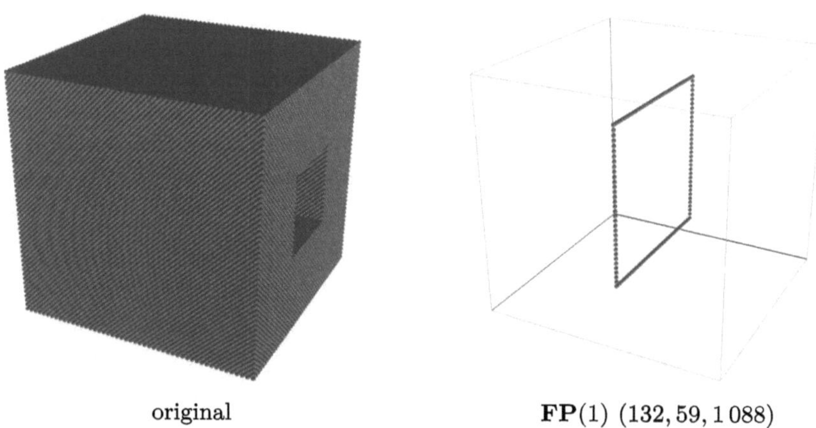

original $\mathbf{FP}(1)$ $(132, 59, 1\,088)$

Fig. 8. A $99 \times 99 \times 99$ image and its centerline produced by algorithm $\mathbf{FP}(1)$. The original image contains $431\,244$ black points.

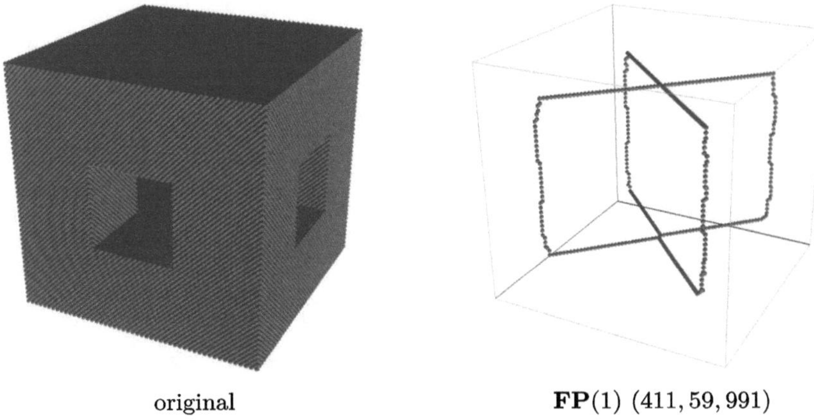

original   FP(1) (411, 59, 991)

Fig. 9. A $99 \times 99 \times 99$ image and its centerline produced by algorithm **FP**(1). The original image contains 395 306 black points.

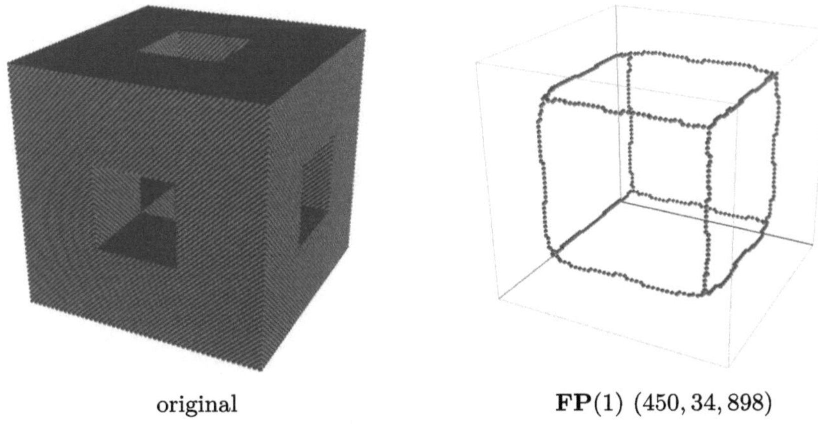

original   FP(1) (450, 34, 898)

Fig. 10. A $99 \times 99 \times 99$ image and its centerline produced by algorithm **FP**(1). The original image contains 359 368 black points.

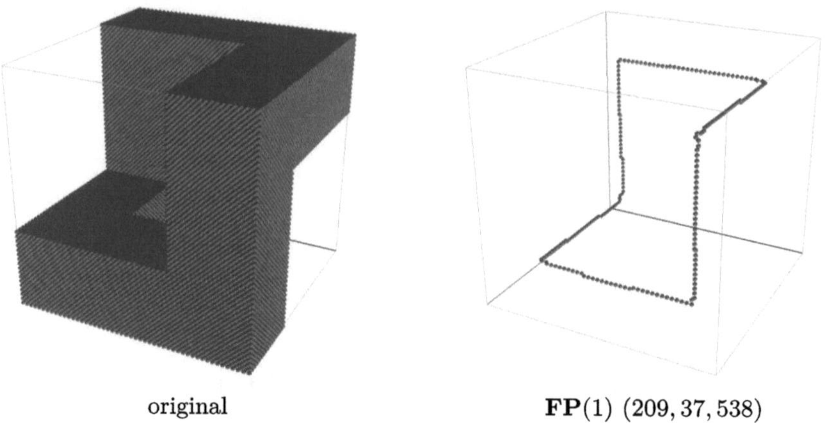

original   FP(1) (209, 37, 538)

Fig. 11. A $99 \times 99 \times 99$ image and its centerline produced by algorithm **FP**(1). The original image contains 215 622 black points.

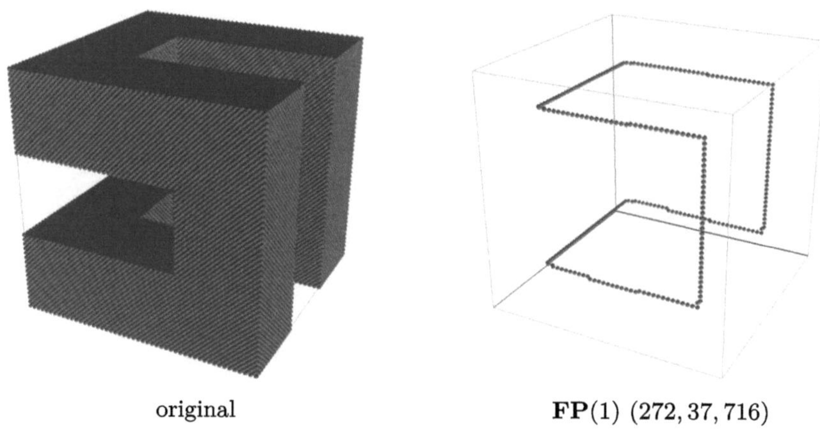

original   FP(1) (272, 37, 716)

Fig. 12. A $99 \times 99 \times 99$ image and its centerline produced by algorithm **FP**(1). The original image contains 287 496 black points.

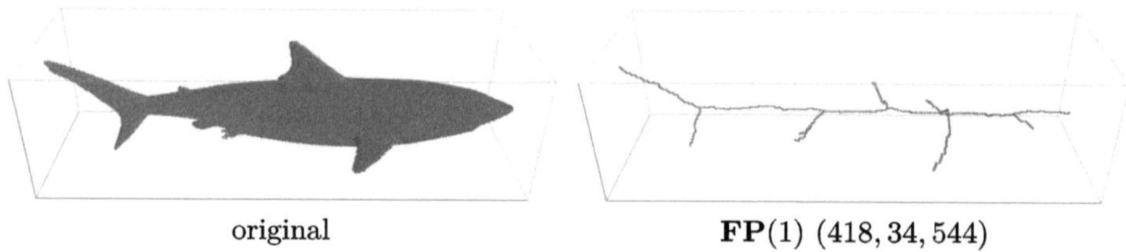

original   FP(1) (418, 34, 544)

Fig. 13. A $105 \times 129 \times 380$ image and its centerline produced by algorithm **FP**(1). The original image contains 193 646 black points.

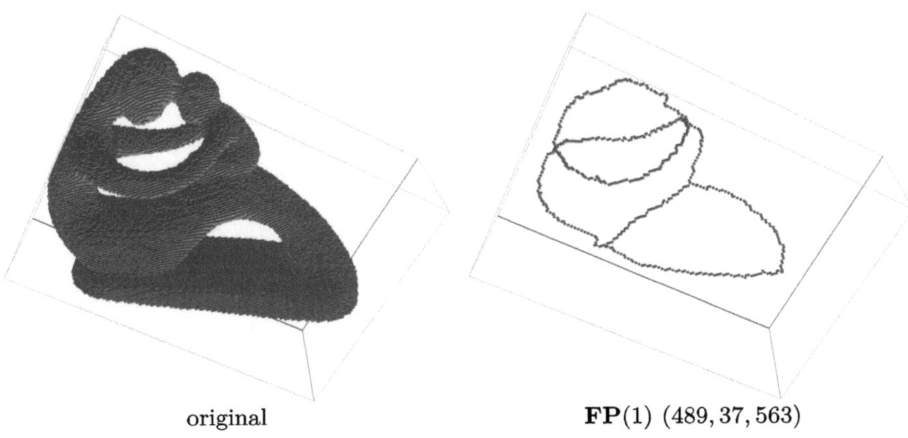

original   **FP**(1) (489, 37, 563)

Fig. 14. A $138 \times 70 \times 189$ image and its centerline produced by algorithm **FP**(1). The original image contains 200 535 black points.

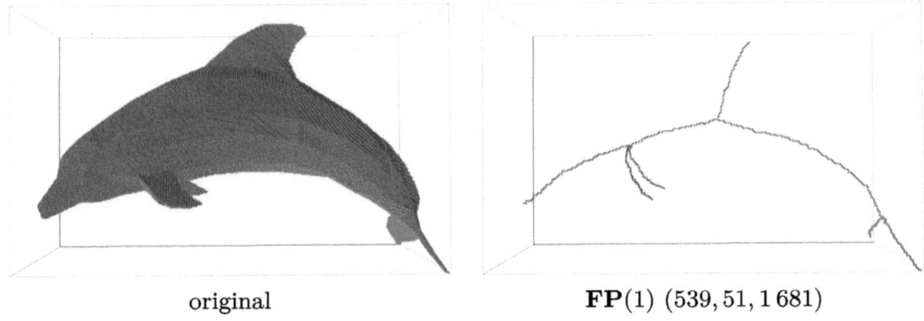

original   **FP**(1) (539, 51, 1 681)

Fig. 15. A $348 \times 130 \times 215$ image and its centerline produced by algorithm **FP**(1). The original image contains 601 457 black points.

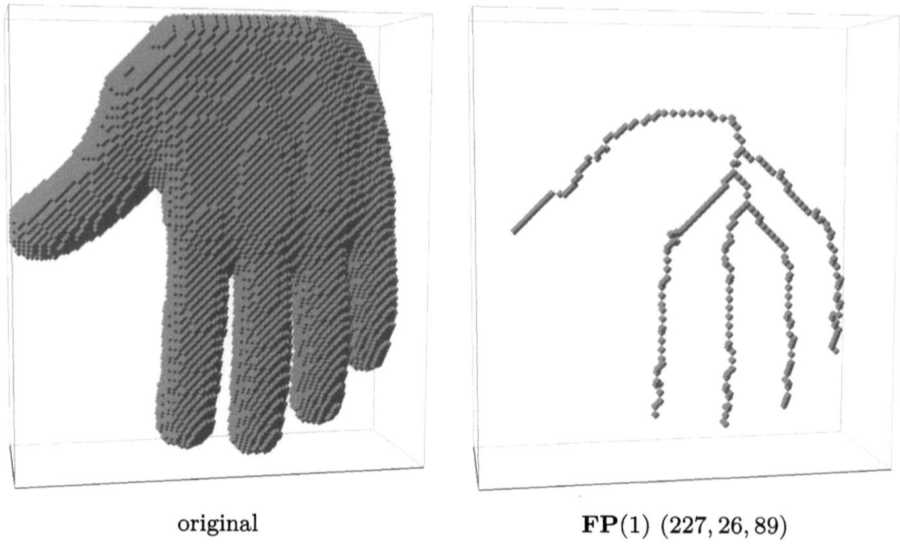

original   **FP**(1) $(227, 26, 89)$

Fig. 16. A $91 \times 101 \times 36$ image and its centerline produced by algorithm **FP**(1). The original image contains 38 106 black points.

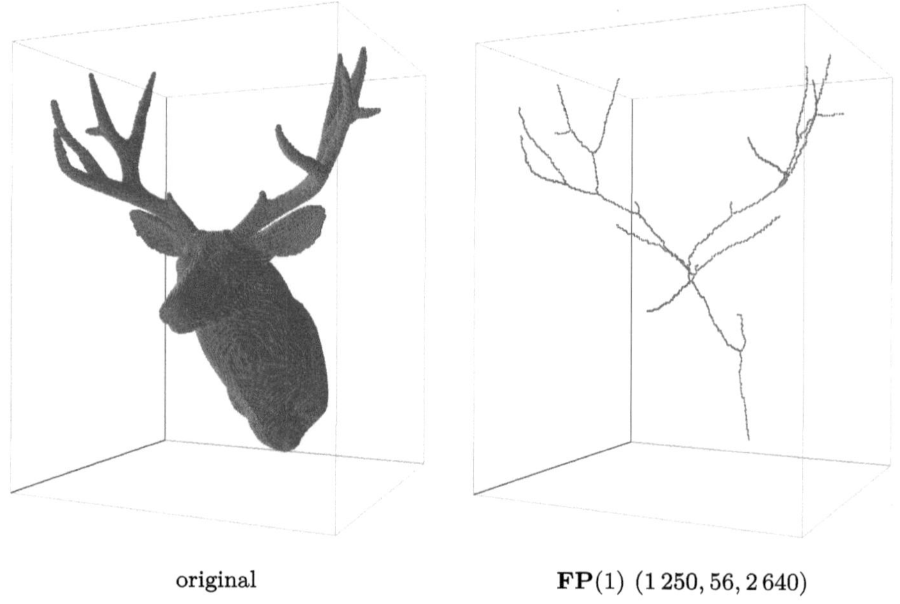

original   **FP**(1) $(1\,250, 56, 2\,640)$

Fig. 17. A $380 \times 287 \times 271$ image and its centerline produced by algorithm **FP**(1). The original image contains 904 114 black points.

7   Conclusions

In this work, the first fully parallel curve-thinning algorithm on $(18, 12)$ pictures of the unusual FCC grid is reported. Our algorithm combines a sufficient condition for topology-preserving reductions with endpoint re-checking, that reviews the emerging endpoints at the iteration level. We proved that the proposed algorithm is topology-preserving and it is capable of producing one voxel thin centerlines for all possible pictures.

As a future work we intend to construct fully parallel 3D curve thinning algorithms for the remaining two types of 'reasonable' (i.e., $(12, 18)$ and $(12, 12)$) pictures of the FCC grid.

Acknowledgements. This research was supported by project no. TKP2021-NVA-09 provided by the Ministry of Culture and Innovation of Hungary from the National Research, Development and Innovation Fund.

References

1. Biswas, R., Largeteau-Skapin, G., Zrour, R., Andres, E.: Rhombic dodecahedron grid – Coordinate system and 3D digital object definitions. In: Proceedings 21st IAPR International Conference on Discrete Geometry for Computer Imagery, pp. 27–37 (2019). https://doi.org/10.1007/978-3-030-14085-4_3
2. Čomić, L., Magillo, P.: Repairing 3D binary images using the FCC grid. J. Math. Imag. Vision **61**(9), 1301–1321 (2019). https://doi.org/10.1007/s10851-019-00904-0
3. Čomić, L., Magillo, P.: On hamiltonian cycles in the FCC grid. Comput. Graph. **89**, 88–93 (2020). https://doi.org/10.1016/j.cag.2020.05.015
4. Čomić, L., Nagy, B.: A combinatorial 3-coordinate system for the face centered cubic grid. In: in 9th International Symposium on Image and Signal Processing and Analysis, pp. 298–303 (2015). https://doi.org/10.1109/ISPA.2015.7306076
5. Čomić, L., Nagy, B.: A topological 4-coordinate system for the face centered cubic grid. Pattern Recogn. Lett. **83**, 67–74 (2016). https://doi.org/10.1016/j.patrec.2016.03.012
6. Cornea, N., Silver, D., Min, P.: Curve-skeleton properties, applications, and algorithms. IEEE Trans. Visual Comput. Graphics **13**, 530–548 (2007). https://doi.org/10.1109/TVCG.2007.1002
7. Edelsbrunner, H., Iglesias-Ham, M., Kurlin, V.: Relaxed disk packing. In: Proceeding 27th Canadian Conference on Computational Geometry, pp. 128–135 (2015). https://doi.org/10.48550/arXiv.1505.03402
8. Gastineau, N., Togni, O.: Coloring of the dth power of the face-centered cubic grid. Discuss. Math. Graph Theory **41**, 1001–1020 (2021). https://doi.org/10.7151/dmgt.2257
9. Gau, C.J., Kong, T.Y.: Minimal nonsimple sets of voxels in binary images on a face-centered cubic grid. Int. J. Pattern Rec. Artifi; Intell. **13**, 485–502 (1999). https://doi.org/10.1142/S021800149900029X
10. Hall, R.W.: Parallel connectivity-preserving thinning algorithms. In: Kong, T.Y., Rosenfeld, A. (eds.) Topological algorithms for digital image processing, pp. 145–179. Elsevier Science, Amsterdam (1996). https://doi.org/10.1016/S0923-0459(96)80014-0
11. Hall, R.W., Kong, T.Y., Rosenfeld, A.: Shrinking binary images. In: Kong, T.Y., Rosenfeld, A. (eds.) Topological algorithms for digital image processing, pp. 31–98. Elsevier Science, Amsterdam (1996). https://doi.org/10.1016/S0923-0459(96)80012-7

12. Karai, G., Kardos, P., Palágyi, K.: Sufficient conditions for topology-preserving parallel reductions on the face-centered cubic grid. J. Math. Imag. Vision **66**, 271–292 (2024). https://doi.org/10.1007/s10851-024-01177-y
13. Kong, T.Y.: On topology preservation in 2-d and 3-d thinning. Int. J. Pattern Recog. Artifi. Intell. **9**, 813–844 (1995). https://doi.org/10.1142/S0218001495000341
14. Kong, T.Y., Rosenfeld, A.: Digital topology: Introduction and survey. Comput. Vision, Graph. Imag. Process. **48**, 357–393 (1989). https://doi.org/10.1016/0734-189X(89)90147-3
15. Koshti, G., et al.: Sphere construction on the FCC grid interpreted as layered hexagonal grids in 3D. In: Proceeding 19th International Workshop on Combinatorial Image Analysis, pp. 82–96 (2018). https://doi.org/10.1007/978-3-030-05288-1_7
16. Magillo, P.: Non-square grids: a new trend in imaging and modeling? Comput. Sci. Rev. **56**, 100695 (2025). https://doi.org/10.1016/j.cosrev.2024.100695
17. Palágyi, K.: A 3d fully parallel surface-thinning algorithm. Theoret. Comput. Sci. **406**, 119–135 (2008). https://doi.org/10.1016/j.tcs.2008.06.041
18. Rácz, G.F., Csébfalvi, B.: Cosine-weighted B-spline interpolation on the face-centered cubic lattice. Comput. Graph. Forum **37**, 503–511 (2018). https://doi.org/10.1111/cgf.13437
19. Saha, P.K., Borgefors, G., Sanniti di Baja, G.: Skeletonization: theory, methods and applications. Academic Press, London (2017). https://doi.org/10.1016/B978-0-08-101291-8.00017-1
20. Strand, R., Stelldinger, P.: Topology preserving marching cubes-like algorithms on the face-centered cubic grid. In: 14th International Conference on Image Analysis and Processing, pp. 781–786 (2007). https://doi.org/10.1109/ICIAP.2007.4362871
21. Vad, V., Csébfalvi, B., Rautek, P., Gröller, E.: Towards an unbiased comparison of CC, BCC, and FCC lattices in terms of prealiasing. Comput. Graph. Forum **33**, 81–90 (2014). https://doi.org/10.1111/cgf.12364

Edge Detection on a Triangular Grid

Hussain M. Abass[1] and Benedek Nagy[1,2]($\boxtimes$)

[1] Department of Mathematics, Eastern Mediterranean University, Famagusta, North
Cyprus, via Mersin-10, Turkey
`nbenedek.inf@gmail.com`
[2] Institute of Mathematics and Informatics, Eszterházy Károly Catholic University,
Eger 3300, Hungary

Abstract. Edge detection is a vital step in image processing, pattern
recognition and computer vision. While edge detection were tackled on
the traditional rectangular and on the hexagonal grids, there is a lack of
studies on the triangular grid. In this paper, we fill this gap, we adapt
various classical edge detection techniques and algorithms into the triangular grid. We start by gradient-based technique, where we introduce filters on the triangular grid corresponding to the Sobel filter on the square
grid. We also discuss the second-derivative operators such as Laplacian
filter as well as some low pass filters, such as Gaussian smoothing, allowing us to derive the Canny and Laplacian of Gaussian filters. We adapt
these techniques keeping in mind the properties, the topology and neighborhood of the triangular grid with some modifications to the filters for
improved results. Some filters give more accurate edges while some others
have nice properties like robustness to noise. By our results, the Laplacian filter gives the best results on denoised images, while the Prewitt-T
and Canny-T filters give good results on noisy images on the triangular
grid.

Keywords: Computer vision · Edge detection · Triangular grid ·
Image processing on a nontraditional grid · Image filters

1 Introduction

Digital image processing and computer vision use discrete algorithms that are
based on digital geometry [9,10]. The geometry that can be used is connected
to the image grid. There are three regular and some other known and frequently
used grids [13,18]. The most used digital geometry is based on the square and on
the rectangular grids. However, they may not be the best options, e.g., because
of their symmetry. There are many algorithms and techniques on the hexagonal
grid, as well, [14]. The hexagonal and the rectangular grids are point lattices, in
the sense that any of their pixels can be translated into any other by mapping
the grid into itself. This pleasant property, however, is not automatic in every
grid. The third regular grid is the triangular grid that is built up by triangle
pixels, commonly referred to as trixels. This latter grid is not a point lattice; the

P. Balázs et al. (Eds.): IWCIA 2025, LNCS 15985, pp. 83–97, 2026.
https://doi.org/10.1007/978-3-032-19347-6_6

two oppositely oriented triangles made the theory a little bit more sophisticated. However, there are various basic image processing techniques and algorithms that are already adapted to this grid: digital distances and distance transform [11,15, 19], mathematical morphology [1,2], digital topology and thinning algorithms [4,5,8,17,20,26], binary tomography [12,22,24], just to mention a few.

On the other hand, edge detection is also a key process in computer vision. It is the basis of various algorithms and applications. To our knowledge, edge detection on the triangular grid has not been considered yet, thus with this paper, we try to fill this gap. Given the unique topology and neighborhood of the triangular grid, it is needed for efficient work with images in this grid. The main difference between the square and triangular grids lies in the neighbors for the tiles; a tile in the square grid has 4-neighbors and 8-neighbors totaling eight neighbors, while in the triangular grid, it has three 1-neighbors, six 2-neighbors and three 3-neighbors totaling twelve neighbors; Fig. 1 shows those neighborhood relations. On the square grid we use the 4- and 8-neighborhood that is based on the number of the given type of neighbors, however, on the triangular grid we use 1-, 2- and 3-neighborhood which is based on the distance between the trixels. The triangular grid has another property which is the orientation of the tiles, there are odd and even tiles. These structural differences may result in some unique properties. In the next section we recall some edge detection filters widely used on the square and hexagonal grids and we also describe briefly the triangular grid. Then in Sect. 3 we present gradient-based edge detection on the triangular grid, while in Sect. 4, second derivative edge detection is presented based both on techniques used on the square and on the hexagonal grids. The independent approach shown in Sect. 5 is based on techniques used on the hexagonal grid. We show real-life examples in Sect. 6. Finally, future works and conclusions close the paper.

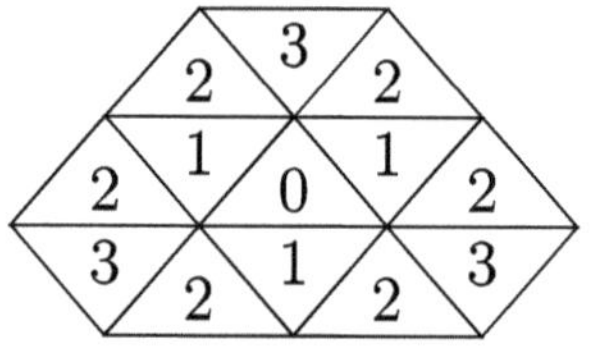

(a) Even trixel neighborhood

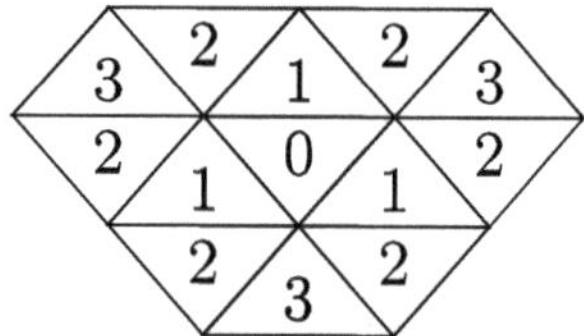

(b) Odd trixel neighborhood

Fig. 1. Neighborhood of a trixel and the types of neighbor relation based on distances of the midpoints.

2   Preliminaries

In this section, first we recall some terminology widely used on the square grid and then we also recall a formal description of the triangular grid including, e.g., coordinate system and neighborhood relations.

First, we recall that a convolution kernel, also known as a filter or an operator [6], is a matrix used for sharpening, smoothing, edge detecting, feature extraction and more.

Definition 1. (Gaussian Smoothing, [25]) *Smoothing is a convolution operation on an image. Gaussian smoothing is weighted smoothing based on the 2-D Gauss function*

$$G(x, y) = \frac{1}{2\pi\sigma^2} \cdot e^{-\frac{x^2+y^2}{2\sigma^2}}.$$

where x and y are the distances from the center of the kernel along the x and y axes, respectively, and σ is the standard deviation of a Gaussian distribution.

2.1   Edge Detection Filters

Gradient Filters. Gradient based edge detectors, or directional filters, utilize the partial derivative along the axes of the grid. It is useful to estimate the edges and their directions [7]. On the square grid, we take the partial-derivative in the x and y directions, independently, then we compute the gradient magnitude using the formula

$$|\mathcal{D}| = \sqrt{\mathcal{D}_x^2 + \mathcal{D}_y^2}. \tag{1}$$

applying appropriate thresholding. The two commonly used gradient filters are Prewitt and the Sobel [14]. In this paper, we use the name of the filters when they are given for images on the traditional square grid, and we use the suffixes -H and -T, when a variant of the filter is adapted to the hexagonal and triangular grids, respectively(Figs. 2 and 12).

$$P_x = \begin{bmatrix} 1 & 0 & -1 \\ 1 & 0 & -1 \\ 1 & 0 & -1 \end{bmatrix} \quad P_y = \begin{bmatrix} 1 & 1 & 1 \\ 0 & 0 & 0 \\ -1 & -1 & -1 \end{bmatrix} \quad S_x = \begin{bmatrix} 1 & 0 & -1 \\ 2 & 0 & -2 \\ 1 & 0 & -1 \end{bmatrix} \quad S_y = \begin{bmatrix} 1 & 2 & 1 \\ 0 & 0 & 0 \\ -1 & -2 & -1 \end{bmatrix}$$

(a) Prewitt's kernel   (b) Sobel's kernel

Fig. 2. Kernels for square grid gradient filters.

There is only one type of widely used neighborhood on the hexagonal grid, as each pair of hexels (hexagonal pixels) that share at least one point on their border share a full side. Hence, only the Prewitt-H among the gradient-based filters exists on the grid [14]. Figure 3 shows the kernels of Prewitt-H.

$$P_1 = \begin{bmatrix} 1 & 1 & 1 \\ 0 & 0 & 0 \\ -1 & -1 & -1 \end{bmatrix} \quad P_2 = \begin{bmatrix} 0 & 1 & 1 \\ -1 & 0 & 1 \\ -1 & -1 & 0 \end{bmatrix} \quad P_3 = \begin{bmatrix} 1 & 1 & 0 \\ 1 & 0 & -1 \\ 0 & -1 & -1 \end{bmatrix}$$

Fig. 3. Kernels of Prewitt-H.

Canny Filter. The Canny filter was designed as an optimal edge detector; minimizing false negatives and maximizing true positive edges in the output. In addition to minimize the distance between the located and actual edge. Finally, to minimize multiple responses to a single edge (see Canny-H in [14]). The Canny filter works by applying Gaussian smoothing as a preprocessing step, then using a gradient filter, finally thinning the edges by non-maximal edge suppression. The Canny filter produces binary edges by applying hysteresis thresholding [6].

Laplacian Filter. The Laplacian filter uses second-order derivatives to sharpen images [6]. Unlike the gradient-based approach, kernels in Laplacian filter are isotropic. The Laplacian filter detects edges nicely and symmetrically with high level of details, however, this makes it very noise-prone. The kernels are generated using the discrete Laplacian operator [6]. Figure 4 shows some kernels of the Laplacian filter.

$$\begin{bmatrix} 1 & 1 & 1 \\ 1 & -8 & 1 \\ 1 & 1 & 1 \end{bmatrix} \quad \begin{bmatrix} 0 & -1 & 0 \\ -1 & 4 & -1 \\ 0 & -1 & 0 \end{bmatrix} \quad \begin{bmatrix} -1 & -1 & -1 \\ -1 & 8 & -1 \\ -1 & -1 & -1 \end{bmatrix}$$

Fig. 4. Kernels of the Laplacian.

Laplacian of Gaussian. Instead of using the maxima in the gradients, an alternative methodology to find the edges in an image is to use the zero crossings of the second derivative [14]. The Laplacian of Gaussian, LoG and LoG-H, for short, much like the Canny filter, applies smoothing before the Laplacian filter. The smoothing can be tuned to make the edge detector sensitive to the different strength of edges.

Since our aim here is to provide similar filters on the triangular grid, now, we recall the description of this grid.

2.2   The Triangular Grid

The triangular grid is built up by same size regular triangle pixels, trixels for short, in edge-to-edge manner. The two orientations are alternate in the grid. A symmetric coordinate system is used in [15] to describe the grid with 0-sum and 1-sum integer triplets. The triangles $\triangle$ are addressed by 0-sum triplets and they are referred to as even trixels; while the $\triangledown$ oriented trixels are addressed by 1-sum integer triplets and they are the odd trixels. The three coordinate axes are in the direction from the middle of the even trixel referred to as the Origin to the directions of the middle of its three closest neighbor odd trixels. In this way, the angle of any two of the axes is 120°. A trixel has three types of neighbors, as the Euclidean distance of the midpoints of trixels sharing at least one point an their boundary can be three different values. According to this, we use the concepts of 1-neighbor, they are the three side neighbors of a trixel, each has the opposite orientation as the original trixel. The six 2-neighbor trixels of a trixel are, in fact, the closest same oriented trixels. The three 3-neighbors are also having opposite orientations than the original trixel. Figure 1 shows the neighbor trixels for both an even and an odd trixel.

3 Gradient-Based Edge Detection

3.1 Sobel Filter

If we put into account the trixel neighborhood relations when designing the kernel for the filter we get the Sobel filter [6]. The idea behind the Sobel filter is to give more weight to pixels closer to the center to achieve some smoothing. Now, we introduce Sobel-T. Since we have three types of neighbors on the triangular grid, Sobel-T is designed with 1-neighbors having the largest weight followed by 2-neighbors and then 3-neighbors. We make the necessary modification to fulfill the zero-sum condition of gradient filters. Figure 5 shows the Sobel-T kernels for the even trixels of the grid, the kernels for odd trixels can be obtained by mirroring the even kernels vertically and multiplying the weights by -1. The output of Sobel-T is a grayscale image, we can apply a threshold, t, which gives us binary edges (see Fig. 13 for an example of a grayscale image result). By applying the approach taken in [14] for computing the gradient magnitude, the hexagonal grid is the dual of the triangular grid [18], we get the formula (2) which give a good approximation of the magnitude of the gradient; however, formula (3) gives better results since it uses the redundancy of the kernels to compute the orthogonal gradient magnitude, since we have $G_y = S_3$ and $G_x = S_1 + S_2$. Similarly, we get the direction of the gradient formula (4). See Fig. 6, thresholding is applied for better visualization.

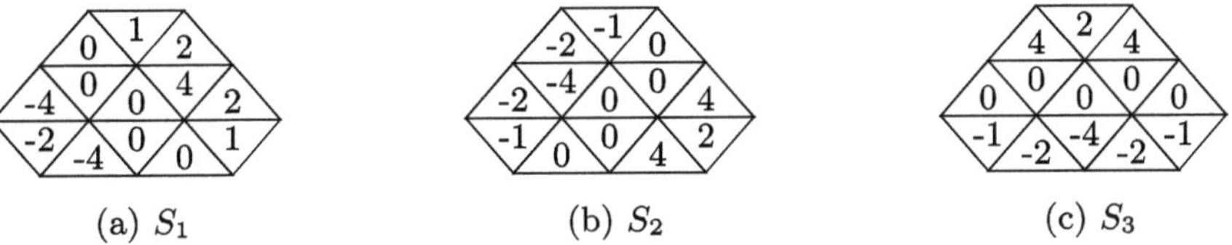

Fig. 5. Convolution kernels of Sobel-T filter.

$$|G| = \sqrt{S_1^2 + S_2^2 + S_1 \cdot S_2}. \tag{2}$$

$$|G| = \sqrt{S_1^2 + S_2^2 + S_3^2 + 2S_1 \cdot S_2}. \tag{3}$$

$$\theta = \tan^{-1}\left(\frac{S_3}{S_1 + S_2}\right). \tag{4}$$

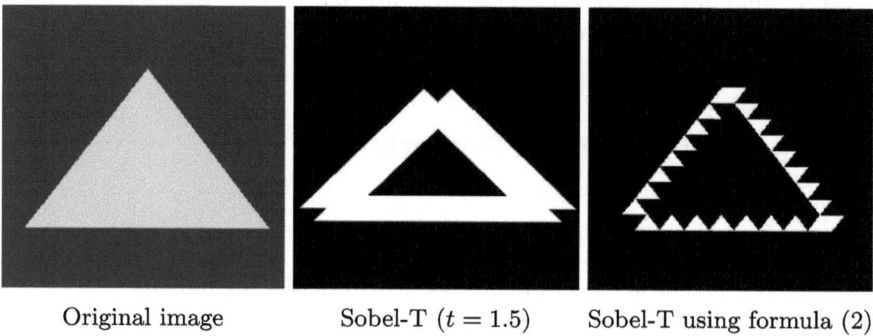

Original image   Sobel-T ($t = 1.5$)   Sobel-T using formula (2)

Fig. 6. Example of Sobel-T on an image.

3.2 Canny Filter

From the continuous coordinate system defined in [3, 21] and the conversion from triangular grid to the Cartesian plane we can generate a Gaussian smoothing kernel for the triangular grid analogously to Definition 1. The Canny-T filter uses the same approach used on the square grid: first, smoothing the image with a Gaussian-T kernel, second, applying the Sobel-T filter, third, applying non-maximum edge suppression, next, double-thresholding the result, and finally, edge tracking by hysteresis. A Gaussian-T kernel in size of a trixel neighborhood of the triangular grid, with $\sigma = 3$, is used for the results in Fig. 7. The Canny-T filter is highly robust to noise; however, it struggles with detecting finer details of an image, specially on images with little to no noise (see Fig. 16).

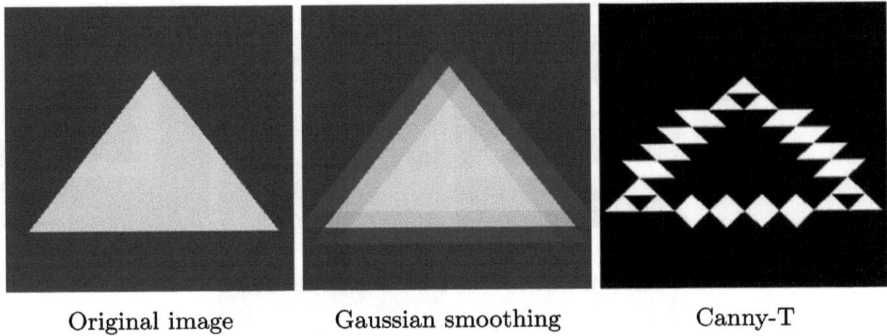

Fig. 7. Results of applying Gaussian-T smoothing and Canny-T filter.

4   Second Derivative Edge Detection

For this section, we discuss application of second-order derivatives and their usage for edge detection.

4.1   Laplacian Filter

For the Laplacian-T filter we mimic the approach taken in [14] for Laplacian-H. We can derive multiple kernels, two has been depicted in Fig. 8. So far the Laplacian-T gives the best results, with clean and accurate edges and corners. It is, however, very noise-sensitive. The different kernels yield exactly the same results: many distributions of the weights putting into account the neighborhood relations of the trixels and fulfilling the zero sum condition gives good results. In Fig. 8, kernel (a) is faster to apply since it only uses 1- and 2- neighbors. Figure 9 shows the result of the Laplacian-T filter.

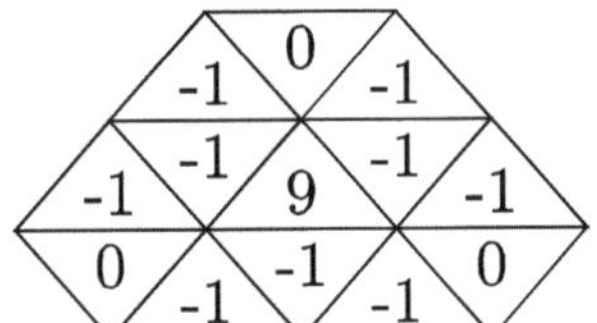
(a) Kernel on 1 and 2-neighbors.

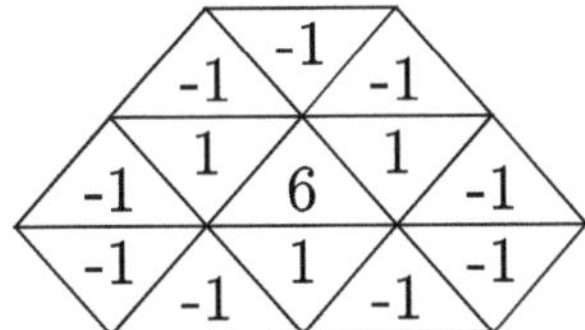
(b) Kernel involving all neighborhood.

Fig. 8. Different kernels of Laplacian-T.

4.2   Laplacian of Gaussian

The Laplacian of Gaussian filter on the triangular grid, denoted LoG-T, is introduced to solve the same problem as LoG: the Laplacian-T is very sensitive to noise and needs noise suppression. LoG-T uses Gaussian-T as defined in Sect. 3.2. Figure 10 shows LoG on primitives shown in Fig. 9.

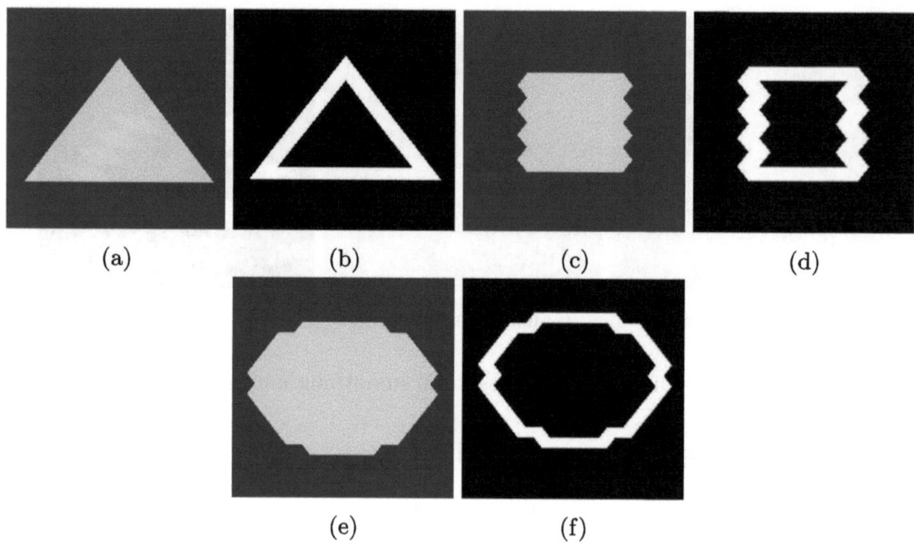

Fig. 9. The Laplacian results (b)(d)(f) on the images (a)(c)(e), respectively.

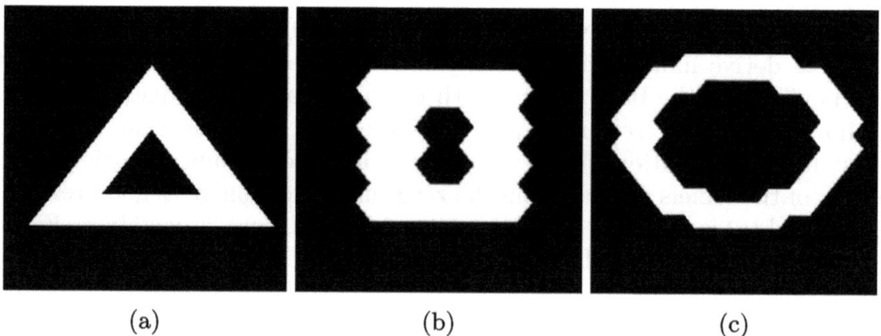

Fig. 10. Results of Laplacian of Gaussian using the same inputs as in Fig. 9.

5   The "Independent" Approach

The triangular grid can also be seen as the union of two hexagonal grids [16], i.e., both the sets of even and odd trixels define hexagonal subgrids with the six 2-neighbors of each trixel. In this section, we present methods that apply the methods for the hexagonal grid [14] separately for the even and odd subgrids of the triangular grid. Similar approach was already used in tomography [23], and in mathematical morphology [1]. In this section, we show that this approach can also be used for edge detection.

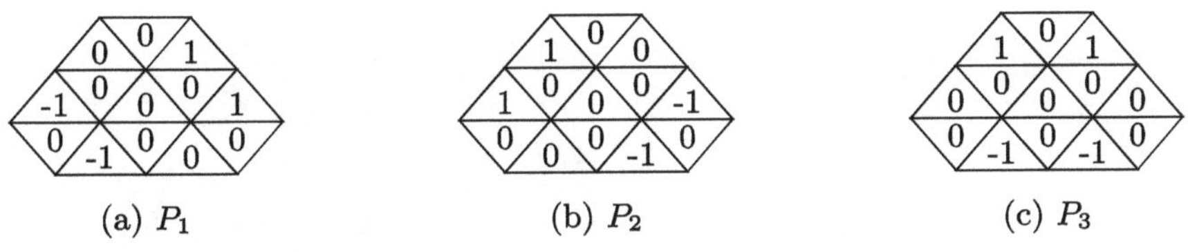

(a) P_1 (b) P_2 (c) P_3

Fig. 11. Prewitt-T filter for the even trixels based on Prewitt-H filter [14].

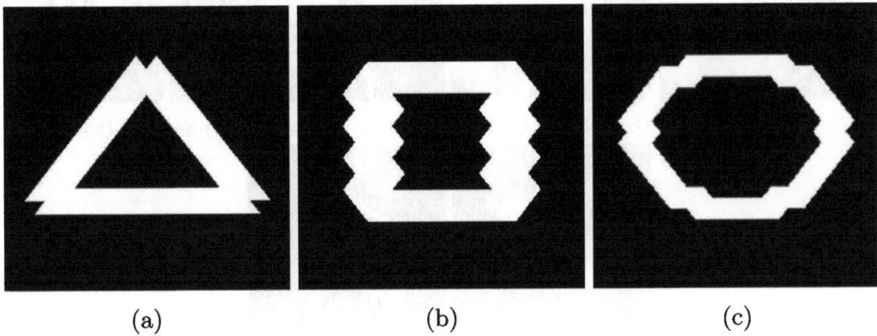

(a)   (b)   (c)

Fig. 12. Results of Prewitt-T filter.

As Fig. 11 shows for the even trixels, the weights for the hexagonal approach are assigned to the closest even neighbors of the given trixel. As odd trixels are not in use for even trixels, their weights are set to be zero. Similarly, by flipping the role of even and odd trixels (and, e.g., by mirroring the figures), the independent Prewitt filter, denoted Prewitt-T, for the odd trixels can also be obtained based on the hexagonal analog.

Using this approach, smoothing kernels can also be applied independently. Thus, we may use any of the filters of the hexagonal grid mentioned in [14] on the triangular grid, in this way.

6   Real-World Examples

In this section, we showcase the results of the filters on various images. The images which the filters are applied on are resampled from the square grid into triangles, this process reduces the level of details of the image having an effect similar to a blur. The image in Fig. 14 contains a medium amount of noise. Hence, the gradient-based filters perform better than the second derivative filters, with Canny-T giving the best results due to noise-suppression. This is also the case with the image in Fig. 15, as it has high amount of noise. However, on an artificially rendered image with little to no noise, Laplacian-T gives the best results as shown in Fig. 16.

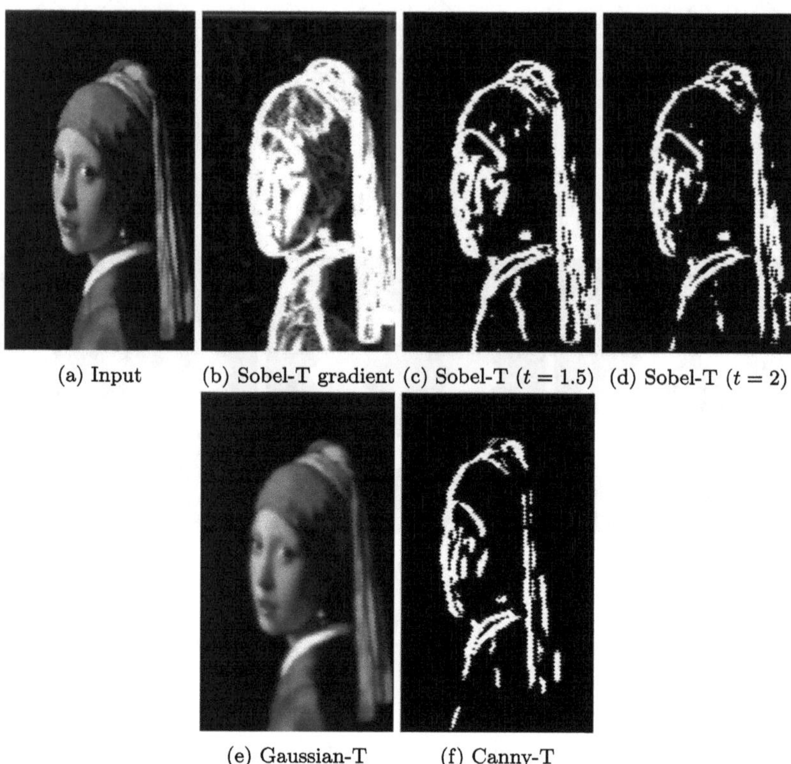

(a) Input   (b) Sobel-T gradient (c) Sobel-T ($t = 1.5$) (d) Sobel-T ($t = 2$)

(e) Gaussian-T   (f) Canny-T

Fig. 13. Applying Sobel-T, Gaussian-T and Canny-T on an image.

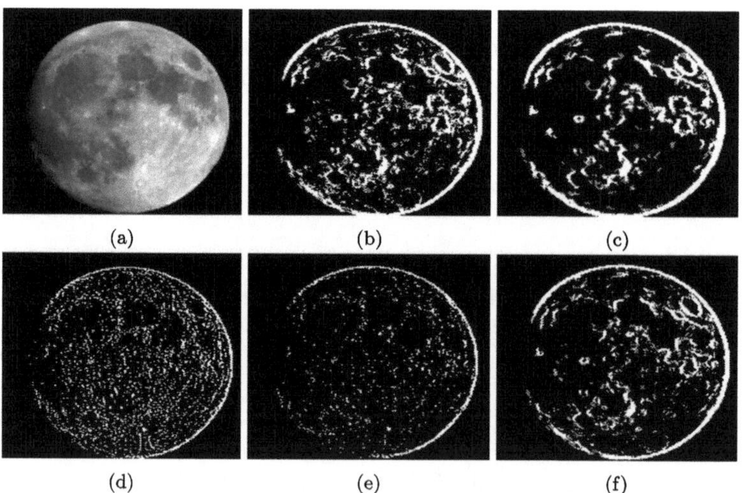

(a)   (b)   (c)

(d)   (e)   (f)

Fig. 14. Results of applying on an image (a), the filters (b) Sobel-T, (c) Canny-T, (d) Laplacian-T, (e) LoG-T, and (f) Prewitt-T.

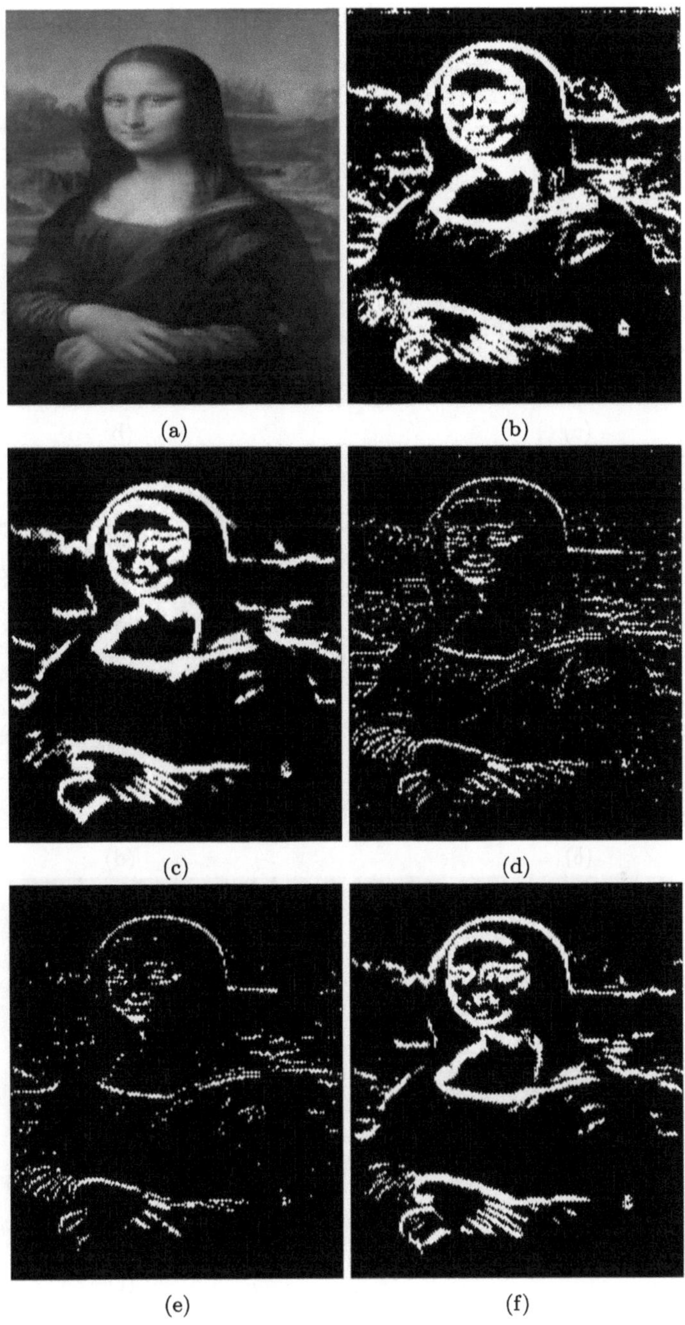

Fig. 15. (a) Image, (b) Sobel-T, (c) Canny-T, (d) Laplacian-T, (e) LoG-T, and (f) Prewitt-T.

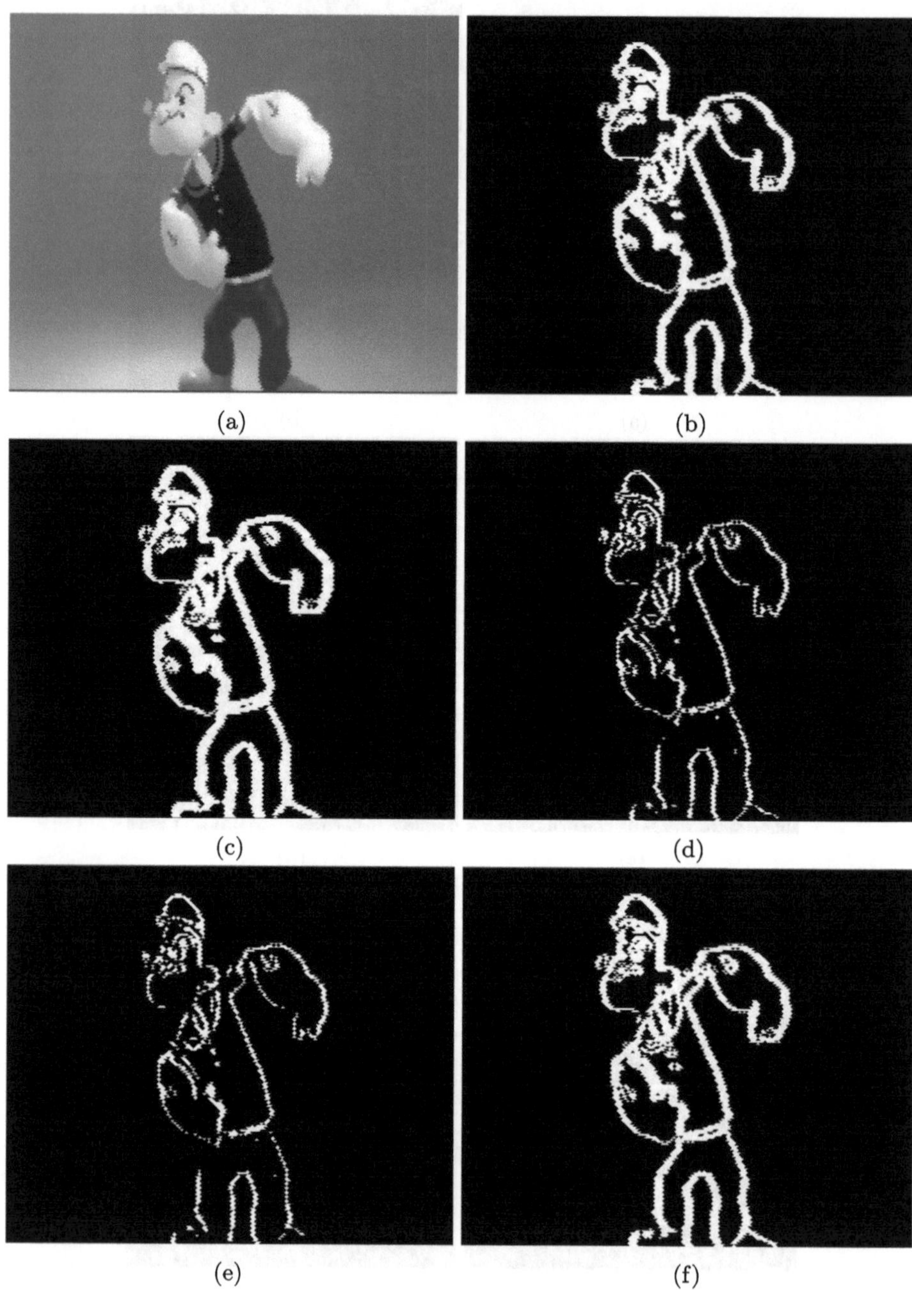

Fig. 16. (a) Image, (b) Sobel-T, (c) Canny-T, (d) Laplacian-T, (e) LoG-T, and (f) Prewitt-T.

7   Conclusion and Future Work

We introduced many of the classical edge detectors such as Sobel, Canny and Laplacian to the triangular grid. In addition to that, we adapted the Gaussian smoothing and and its application in edge detection is shown. Further, we have shown that hexagonal filters works well with the triangular grid, we have generalized the Prewitt filter used on the hexagonal grid to the triangular grid based on the independent approach, i.e., considering the triangular grid as a disjoint union of two hexagonal grids.

We found some high performance kernels with good edge detection, the Laplacian-T filter for example; and some kernels with less edge accuracy but more noise robustness, for example, the Canny-T filter. Some filters give information about the direction of the edges, Prewitt-T, Sobel-T and Canny-T. In our work we have successfully adapted methods from the traditional square and hexagonal grids, as well as, a way to apply hexagonal filters directly on the triangular grid.

We introduced smoothing, specifically Gaussian, to the triangular grid, however, we only used 12 neighbors as the kernel size. More work could be done by testing larger kernels and their effects on the Canny-T and LoG-T.

We recommend the Laplacian-T filter for good results on a denoised or artificially drawn image, otherwise, the Prewitt-T or Canny-T filter can be used for the best results.

References

1. Abdalla, M., Nagy, B.: Dilation and erosion on the triangular tessellation: An independent approach. IEEE Access **6**, 23108–23119 (2018). https://doi.org/10.1109/ACCESS.2018.2827566
2. Abdalla, M., Nagy, B.: Mathematical morphology on the triangular grid: The strict approach. SIAM J. Imaging Sci. **13**(3), 1367–1385 (2020). https://doi.org/10.1137/19M128017X
3. Abuhmaidan, K., Aldwairi, M., Nagy, B.: Vector arithmetic in the triangular grid. Entropy **23**(3), 373 (3 2021). https://doi.org/10.3390/e23030373
4. Comic, L., Blesic, A.: On the computation of the Euler characteristic of binary images in the triangular grid. In: Vento, M., Percannella, G. (eds.) Computer Analysis of Images and Patterns - 18th International Conference, CAIP 2019, Salerno, Italy, September 3-5, 2019, Proceedings, Part II. Lecture Notes in Computer Science, vol. 11679, pp. 556–567. Springer (2019). https://doi.org/10.1007/978-3-030-29891-3_49
5. Deutsch, E.S.: Thinning algorithms on rectangular, hexagonal, and triangular arrays. Communications of the ACM **15**(9), 827–837 (nov 1972). https://doi.org/10.1145/361573.361583
6. Gonzalez, R.C., Woods, R.E.: Digital Image Processing, Global Edition. Pearson Higher Education (10 2017)
7. Jähne, B.: Digital Image Processing. Springer-Verlag, Berlin Heidelberg (2005)
8. Kardos, P., Palágyi, K.: Topology preservation on the triangular grid. Ann. Math. Artif. Intell. , 53–68 (2014). https://doi.org/10.1007/s10472-014-9426-6

9. Kiselman, C.: Elements of Digital Geometry, Mathematical Morphology, and Discrete Optimization. World Scientific, Singapore (2022). https://doi.org/10.1142/12584

10. Klette, R., Rosenfeld, A.: Digital Geometry: Geometric Methods for Digital Picture Analysis. Elsevier (2004)

11. Kovács, G., Nagy, B., Vizvári, B.: Chamfer distances on the isometric grid: a structural description of minimal distances based on linear programming approach. J. Comb. Optim. **38**(3), 867–886 (2019). https://doi.org/10.1007/S10878-019-00425-X

12. Lukic, T., Nagy, B.: Deterministic discrete tomography reconstruction by energy minimization method on the triangular grid. Pattern Recognit. Lett. **49**, 11–16 (2014). https://doi.org/10.1016/J.PATREC.2014.05.014

13. Magillo, P.: Non-square grids: A new trend in imaging and modeling? Comput. Sci. Rev. **56**, 100695 (2025). https://doi.org/10.1016/J.COSREV.2024.100695

14. Middleton, L., Sivaswamy, J.: Hexagonal Image Processing: A Practical Approach. Springer-Verlag, London Limited (2005)

15. Nagy, B.: Metrics based on neighbourhood sequences in triangular grids. Pure Mathematics and Applications **13**, 259–274 (1 2002), https://econpapers.repec.org/article/cmtpumath/puma2002v013pp0259-0274.htm

16. Nagy, B.: Generalised triangular grids in digital geometry. Acta Mathematica Academiae Paedagogicae Nyíregyháziensis **20**, 63–78 (01 2004)

17. Nagy, B.: Cellular topology on the triangular grid. In: Barneva, R.P., Brimkov, V.E., Aggarwal, J.K. (eds.) Combinatorial Image Analysis - 15th International Workshop, IWCIA 2012, Austin, TX, USA, November 28-30, 2012. Proceedings. Lecture Notes in Computer Science, vol. 7655, pp. 143–153. Springer (2012). https://doi.org/10.1007/978-3-642-34732-0_11

18. Nagy, B.: Non-traditional 2d grids in combinatorial imaging - advances and challenges. In: Barneva, R.P., Brimkov, V.E., Nordo, G. (eds.) Combinatorial Image Analysis - 21st International Workshop, IWCIA 2022, Messina, Italy, July 13-15, 2022, Proceedings. Lecture Notes in Computer Science, vol. 13348, pp. 3–27. Springer (2022). https://doi.org/10.1007/978-3-031-23612-9_1

19. Nagy, B.: Weighted distances and distance transforms on the triangular tiling. Trans. GIS **27**(7), 2042–2098 (2023). https://doi.org/10.1111/TGIS.13112

20. Nagy, B.: A Khalimsky-like topology on the triangular grid. In: Brunetti, S., Frosini, A., Rinaldi, S. (eds.) Discrete Geometry and Mathematical Morphology - Third International Joint Conference, DGMM 2024, Florence, Italy, April 15-18, 2024, Proceedings. Lecture Notes in Computer Science, vol. 14605, pp. 150–162. Springer (2024). https://doi.org/10.1007/978-3-031-57793-2_12

21. Nagy, B., Abuhmaidan, K.: A continuous coordinate system for the plane by triangular symmetry. Symmetry **11**(2), 191 (2019). https://doi.org/10.3390/SYM11020191

22. Nagy, B., Lukić, T.: Dense Projection Tomography on the Triangular Tiling. Fundamenta Informaticae **145**(2), 125–141 (nov 2016). https://doi.org/10.3233/FI-2016-1350

23. Nagy, B., Lukic, T.: Binary tomography on triangular grid involving hexagonal grid approach. In: Barneva, R.P., Brimkov, V.E., Tavares, J.M.R.S. (eds.) Combinatorial Image Analysis - 19th International Workshop, IWCIA 2018, Porto, Portugal, November 22-24, 2018, Proceedings. Lecture Notes in Computer Science, vol. 11255, pp. 68–81. Springer (2018). https://doi.org/10.1007/978-3-030-05288-1_6

24. Nagy, B., Moisi, E.V.: Memetic algorithms for reconstruction of binary images on triangular grids with 3 and 6 projections. Appl. Soft Comput. **52**, 549–565 (2017). https://doi.org/10.1016/J.ASOC.2016.10.014
25. Parker, J.: Algorithms for Image Processing and Computer Vision, 2nd edn. Wiley Publishing, Inc (2011)
26. Wiederhold, P., Morales, S.: Thinning on Quadratic, Triangular, and Hexagonal Cell Complexes. In: Brimkov, V.E., Barneva, R.P., Hauptman, H.A. (eds.) IWCIA 2008. LNCS, vol. 4958, pp. 13–25. Springer, Heidelberg (2008). https://doi.org/10.1007/978-3-540-78275-9_2

Morse Sequences on Stacks and Flooding Sequences

Gilles Bertrand[(✉)]

Univ Gustave Eiffel, CNRS, LIGM, 77454 Marne-la-Vallée, France
`gilles.bertrand@esiee.fr`

Abstract. This paper builds upon the framework of *Morse sequences*, a simple and effective approach to discrete Morse theory. A Morse sequence on a simplicial complex consists of a sequence of nested subcomplexes generated by expansions and fillings—two operations originally introduced by Whitehead. Expansions preserve homotopy, while fillings introduce critical simplexes that capture essential topological features.We extend the notion of Morse sequences to *stacks*, which are monotonic functions defined on simplicial complexes,and define *Morse sequences on stacks* as those whose expansions preserve the homotopy of all sublevel sets. This extension leads to a generalization of the fundamental collapse theorem to weighted simplicial complexes. Within this framework, we focus on a refined class of sequences called *flooding sequences*, which exhibit an ordering behavior similar to that of classical watershed algorithms. Although not every Morse sequence on a stack is a flooding sequence, we show that the gradient vector field associated with any Morse sequence can be recovered through a flooding sequence.

Finally, we present algorithmic schemes for computing flooding sequences using cosimplicial complexes.

Keywords: Discrete Morse theory · Expansions and collapses · Fillings and perforations · Simplicial complex

1 Introduction

This work is set within the framework of *Morse sequences* [5] offering a simple approach to discrete Morse theory [13]. If L is a subcompex of a simplicial complex K, a Morse sequence from L to K is a sequence $\overrightarrow{W} = \langle L = K_0, \ldots, K_k = K \rangle$ of nested simplicial complexes such that, for each $i \in [1, k]$, K_i is either an *expansion* or a *filling* of K_{i-1}. These two elementary operations were introduced in [23]: expansions (the inverse of collapses) preserve homotopy, while fillings introduce critical simplices. The set of all expansions of the sequence induces a gradient vector field between L and K, while the set of all critical simplexes captures fundamental homological properties of the pair (L, K).

In this paper, we extend these notions by exploring Morse sequences constructed on stacks defined on simplicial complexes (Sect. 3). A *stack* on a simplicial complex K is a monotonic function F assigning a value $F(\nu)$ to each

P. Balázs et al. (Eds.): IWCIA 2025, LNCS 15985, pp. 98–113, 2026.
https://doi.org/10.1007/978-3-032-19347-6_7

simplex $\nu \in K$. A *Morse sequence on F* is a Morse sequence on K in which the expansions preserve the homotopy of all sublevel sets of F. This naturally generalizes the fundamental collapse theorem of discrete Morse theory to weighted simplicial complexes.

We introduce and study a refined class of Morse sequences on stacks, called *flooding sequences* (Sect. 4). These sequences satisfy an ordering condition analogous to those found in some classical watershed algorithms. A Morse sequence on a stack is not necessarily a flooding sequence. Nevertheless, we prove that the gradient vector field of an arbitrary Morse sequence on a stack F can be obtained by computing a flooding sequence on F.

Thanks to the notion of a *cosimplicial complex* (a complex which is the difference of two simplicial complexes), we derive simple schemes for computing flooding sequences (Sect. 5 and 6). In particular, we consider a *maximal increasing scheme* (building $\overrightarrow{W}$ from L to K by prioritizing expansions), and a *minimal decreasing scheme* (building $\overrightarrow{W}$ from K to L by prioritizing collapses).

These methods follow a propagation paradigm aimed at minimizing the number of critical simplexes. We show that they yield efficient algorithms for computing flooding sequences.

2 Basic Definitions

2.1 Simplicial and Cosimplicial Complexes

In this section, we introduce *cosimplicial complexes*, which generalize the classical notion of simplicial complexes. In the context of combinatorial dynamics, these structures are also referred to as *multivectors* [11]. In this paper, we employ cosimplicial complexes to construct the sequences of simplicial complexes that form Morse sequences.

By a *simplex*, we mean a finite non-empty set. The *dimension* of a simplex ν, written $dim(\nu)$, is the number of its elements minus one.

Definition 1. *Let S be a finite set of simplexes. The set S is a* cosimplicial complex *if, for any $\sigma, \tau \in S$, we have $\nu \in S$ whenever $\sigma \subseteq \nu \subseteq \tau$.*

A simplex of a cosimplicial complex S is a *face of S*. A face of S is a *facet of S* if it is maximal for inclusion, and a *cofacet of S* if it is minimal for inclusion.

A finite set K of simplexes is a *simplicial complex* if, for each $\tau \in K$, we have $\sigma \in K$ whenever $\sigma \subseteq \tau$ and $\sigma \neq \emptyset$. Thus, each simplicial complex is also a cosimplicial complex.

Let K be a simplicial complex and let $S \subseteq K$. We say that S is *closed for K* if S is a simplicial complex. We say that S is *open for K* if, for each $\sigma \in S$, we have $\tau \in S$ whenever $\sigma \subseteq \tau$ and $\tau \in K$. Therefore, a set $S \subseteq K$ is closed for K if and only if $K \setminus S$ is open for K. The family $\mathcal{O}_K$ composed of all sets which are open for K satisfies the axioms of a topology on K. So we obtain the structure of a finite topological space, see [1,2]. We easily get the following.

Proposition 1. *Let K be a simplicial complex and let $S \subseteq K$. The set S is a cosimplicial complex if and only if S is the intersection of an open set for K and a closed set for K.*

In general topology, a subset which satisfies the condition of Proposition 1 is said to be *locally closed*, see [9], Ch.1, §3, no3.

Let R and T be two subsets of K. We have $S = R \cap T$ if and only if $S = R \setminus R'$, with $R' = K \setminus T$. Thus, by Proposition 1 and by the properties of the sets which are open or closed for K, we obtain the following.

Proposition 2. *Let K be a simplicial complex and let $S \subseteq K$.*

- *S is a cosimplicial complex iff S is the difference of two open sets for K.*
- *S is a cosimplicial complex iff S is the difference of two closed sets for K.*

Since a closed set for K is a simplicial complex in its own right, we derive:

Proposition 3. *A set S is a cosimplicial complex if and only if S is the difference of two simplicial complexes.*

2.2 An Ambient Space

In [17], an *open simplicial complex* is defined precisely as the difference of two simplicial complexes. Consequently, by Proposition 3, cosimplicial complexes and open simplicial complexes are equivalent notions. In this paper, we will take advantage of the local characterization of Definition 1 for deriving some operators which act directly on a cosimplicial complex.

Let S be a finite set of simplexes. Let $\overline{S}$ be the set of simplexes such that $\sigma \in \overline{S}$ if and only if there exists $\tau \in S$ with $\sigma \subseteq \tau$. The set $\overline{S}$ is the *closure of S*. In the sequel of this paper, we write $\underline{S} = \overline{S} \setminus S$.

We have $S \subseteq \overline{S}$ and we observe that $\overline{S}$ is a simplicial complex, this complex is the smallest simplicial complex that contains S. It follows that S is a simplicial complex if and only if $\overline{S} = S$. Also, the set $\underline{S}$ is the smallest set such that $S \cup \underline{S}$ is a simplicial complex.

By the previous remarks, the simplicial complex $\overline{S}$ may be considered as "an ambient topological space for S". As a direct consequence of the definition of a cosimplicial complex we obtain:

Proposition 4. *Let S be a finite set of simplexes.*

1) The set S is a cosimplicial complex if and only if S is open for $\overline{S}$.

2) The set S is a cosimplicial complex if and only if $\underline{S}$ is a simplicial complex.

If S is a cosimplicial complex, then the sets $L = \underline{S}$ and $K = \overline{S}$ are two simplicial complexes such that $L \subseteq K$ and $S = K \setminus L$. The following proposition generalizes this situation.

Proposition 5. *Let (L, K) be a pair of simplicial complexes such that $L \subseteq K$. The cosimplicial complex $S = K \setminus L$ is such that $L \cup \overline{S} = K$ and $L \cap \overline{S} = \underline{S}$.*

Let S be a cosimplicial complex and let $\nu \in S$. We write:
- $\partial(\nu, S) = \{\eta \in S \mid \eta \subseteq \nu$ and $dim(\eta) = dim(\nu) - 1\}$, and
- $\delta(\nu, S) = \{\mu \in S \mid \nu \subseteq \mu$ and $dim(\mu) = dim(\nu) + 1\}$;
$\partial(\nu, S)$ and $\delta(\nu, S)$ are, respectively, the *boundary* and the *coboundary of ν in S.*
We observe that:
- the face ν is a facet of S if and only if $\delta(\nu, S) = \emptyset$,
- the face ν is a cofacet of S if and only if $\partial(\nu, S) = \emptyset$.

2.3   Stacks and Filtrations

With the notion of a stack, we introduce some weights on a cosimplicial or a simplicial complex. A stack is just a monotone function that assigns values to the simplices of the complex.

Let F be a map from a cosimplicial complex S to $\mathbb{Z}$. We say that F is a *stack on S* if we have $F(\sigma) \leq F(\tau)$ whenever $\sigma, \tau \in S$ and $\sigma \subseteq \tau$.

Let F be a map from a cosimplicial complex S to $\mathbb{Z}$. For any $\lambda \in \mathbb{Z}$, we write:
$$F_\lambda = \{\nu \in S \mid F(\nu) \leq \lambda\} \text{ and } F[\lambda] = \{\nu \in S \mid F(\nu) = \lambda\},$$
F_λ and $F[\lambda]$ are, respectively, the *cut* and *the section of F at level λ*.

If F is a stack on a cosimplicial complex S, then the indexed family $(F_\lambda)_{\lambda \in \mathbb{Z}}$ is a *filtration on S*. That is we have $F_\lambda \subseteq F_{\lambda'}$ whenever $\lambda \leq \lambda'$. Also if F is a stack on a simplicial complex, then any cut of F is a simplicial complex. Furthermore:

Proposition 6. *If F is a stack on a cosimplicial complex, then any cut of F and any section of F is a cosimplicial complex.*

3   Morse Sequences on Stacks

We recall the definitions of simplicial collapses and simplicial expansions [23].
Let K be a simplicial complex and let $\sigma, \tau \in K$. The couple (σ, τ) is a *free pair for K*, if τ is the only face of K that contains σ. If (σ, τ) is a free pair for K, then the simplicial complex $L = K \setminus \{\sigma, \tau\}$ is *an elementary collapse of K*, and K is *an elementary expansion of L*. We say that K *collapses onto L*, or that L *expands onto K*, if there exists a sequence $\langle K = K_0, \ldots, K_k = L \rangle$, such that K_i is an elementary collapse of K_{i-1}, $i \in [1, k]$.

We recall also the definitions of perforations and fillings [23].
Let K, L be simplicial complexes. If $\nu \in K$ is a facet of K and if $L = K \setminus \{\nu\}$, we say that L is *an elementary perforation of K*, and that K is *an elementary filling of L*.

We now introduce the notion of a "Morse sequence" by simply considering expansions and fillings of a simplicial complex [5].

Definition 2. *Let $L \subseteq K$ be two simplicial complexes. A* Morse sequence *(from L to K) is a sequence $\overrightarrow{W} = \langle L = K_0, \ldots, K_k = K \rangle$ of simplicial complexes such that, for each $i \in [1, k]$, K_i is either an elementary expansion or an elementary filling of K_{i-1}. If $L = \emptyset$, we say that $\overrightarrow{W}$ is a* Morse sequence on K*. Thus, the sequence $\overrightarrow{W} = \langle K \rangle$ is trivially a Morse sequence on K.*

Note that a Morse sequence $\overrightarrow{W} = \langle K_0, \ldots, K_k \rangle$ is a filtration, that is, for each $i \in [0, k-1]$, we have $K_i \subseteq K_{i+1}$; see [12,22].

If $\overrightarrow{W} = \langle \emptyset = K_0, \ldots, K_k = K \rangle$ is a Morse sequence on K, with $k \geq 1$, we also note that K_1 is necessarily a filling of $\emptyset$.

Let $\overrightarrow{W} = \langle L = K_0, \ldots, K_k = K \rangle$ be a Morse sequence from L to K. If $k \geq 1$, we write $\diamond\overrightarrow{W}$ for the sequence $\diamond\overrightarrow{W} = \langle \kappa_1, \ldots, \kappa_k \rangle$ such that, for each $i \in [1, k]$:

- If K_i is an elementary filling of K_{i-1}, then κ_i is the simplex such that $K_i = K_{i-1} \cup \{\kappa_i\}$; we say that K_i and the face κ_i are *critical for* $\overrightarrow{W}$.
- If K_i is an elementary expansion of K_{i-1}, then κ_i is the free pair (σ, τ) such that $K_i = K_{i-1} \cup \{\sigma, \tau\}$; we say that K_i, the pair κ_i, and the faces σ, τ, are *regular for* $\overrightarrow{W}$.

If $k = 0$, we write $\diamond\overrightarrow{W} = \langle \rangle$. That is, $\diamond\overrightarrow{W}$ is the empty sequence.

We say that $\diamond\overrightarrow{W}$ is a *simplex-wise (Morse) sequence (from L to K)*. Note that $\diamond\overrightarrow{W}$ is a sequence of faces and pairs.

Definition 3. *Let $\overrightarrow{W}$ be a Morse sequence. The* gradient vector field *of $\overrightarrow{W}$ is the set composed of all regular pairs for $\overrightarrow{W}$. We say that two Morse sequences $\overrightarrow{V}$ and $\overrightarrow{W}$ from L to K are* equivalent *if they have the same gradient vector field.*

Let us consider the simplicial complex K depicted Fig. 1 (a). For convenience, we will describe a simplex by a concatenation of its vertices.

Let $\overrightarrow{W} = \langle \emptyset = K_0, \ldots, K_7 = K \rangle$ be the sequence such that $K_1 = \{a\}$, $K_2 = K_1 \cup \{b, ab\}$, $K_3 = K_2 \cup \{c, bc\}$, $K_4 = K_3 \cup \{d, cd\}$, $K_5 = K_4 \cup \{e, ed\}$, $K_6 = K_5 \cup \{be\}$, $K_7 = K_6 \cup \{bd, bde\}$. It can be seen that $\overrightarrow{W}$ is a Morse sequence on K. The corresponding simplex-wise sequence $\diamond\overrightarrow{W}$ is such that $\diamond\overrightarrow{W} = \langle a, (b, ab), (c, bc), (d, cd), (e, ed), be, (bd, bde) \rangle$. In Fig. 1 (b) the gradient vector field of $\overrightarrow{W}$ is given by arrows, the two critical faces a and be of $\overrightarrow{W}$ are in red.

Now, we extend the notion of a Morse sequence for an arbitrary stack F.
In the sequel of this paper, L and K will denote simplicial complexes.

Let F be a stack on K and let L be a subcomplex of K. Let $\kappa = (\sigma, \tau)$ be a free pair for L. If $F(\sigma) = F(\tau)$, then we say that $L' = L \setminus \{\sigma, \tau\}$ is an *(elementary) F-collapse of L* and L is an *(elementary) F-expansion of L'*. We write $F(\kappa) = F(\sigma) = F(\tau)$.

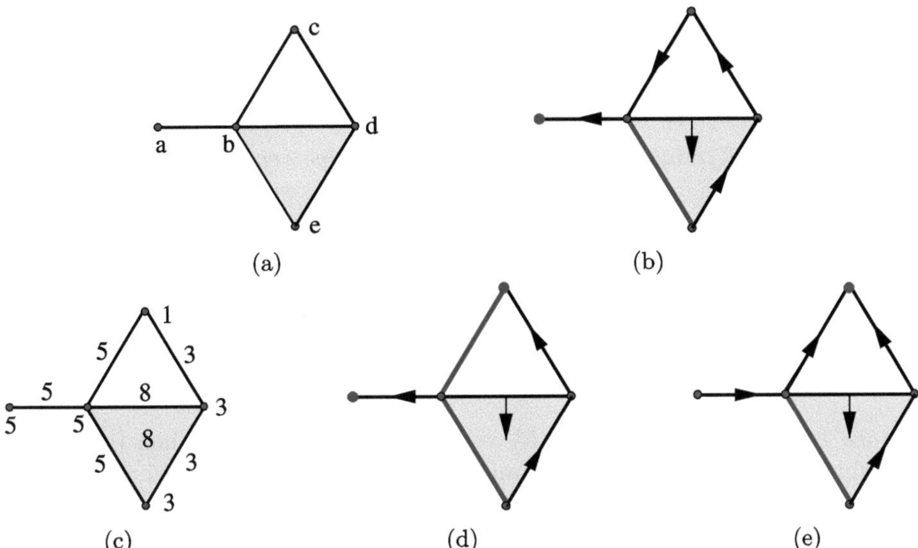

Fig. 1. (a) A simplicial complex K. (b) The gradient vector field of a Morse sequence $\overrightarrow{W}$ on K. (c) A stack F on K. (d) The gradient vector field of a Morse sequence $\overrightarrow{W}'$ on F. (e) The gradient vector field of a flooding sequence $\overrightarrow{W}''$ on F. See text for the description of the sequences $\overrightarrow{W}$, $\overrightarrow{W}'$, and $\overrightarrow{W}''$. (Color figure online)

Definition 4. *Let $\overrightarrow{W}$ be a Morse sequence from L to K and let F be a stack on K. We say that $\overrightarrow{W}$ is a (Morse) sequence on F or an F-sequence (from L to K), if we have $F(\sigma) = F(\tau)$ whenever (σ, τ) is a regular pair for $\overrightarrow{W}$.*

An example of a stack F on a complex K is given Fig. 1 (c). Let us consider again the Morse sequence $\overrightarrow{W}$ and its gradient vector field given in (b). We see that $\overrightarrow{W}$ does not satisfy the condition of Definition 4 for the stack F. Thus, the sequence $\overrightarrow{W}$ is not an F-sequence. Now let us consider the sequence $\diamond\overrightarrow{W}' = \langle a, (b, ab), c, bc, (d, cd), (e, ed), be, (bd, bde) \rangle$. This sequence is a simplex-wise Morse sequence on K. Furthermore the corresponding sequence $\overrightarrow{W}'$ is an F-sequence on K. The gradient vector field of $\overrightarrow{W}'$ is depicted in (d), the four critical faces a, c, bc, and be of $\overrightarrow{W}'$ are in red.

Let F be a stack on K and L be a subcomplex of K. If $\lambda \in \mathbb{Z}$, the set $F_\lambda \cap L$ is the cut of L at level λ which is induced by F. Note that $F_\lambda \cap K = F_\lambda$.
Let (σ, τ) be a free pair for L such that $F(\sigma) = F(\tau)$. Then:
- For each $\lambda < F(\sigma)$, we have $\sigma \notin F_\lambda \cap L$ and $\tau \notin F_\lambda \cap L$,
- For each $\lambda \geq F(\sigma)$, one can verify that the pair (σ, τ) is a free pair for $F_\lambda \cap L$.
Now, let $\overrightarrow{W} = \langle L = K_0, ..., K_k = K \rangle$ be an F-sequence. Suppose K_i is an elementary expansion of K_{i-1}. Then, by the above observation, this elementary

expansion is a (possibly trivial) expansion for all cuts of K_{i-1}.
By induction, we have:

Theorem 1. *Let F be a stack on K and $\overrightarrow{W} = \langle L = K_0, \ldots, K_k = K \rangle$ be an F-sequence. Let K_i and K_j, $j > i$, be two consecutive critical complexes for $\overrightarrow{W}$. That is, $K_{i+1}, \ldots, K_{j-1}$ are regular complexes for $\overrightarrow{W}$. Then, for each $\lambda \in \mathbb{Z}$, the complex $F_\lambda \cap K_{j-1}$ collapses onto $F_\lambda \cap K_i$.*

This last property may be seen as an extension to stacks of a fundamental theorem, called *the collapse theorem*, which makes the link between the basic definitions of discrete Morse theory and discrete homotopy (See Theorem 3.3 of [13] and Theorem 4.27 of [22]).

Remark 1. Let $\overrightarrow{W} = \langle \emptyset = K_0, \ldots, K_k = K \rangle$ be an F-sequence. Then $\overrightarrow{W}$ induces a "double filtration": the indexed family $(K_i)_{i \in [0,k]}$ is a filtration where each K_i induces the filtration $(F_\lambda \cap K_i)_{\lambda \in \mathbb{Z}}$. Also, $\overrightarrow{W}$ induces the sequence $\langle F_0, \ldots, F_k \rangle$ where each F_i is the stack on K_i which is the restriction of F to K_i.

If S is a set, we write $\mathbb{1}_S$ for the constant function $S \to \mathbb{Z}$ such that, for each $x \in S$, we have $\mathbb{1}_S(x) = 1$. We observe that a sequence $\overrightarrow{W}$ is a Morse sequence from L to K if and only if $\overrightarrow{W}$ is an F-sequence from L to K for $F = \mathbb{1}_K$.

It is well known that discrete Morse theory can be formulated from the perspective of gradient vector fields [18]. In the remainder of this section, we present a natural extension of this classical notion to stacks. We then emphasize that a Morse sequence is in no way a restrictive notion with respect to this point of view. That is, any gradient vector field on a stack F is the gradient vector field of some Morse sequence on F.

Let F be a stack on K and let V be a set of pairs (σ, τ), with $\sigma, \tau \in K$, such that $\sigma \in \partial \tau$ and $F(\sigma) = F(\tau)$. We say that V is a *(discrete) vector field on F* if each simplex of K is in at most one pair of V. A simplex $\nu \in K$ is said to be *critical for V* if it does not belong to any pair in V.

Let V be a vector field on F. A *(p-)gradient path in V (from σ_0 to σ_k)* is a sequence $\pi = \langle \sigma_0, \tau_0, \sigma_1, \tau_1, \ldots, \sigma_{k-1}, \tau_{k-1}, \sigma_k \rangle$, with $k \geq 0$, composed of faces such that, for all $i \in [0, k-1]$, $dim(\sigma_i) = dim(\sigma_k) = p$, (σ_i, τ_i) is in V, $\sigma_{i+1} \subset \tau_i$, and $\sigma_{i+1} \neq \sigma_i$. This sequence π is said to be *trivial* if $k = 0$, that is, if $\pi = \langle \sigma_0 \rangle$; otherwise, if $k \geq 1$, we say that π is *non-trivial*. Also, the sequence π is *closed* if $\sigma_0 = \sigma_k$. We say that a vector field V on F is *a gradient vector field* if V is *acyclic*, that is if V contains no non-trivial closed p-gradient path.

Note that, if F is the constant map $\mathbb{1}_K$, then the above definitions correspond exactly to the classical ones. The proof of the following result is a direct extension of the proof given in [5] (Theorem 51).

Theorem 2. *Let F be a stack on K. A vector field V on F is a gradient vector field if and only if V is the gradient vector field of a Morse sequence on F.*

4   Flooding Sequences

We introduce the following refinement of an F-sequence. A flooding sequence corresponds to a process similar to the one often used for computing the watershed transform, which is a key tool for image segmentation in Mathematical Morphology [7,8].

Definition 5. *Let F be a stack on K and let $\overrightarrow{W} = \langle \emptyset = K_0, ..., K_k = K \rangle$ be an F-sequence. We say that $\overrightarrow{W}$ is a flooding sequence (on F), if we have $F(\sigma) \le F(\tau)$ whenever $\sigma \in K_i$, $\tau \in K_j \setminus K_i$, and $i < j$.*

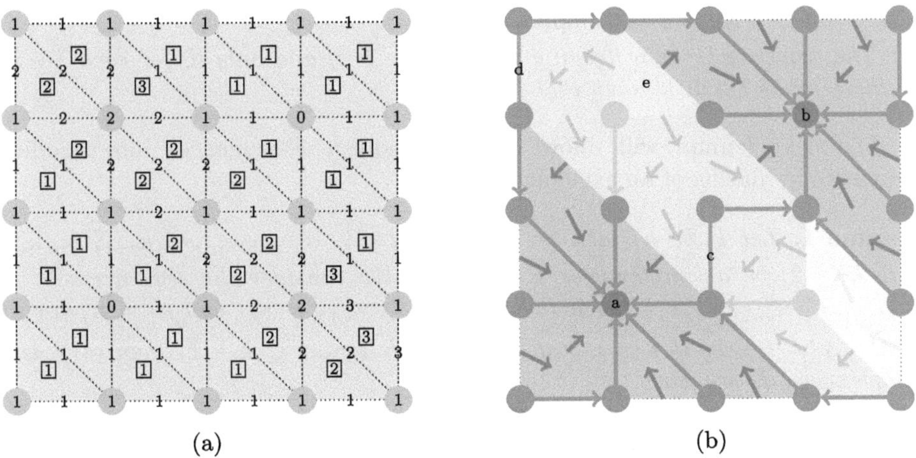

<table>
<tr><td align="center">(a)</td><td align="center">(b)</td></tr>
</table>

Fig. 2. A flooding sequence. (a) a simplicial stack F on a triangulation K of a square. (b) A flooding sequence on F. (Color figure online)

In the example of Fig. 1, it can be seen that the sequence $\overrightarrow{W'}$ is not a flooding sequence on F. The first element of $\overrightarrow{W'}$ is the face a. In a flooding sequence, the face c should appear before a since $F(a) = 5$ and $F(c) = 1$. Now, let us consider the sequence $\diamond\overrightarrow{W''} = \langle c, (d, cd), (e, ed), (b, bc), (a, ab), be, (bd, bde) \rangle$. The corresponding sequence $\overrightarrow{W''}$ is a flooding sequence on F. The gradient vector field of $\overrightarrow{W''}$ is given in (e), the two critical faces c and be of $\overrightarrow{W''}$ are in red.

Figure 2 illustrates an example of a flooding sequence on a less simple complex. This sequence $\overrightarrow{W}$ can begin from two possible points, both at level 0, as shown in Fig. 2.a. If $\overrightarrow{W}$ begins with **a**, then the second element of $\overrightarrow{W}$ must be **b**, both of them are critical for $\overrightarrow{W}$. Then the sequence must continue with a subsequence composed of all simplexes at level 1. This subsequence is illustrated in blue, it turns out that it is composed solely of regular elements. The next subsequence composed of all simplexes at level 2 is illustrated in green, it contains two critical simplexes $\mathbf{c}$ and $\mathbf{d}$. The last subsequence composed of all simplexes at level 3 is illustrated in yellow, it contains one critical simplex $\mathbf{e}$. See also Fig. 1 in [6] which gives, on the same complex, two examples of F-sequences that are not flooding sequences.

Let F be a stack on K and let L be a subcomplex of K. We say that L is *a subcomplex for F* if, for all $\sigma \in K$ and all $\tau \in L$, we have $\sigma \in L$ whenever $F(\sigma) < F(\tau)$. The following provides two characterizations of flooding sequences.

Proposition 7. *Let F be a stack on K and let $\overrightarrow{W} = \langle \emptyset = K_0, ..., K_k = K \rangle$ be an F-sequence.*

1) *The sequence $\overrightarrow{W}$ is a flooding sequence on F if and only if, for each $i \in [0, k]$, the complex K_i is a subcomplex for F.*
2) *The sequence $\overrightarrow{W}$ is a flooding sequence on F if and only if, for each $\lambda \in \mathbb{Z}$, there exists $i \in [0, k]$ such that $F_\lambda = K_i$.*

The next lemma will allow us to transform F-sequences into flooding sequences by means of an exchange property.

Lemma 1. *Let F be a stack on K, and let $\diamond \overrightarrow{W} = \langle \kappa_1, \ldots, \kappa_i, \kappa_{i+1}, \ldots, \kappa_k \rangle$, with $k \geq 2$, be a simplex-wise F-sequence. We consider the sequence $\diamond \overrightarrow{W}' = \langle \kappa_1', \ldots, \kappa_i', \kappa_{i+1}', \ldots, \kappa_k' \rangle$, such that $\kappa_i' = \kappa_{i+1}$, $\kappa_{i+1}' = \kappa_i$, and $\kappa_j' = \kappa_j$ for all $j \neq i$, $j \neq i+1$. If $F(\kappa_i) > F(\kappa_{i+1})$, with $1 \leq i \leq k-1$, then $\diamond \overrightarrow{W}'$ is a simplex-wise F-sequence.*

Proof. We write $\overrightarrow{W} = \langle \emptyset = K_0, ..., K_k = K \rangle$ for the Morse sequence corresponding to $\diamond \overrightarrow{W}$, and $\overrightarrow{W}' = \langle \emptyset = K_0', ..., K_k' = K \rangle$ for the one corresponding to $\diamond \overrightarrow{W}'$. We consider the different cases for κ_i and κ_{i+1}:

1. Suppose κ_i is a critical simplex; that is, K_i a filling of K_{i-1}.
   (a) Suppose κ_{i+1} is a critical simplex; that is K_{i+1} is a filling of K_i.
       - If $\kappa_i \notin \partial(\kappa_{i+1})$, then we see that $K_i' = K_{i-1} \cup \{\kappa_{i+1}\}$ is a filling of K_{i-1}, and $K_{i+1} = K_i' \cup \{\kappa_i\}$ is a filling of K_i'. Thus $\diamond \overrightarrow{W}'$ is a simplex-wise F-sequence.
       - If $\kappa_i \in \partial(\kappa_{i+1})$, then $\overrightarrow{W}'$ is not a filtration. But, in this case, we have $F(\kappa_i) \leq F(\kappa_{i+1})$ since F is a stack on K.
   (b) Suppose $\kappa_{i+1} = (\sigma, \tau)$; that is K_{i+1} is an expansion of K_i. The pair (σ, τ) is a free pair for F, we have $F(\kappa_{i+1}) = F(\sigma) = F(\tau)$.
       - If $\kappa_i \notin \partial(\tau)$, then $K_i' = K_{i-1} \cup \{\sigma, \tau\}$ is an expansion of K_{i-1}, and $K_{i+1} = K_i' \cup \{\kappa_i\}$ is a filling of K_i'. Thus $\diamond \overrightarrow{W}'$ is a simplex-wise F-sequence.
       - If $\kappa_i \in \partial(\tau)$, we have $F(\kappa_i) \leq F(\tau) = F(\kappa_{i+1})$ since F is a stack on K.
2. Suppose $\kappa_i = (\sigma, \tau)$; that is K_i is an expansion of K_{i-1}.

(a) Suppose κ_{i+1} is a critical simplex; that is K_{i+1} is a filling of K_i. We observe that we cannot have $\tau \in \partial(\kappa_{i+1})$, otherwise there would exist a face $\nu \in K_i$, with $\nu \in \partial(\kappa_{i+1})$, $\nu \neq \tau$, such that σ is a proper face of ν.
   - If $\sigma \notin \partial(\kappa_{i+1})$, then $\diamond\overrightarrow{W}'$ is a simplex-wise F-sequence.
   - If $\sigma \in \partial(\kappa_{i+1})$, then $F(\kappa_i) = F(\sigma) \leq F(\kappa_{i+1})$ since F is a stack on K.
(b) Suppose $\kappa_{i+1} = (\sigma', \tau')$; that is K_{i+1} is an expansion of K_i.
   - If $\sigma \notin \partial(\tau')$ and $\tau \notin \partial(\tau')$, then $\diamond\overrightarrow{W}'$ is a simplex-wise F-sequence.
   - If $\sigma \in \partial(\tau')$ or $\tau \in \partial(\tau')$, we have $F(\kappa_i) \leq F(\tau') = F(\kappa_{i+1})$ since F is a stack on K. $\qquad\square$

An F-sequence is not necessary a flooding sequence. Nevertheless we have the following result which indicates that the gradient vector field of an arbitrary Morse sequence on F may be obtained by computing a flooding sequence on F.

Theorem 3. *Let F be a stack on K, and let $\overrightarrow{W}$ be an F-sequence from $\emptyset$ to K. Then, there exists a flooding sequence on F which is equivalent to $\overrightarrow{W}$.*

Proof. If $\overrightarrow{W}$ is not a flooding sequence, then there exist two elements κ_i and κ_{i+1} in $\diamond\overrightarrow{W}$ such that $F(\kappa_i) > F(\kappa_{i+1})$. By Lemma 1, we can swap these elements. Note that this swap corresponds exactly to the swap of the bubble sorting algorithm [10]. By repeating this operation, we obtain a sorted sequence $\overrightarrow{V}$ which is equivalent to $\overrightarrow{W}$, and which is a flooding sequence.$\square$ $\qquad\square$

Remark 2. By Lemma 1, a flooding sequence may be obtained from an arbitrary F-sequence from $\emptyset$ to K with a sorting algorithm. In order to effectively obtain a Morse sequence, it can be seen that it is sufficient for the sorting algorithm to be *stable*. That is, equal elements must appear in the same order in the sorted sequence as they do in the original F-sequence [10].

We conclude this section by illustrating Theorem 3 with the sequence $\overrightarrow{W}'$ given Fig. 1. We have $\diamond\overrightarrow{W}' = \langle a, (b, ab), c, bc, (d, cd), (e, ed), be, (bd, bde)\rangle$. By iteratively applying Lemma 1, we see that we can derive the simplex-wise sequence $\diamond\overrightarrow{W}''' = \langle c, (d, cd), (e, ed), a, (b, ab), bc, be, (bd, bde)\rangle$: the corresponding sequence $\overrightarrow{W}'''$ is an F-sequence which is equivalent to $\overrightarrow{W}'$. Furthermore, $\overrightarrow{W}'''$ is a flooding sequence on F.

5   Flooding Sequences and Cosimplicial Complexes

Let F be a stack on K and let $\overrightarrow{W} = \langle \emptyset = K_0, ..., K_k = K \rangle$ be an F-sequence. Let $\diamond\overrightarrow{W} = \langle \kappa_1, \ldots, \kappa_k \rangle$. For each $\lambda \in \mathbb{Z}$ we write $\diamond\overrightarrow{W}[\lambda]$ for the subsequence of $\diamond\overrightarrow{W}$ composed of all κ_i such that $F(\kappa_i) = \lambda$. Recall that a subsequence is a *substring* if it a sequence of contiguous elements of the original sequence. If the sequence $\overrightarrow{W}$ is a flooding sequence then, for each $\lambda \in \mathbb{Z}$, the sequence $\diamond\overrightarrow{W}[\lambda]$ is a substring of $\diamond\overrightarrow{W}$ which is a simplex-wise Morse sequence from $F_{\lambda-1}$ to F_λ. The set $S = F_\lambda \setminus F_{\lambda-1}$ is a cosimplicial complex. From Proposition 5, we obtain:

Proposition 8. *Let (L, K) be a pair of simplicial complexes such that $L \subseteq K$, and let $S = K \setminus L$. A sequence $\vec{W}$ is a Morse sequence from L to K if and only if $\vec{W}$ is a Morse sequence from $\underline{S}$ to $\overline{S}$.*

Let S be a cosimplicial complex. If $\vec{V}$ is a Morse sequence from $\underline{S}$ to $\overline{S}$, then we say that $\vec{V}$ and $\diamond\vec{V}$ are *Morse sequences on* S.

If F is a stack on K, recall that $F[\lambda] = \{\nu \in K \mid F(\nu) = \lambda\}$; we have $F[\lambda] = F_\lambda \setminus F_{\lambda-1}$. We derive the following result from the previous observations.

Proposition 9. *Let F be a stack on K and let $\vec{W} = \langle \emptyset = K_0, ..., K_k = K \rangle$ be an F-sequence. Let $[0, H]$ be the range of F. The sequence $\vec{W}$ is a flooding sequence if and only if $\diamond\vec{W}$ is a concatenation of the form $\diamond\vec{W} = \diamond\vec{W}[0] \ ... \ \diamond\vec{W}[\lambda]$ $... \ \diamond\vec{W}[H]$ where, for each $\lambda \in [0, H]$, $\diamond\vec{W}[\lambda]$ is a Morse sequence on $F[\lambda]$.*

Definition 6. *Let S be a cosimplicial complex and let $\sigma, \tau, \nu \in S$, with $\sigma \in \partial\tau$.*

- *If $\delta(\sigma, S) = \{\tau\}$, we say that $S' = S \setminus \{\sigma, \tau\}$ is a reduction of S,*
- *If $\delta(\nu, S) = \emptyset$, we say that $S' = S \setminus \{\nu\}$ is a perforation of S,*
- *If $\partial(\tau, S) = \{\sigma\}$, we say that $S' = S \setminus \{\sigma, \tau\}$ is a coreduction of S,*
- *If $\partial(\nu, S) = \emptyset$, we say that $S' = S \setminus \{\nu\}$ is a coperforation of S.*

In each of the above four cases, we say that S' is a contraction *of S.*

Reductions and coreductions have been introduced in [19] in the context of $\mathcal{S}$-complexes. In our context, the following property is crucial for an efficient computation of Morse sequences.

Proposition 10. *Let S be a cosimplicial complex and let $\sigma, \tau, \nu \in S$.*

- *$\overline{S} \setminus \{\sigma, \tau\}$ is a collapse of $\overline{S}$ iff $S \setminus \{\sigma, \tau\}$ is a reduction of S.*
- *$\overline{S} \setminus \{\nu\}$ is a perforation of $\overline{S}$ iff $S \setminus \{\nu\}$ is a perforation of S.*
- *$\underline{S} \cup \{\sigma, \tau\}$ is an expansion of $\underline{S}$ iff $S \setminus \{\sigma, \tau\}$ is a coreduction of S.*
- *$\underline{S} \cup \{\nu\}$ is a filling of $\underline{S}$ iff $S \setminus \{\nu\}$ is a coperforation of S.*

We note that a contraction of a cosimplicial complex is also a cosimplicial complex. Thus, the following definition makes sense.

Let S be a cosimplicial complex. A *Morse cosequence on* S is a sequence $\vec{U} = \langle S = S_0, \ldots, S_k = \emptyset \rangle$ such that, for each $i \in [1, k]$, S_i is a contraction of S_{i-1}. The sequence is increasing (resp. decreasing) if each S_i is either a coreduction or a coperforation of S_{i-1} (resp. a reduction or a perforation of S_{i-1}).

Let $L \subseteq K$ be two simplicial complexes. Let $\vec{W} = \langle L = K_0, \ldots, K_k = K \rangle$ be a sequence of nested complexes. From Propositions 8 and 10, we obtain:
Let $\vec{U} = \langle S_0, \ldots, S_k \rangle$ with $S_i = K \setminus K_i$. Then $\vec{W}$ is a Morse sequence if and only if $\vec{U}$ is an increasing Morse cosequence on $K \setminus L$.
Let $\vec{U} = \langle S_0, \ldots, S_k \rangle$ with $S_i = K_{k-i} \setminus L$. Then $\vec{W}$ is a Morse sequence if and only if $\vec{U}$ is a decreasing Morse cosequence on $K \setminus L$.

6   Computing a Flooding Sequence

We have seen that a Morse sequence $\overrightarrow{W}$ from L to K may be obtained by computing a Morse sequence $\diamond \overrightarrow{V}$ on the set $S = K \setminus L$. The sequence $\diamond \overrightarrow{V}$ can be constructed from the left to the right through iterative coreductions and coperforations of the set S. Alternatively, $\diamond \overrightarrow{V}$ can be constructed from the right to the left through iterative reductions and perforations of S.

Scheme 1 allows us to compute a Morse sequence $\diamond \overrightarrow{V}$ on an arbitrary cosimplicial complex S by considering the two previous constructions. At each step of the scheme, the set S' is a contraction of S.

Scheme 1: Sequence(S)

Data: A cosimplicial complex S
Result: Sequence(S), which is a simplex-wise Morse sequence on S.
1 $\overrightarrow{L} := \langle \rangle$; $\overrightarrow{R} := \langle \rangle$;
2 **while** $S \neq \emptyset$ **do**
3 $\quad$ Extract κ such that:
4 $\quad\quad$ - either $\kappa = (\sigma, \tau)$, with $\sigma \subset \tau$, and $S' = S \setminus \{\sigma, \tau\}$ is a reduction or a
$\quad\quad\quad$ coreduction of S,
5 $\quad\quad$ - or $\kappa = \nu$ and $S' = S \setminus \{\nu\}$ is a perforation or a coperforation of S.
6 $\quad$ **if** S' *is a coreduction or a coperforation of S* **then** $\overrightarrow{L} := \overrightarrow{L} \cdot \kappa$;
7 $\quad$ **if** S' *is a reduction or a perforation of S* **then** $\overrightarrow{R} := \kappa \cdot \overrightarrow{R}$;
8 $\quad$ $S := S'$
9 $\diamond \overrightarrow{V} := \overrightarrow{L} \cdot \overrightarrow{R}$;
10 **return** $\diamond \overrightarrow{V}$

We consider three cases of Scheme 1:

1. If $\diamond \overrightarrow{V}$ is constructed using only coreductions and coperforations, then the scheme is an *increasing scheme*. We say that $\diamond \overrightarrow{V}$ is a *maximal sequence* if we make a coperforation only when no coreduction can be made.
2. If $\diamond \overrightarrow{V}$ is constructed using only reductions and perforations, then the scheme is a *decreasing scheme*. We say that $\diamond \overrightarrow{V}$ is a *minimal sequence* if we make a perforation only when no reduction can be made.
3. Otherwise we say that the scheme is an *interleaved scheme*. We say that $\diamond \overrightarrow{V}$ is a *min/max sequence* if we make a perforation or a coperforation only when no reduction and no coreduction can be made.

Let us illustrate these three schemes with the complex K of Fig. 1 (a).

1. The Morse sequence $\overrightarrow{W}$ corresponding to Fig. 1 (b) is such that:
$$\diamond \overrightarrow{W} = \langle a, (b, ab), (c, bc), (d, cd), (e, ed), be, (bd, bde) \rangle.$$

This sequence is a maximal sequence. It can be built from the left to the right with Scheme 1. Under this ordering, no coreduction can be made when the two coperforations a and be are done.

2. Let $\overrightarrow{W}'''$ be the Morse sequence such that:
$$\diamond\overrightarrow{W}''' = \langle b, (c, bc), (d, cd), (e, ed), be, (a, ab), (bd, bde)\rangle.$$
This sequence is a minimal sequence. It can be built from the right to the left with Scheme 1. Under this ordering, no reduction can be made when the two perforations be and b are done. Note that $\overrightarrow{W}$ is not a minimal sequence and $\overrightarrow{W}'''$ is not a maximal sequence.

3. For a min/max sequence, at the second step of Scheme 1, we must have either $\overrightarrow{L} = \langle\rangle$ and $\overrightarrow{R} = \langle(a, ab), (bd, bde)\rangle$ or $\overrightarrow{L} = \langle\rangle$ and $\overrightarrow{R} = \langle(bd, bde), (a, ab)\rangle$. At the third step, we can choose to make the coperforation b. In this case, we must make three coreductions at the following steps. Thus, at line 9 of Scheme 1, we may obtain:
$$\overrightarrow{L} = \langle b, (c, bc), (d, cd), (e, ed), be\rangle \text{ and } \overrightarrow{R} = \langle(a, ab), (bd, bde)\rangle \qquad .$$
We retrieve the sequence $\overrightarrow{W}'''$ after the concatenation of $\overrightarrow{L}$ and $\overrightarrow{R}$.

The purpose of maximal, minimal, and min/max sequences is to try to minimize the number of critical simplexes. This problem is, in general, NP-hard [16]. Therefore, these sequences do not, in general, yield optimal results. Note also that different numbers of critical simplexes may be obtained by these three types of sequences, see Fig. 11 of [14] and Fig. 4 of [4] for examples which illustrate this fact.

The computational models of increasing and decreasing schemes are classical models for extracting gradient vector fields, see [3, 14, 15, 19]. Also the interleaved model is introduced in [14]. A distinctive feature of our approach is the use of these schemes to compute Morse sequences. A Morse sequence on a complex K not only defines a gradient vector field on K, but also provides a specific structure on K.

With Proposition 9, we derive Scheme 2 which is a direct extension of Scheme 1 for computing flooding sequences. Given a stack F on K, we consider the different sections $F[\lambda]$ of F at level λ. For each λ, a Morse sequence $\diamond\overrightarrow{W}[\lambda]$ on $F[\lambda]$ is computed thanks to Scheme 1. Then, the sequences $\diamond\overrightarrow{W}[\lambda]$ are simply concatenated in order to obtain a flooding sequence on F.

Scheme 2: Flood(F)

Data: A simplicial complex K and a stack F on K; the range of F is $[0, H]$.
We set $F[\lambda] := \{\nu \in K \mid F(\nu) = \lambda\}$.
Result: **Flood**(F), which is a simplex-wise flooding sequence on F.

1 **for** $\lambda \in [0, H]$ **do in parallel**
2 $\quad \diamond\overrightarrow{W}[\lambda] := $ **Sequence**$(F[\lambda])$

3 $\diamond\overrightarrow{W} := \langle\rangle$; $\lambda := 0$;
4 **while** $\lambda \leq H$ **do**
5 $\quad \diamond\overrightarrow{W} := \diamond\overrightarrow{W} \cdot \diamond\overrightarrow{W}[\lambda]$;
6 $\quad \lambda := \lambda + 1$

7 **return** $\diamond\overrightarrow{W}$

With the three above instances of Scheme 1, we obtain three instances of Scheme 2 which allow to compute maximal, minimal, and min/max flooding sequences.

Let us consider again the stack F given Fig. 1 (c). We have:

$$F[1] = \{c\},\ F[3] = \{d, e, de, cd\},\ F[5] = \{a, b, ab, bc, be\},\ F[8] = \{bd, bde\}.$$

With Scheme 1, we may obtain the following maximal sequences:

- **Sequence**$(F[1]) = \langle c \rangle$,
- **Sequence**$(F[3]) = \langle (d, cd), (e, ed) \rangle$,
- **Sequence**$(F[5]) = \langle (b, bc), (a, ab), be \rangle$,
- **Sequence**$(F[8]) = \langle (bd, bde) \rangle$.

Thus, with Scheme 2, we may obtain precisely the flooding sequence $\diamond\overrightarrow{W}''$ which is illustrated Fig. 1 (e). Observe that, in this simple case, we obtain an *optimal F-sequence*, that is, an F-sequence with the smallest number of critical faces. Observe also that, by Theorem 3, an optimal F-sequence may always be obtained by considering solely flooding sequences.

With Scheme 1, it can be seen that a maximal, minimal, or min/max sequence $\diamond\overrightarrow{W}[\lambda]$ of Scheme 2 may be obtained with a time complexity $\mathcal{O}(dN_\lambda)$, where d is the dimension of the complex K, and N_λ is the number of simplexes of $F[\lambda]$, see [6]. Therefore, since the sets $F[\lambda]$ are disjoint, all the sequences $\diamond\overrightarrow{W}[\lambda]$ can be computed with a time complexity $\mathcal{O}(dN)$, where N is the number of simplexes of K. Now, if each sequence is stored as a list, and if the range of F is $[0, H]$, it can be seen that the concatenations may be done with complexity $\mathcal{O}(H)$. Since H is bounded by N, we obtain $\mathcal{O}(dN)$ for computing a flooding sequence with a sequential algorithm.

Also the computations of all sequences may be done in parallel. In the best case, that is, if the values of the simplexes are equally distributed, we achieve $\mathcal{O}(\frac{dN}{H} + H)$ for computing the flooding sequence.

7   Conclusion

This work extends the framework of Morse sequences to stacks on simplicial complexes, introducing a flexible and expressive model for discrete Morse theory. By incorporating stack-based constraints, we define flooding sequences—a refined version of Morse sequences that respect monotonic ordering, reflecting processes commonly observed in watershed algorithms.

We have shown that the gradient vector field arising from any Morse sequence on a stack can be recovered by computing an equivalent flooding sequence, underscoring the generality of the flooding model. We also proposed general construction schemes, based on cosimplicial complexes, that provide efficient methods for computing flooding sequences. It should be noted that all the notions introduced in the paper may be directly adapted to more general complexes, for example to cubical complexes.

Importantly, a Morse sequence encodes richer structural information about a complex than a gradient field alone. In particular, the Morse complex of an object can be easily derived via the reference map introduced in [5], requiring only a linear scan of the sequence.

It is also worth emphasizing that a stack is not necessarily an injective function—different faces in the complex may share the same value. This contrasts with some approaches that require distinct values for each face [14,20]. Allowing arbitrary stacks makes the model more suitable for real-world data, where such constraints are typically not satisfied [21].

Overall, the Morse sequence framework opens promising directions for future research, including extensions to persistent homology as well as applications in image segmentation and topological data analysis.

Acknowledgements. The author wishes to express his thanks to Laurent Najman for his active interest in this paper and for stimulating conversations.

References

1. Alexandrov, P.S.: Diskrete Räume. Mathematiceskii Sbornik (N.S.) **2**, 501–518 (1937)
2. Barmak, J.: Algebraic Topology of Finite Topological Spaces and Applications, Lecture Notes in Mathematics, vol. 2032. Springer Verlag (2011)
3. Benedetti, B., Lutz, F.H.: Random discrete Morse theory and a new library of triangulations. Exp. Math. **23**(1), 66–94 (2014)
4. Bertrand, G.: Morse sequences. In: Brunetti, S., Frosini, A., Rinaldi, S. (eds.) International Conference on Discrete Geometry and Mathematical Morphology, Vol. 14605, pp. 377–389. Springer, Cham (2024). https://doi.org/10.1007/978-3-031-57793-2_29
5. Bertrand, G.: Morse sequences: A simple approach to discrete Morse theory. J. Math. Imaging Vision **67**(16), 1–22 (2025)

6. Bertrand, G., Najman, L.: Computing gradient vector fields with Morse sequences, arXiv:2504.07526 (2025)
7. Beucher, S., Lantuéjoul, C.: Use of watersheds in contour detection. International Workshop on Image Processing, CCETT/IRISA (1979)
8. Beucher, S., Meyer, F.: The morphological approach to segmentation: the watershed. Transformation **34**, 433–481 (1993)
9. Bourbaki, N.: General Topology: Chapters 1–4. Elements of Mathematics, Springer, Berlin Heidelberg (2013)
10. Cormen, T.H., Leiserson, C.E., Rivest, R.L., Stein, C.: Introduction to Algorithms, Problem 2-2, p. 40. The MIT Press, 2nd edn. (2001)
11. Dey, T., Mrozek, M., Slechta, R.: Persistence of Conley-Morse graphs in combinatorial dynamical systems. SIAM J. Appl. Dyn. Syst. **21**, 817–839 (2022)
12. Edelsbrunner, H., Morozov, D.: Persistent homology: theory and practice. In: European Congress of Mathematics. pp. 31–50. European Mathematical Society (EMS) (2013)
13. Forman, R.: Morse theory for cell complexes. Adv. Math. **134**, 90–145 (1998)
14. Fugacci, U., Iuricich, F., De Floriani, L.: Computing discrete Morse complexes from simplicial complexes. Graph. Models **103**, 101023 (2019)
15. Harker, S., Mischaikow, K., Mrozek, M., Nanda, V.: Discrete Morse theoretic algorithms for computing homology of complexes and maps. Found. Comput. Math. **14**, 151–184 (2014)
16. Joswig, M., Pfetsch, M.E.: Computing optimal Morse matchings. SIAM J. Discrete Math. **20**, 11–25 (2006)
17. Knudson, K.P., Scoville, N.A.: Discrete Morse theory for open complexes, arXiv:2402.12116 (2024)
18. Kozlov, D.: Organized Collapse: An Introduction to Discrete Morse Theory. Graduate Studies in Mathematics, American Mathematical Society (2020)
19. Mrozek, M., Batko, B.: Coreduction homology algorithm. Discrete & Comput. Geom. **41**(1), 96–118 (2009)
20. Robins, V., Wood, P.J., Sheppard, A.P.: Theory and algorithms for constructing discrete Morse complexes from grayscale digital images. IEEE Trans. Pattern Anal. Mach. Intell. **33**(8), 1646–1658 (2011)
21. Rocca, L., Iuricich, F., Puppo, E.: Disambiguating flat spots in discrete scalar fields. Graph. Models **141**, 101299 (2025)
22. Scoville, N.A.: Discrete Morse Theory, vol. 90. American Mathematical Soc. (2019)
23. Whitehead, J.H.C.: Simplicial spaces, nuclei and m-groups. Proc. Lond. Math. Soc. **2**(1), 243–327 (1939)

Characterizing Translations on Octagonal-Square Grids, Including the Khalimsky Grid

Benedek Nagy[1,2]([envelope]) [ORCID] and Ede Troll[2] [ORCID]

[1] Eastern Mediterranean University, Department of Mathematics, via Mersin-10, Famagusta, North Cyprus, Turkey
[2] Institute of Mathematics and Informatics, Eszterházy Károly Catholic University, 3300 Eger, Hungary
nbenedek.inf@gmail.com, troll.ede@uni-eszterhazy.hu

Abstract. Translations are transformations that are expected to be bijective and preserve connectedness, shape, perimeter, and area. In digital geometry, as we show here, this is not always obvious; we may need special conditions on the translation vector to have translations with these desired properties. In this paper, we use the Khalimsky grid, i.e., the semi-regular grid induced by the Khalimsky topology. This grid, also known as the truncated quadrille or truncated square tiling, is built up by regular octagons and squares with equal side length. We also generalize the case when the side lengths vary, but the neighborhood structure and, thus, the topology of the grid remains the same.

Keywords: Bijective and non-bijective transformations · Digital continuity · Nontraditional grids · Khalimsky topology · Semi-regular grids · Jordan curves

1 Introduction

In geometry, as well as in (digital) image processing and (computer) graphics, objects such as polygons and shapes are transformed. Thus, in general, both in mathematics and computer science, transformations play central roles. Some of the geometric transformations do not modify the shape. These isomorphic transformations are well known; they consist of translations, rotations, mirroring, and their combinations. In this paper, we restrict our analysis to translations. They, on the one hand, are the simplest geometric transformations. On the other hand, if we change the underlying geometry from Euclidean to digital geometry, they could exhibit some interesting behavior, as we demonstrate here.

In digital geometry, we deal with objects or shapes that are defined on a grid, i.e., typically a periodic tessellation of the plane or space, depending on the dimension in which we work. The digital geometries differ from the Euclidean in various ways, e.g., in them, the neighborhood relation (also known as adjacency)

P. Balázs et al. (Eds.): IWCIA 2025, LNCS 15985, pp. 114–128, 2026.
https://doi.org/10.1007/978-3-032-19347-6_8

plays a significant role. The properties of a digital geometry (and so of digital shapes) highly depend on the underlying grid, i.e., the image grid and its (neighborhood) structure. This is also true for the properties of the digital variants of the geometric transformations. Concentrating on translations, their digital versions exhibit the usual nice properties on point lattices, i.e., in two dimensions, specifically on rectangular and hexagonal grids. Thus, in both the square and hexagonal grids, translation is always bijective and preserves the structure and size of the object or image. On a third regular grid, the triangular grid, however, translations behave already in a more sophisticated way [1,2]: some of them are bijective and preserve shape (called strongly bijective), some of them are bijective, but do not preserve neighborhoods (called weakly bijective), and there are some non-bijective ones. This is caused by the fact that the triangular grid is not a point lattice; it is not closed under translations by each grid vector due to the two opposite orientations of the pixels. The characterization of these translations is based on a continuous coordinate system that reflects the triangular symmetry [17].

Nontraditional grids have various advantages over the traditional grids [13, 14]. There are eleven Archimedean grids apart from the three regular grids. Those tessellations are considered here that contain only regular polygons with the same side length; they are side-to-side, and the corners are indistinguishable, i.e., at every grid point where the edges (sides) meet, exactly the same number and type of polygons meet. These eight grids are also referred to as semi-regular grids, and they may be denoted by the sequence of the numbers of sides of the polygons that meet at a corner, e.g., T(4,8,8) and T(3,6,3,6).

There are some topological problems with the square and triangular grids. Thus, some alternatives are proposed by using various neighborhoods alternately. These solutions lead to the Khalimsky grid [9,10] and a grid built up by triangles and non-regular enneagons [15], respectively. The Khalimsky grid, in this context, can be seen as a semi-regular grid [4,14] having regular octagons and squares with the same side length, alternately. It is the truncated quadrille tiling, and it is denoted by T(4,8,8). Another solution to the topological problem is to work with abstract cell complexes, addressing and including lower dimensional parts of the grid into the studies. Such a solution is shown in [11,12] for the square grid and in [16,19] for the triangular grid.

In this paper, we consider translations (of objects/shapes) on the semi-regular grid T(4,8,8) and in related other grids. We characterize the translations by the so-called fractional part of the translation vector and show examples for three distinct classes. We also demonstrate that the effect we highlight here is due to the two types of neighborhoods considered in the Khalimsky topology for the square grid; thus, translations of a Jordan curve may destroy the property of being a Jordan curve.

2   Preliminaries and Description of the Grid

In this section, we briefly recall, first, the topological problem on the square grid that is resolved by using the Khalimsky topology [5,6]. Thus, let us consider the square grid with its usual representation, i.e., $\mathbb{Z}^2$.

On the one hand, it is known that the two diagonals of the chessboard are colored with different colors, i.e., the line segments represented by them cross each other without having a common square.

On the other hand, if for every square only the four edge neighbors are used, we may have a simple closed curve, e.g., with pixels $(0,0),(1,0),(2,0),(2,1),$ $(3,1),(3,2),(3,3),(2,3),(1,3),(1,2),(0,2),(0,1),(0,0)$. This is a closed digital curve, and it is easy to check that for each of the pixels in the curve, there are exactly two of its neighbors that are also in the curve; thus, it is a simple closed curve. Now, there are exactly two squares inside the curve, $(1,1)$ and $(2,2)$. However, these two squares are not neighbors; thus, the interior is disconnected.

This paradox is also connected to the fact that the Jordan curve theorem does not hold on the square grid with any of the two usual neighborhood relations [7].

In the Khalimsky topology, the 4- and 8-neighborhoods are used alternately on the grid. This topology is nicely depicted by the semi-regular grid T(4,8,8). In this grid, at every corner of a tile, exactly three tiles meet in the order of square-octagon-octagon, hence the notation.

In the next part, we describe the semi-regular grid T(4,8,8), also known as the truncated square tiling, to be able to do mathematics on it in the forthcoming sections. Instead of the original $\mathbb{Z}^2$, we assign coordinates to the points using either even or odd pairs; in this way, the grid is represented by a 45° rotation from its original (this type of description may be referred to as the diagonal square grid [8] and the 2-dimensional diamond grid [3]).

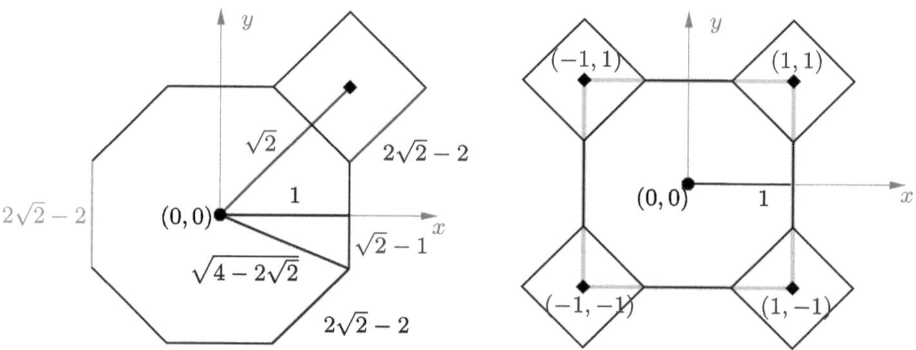

Fig. 1. Left: the size of the tiles, with the coordinate axes. Right: a unit square (highlighted with green borders) in the middle of the coordinate system. (Color figure online)

However, here, instead of one of the original adjacencies of the square grid, we consider the points with the Khalimsky topology. The grid is built by polygons shown in Fig. 1 (left), and it tessellates the plane. The square area highlighted by a green border in Fig. 1 (right) plays an essential role for us, as we refer to it as the unit square. By the tessellation, an infinite grid can be obtained, and the pixels (tiles, polygons) will be referred to with the coordinates of their center, as shown in Fig. 2 (left). We use the notation $\mathbb{K}$ to denote the grid, i.e., the set of its pixels. Observe that in $\mathbb{K}$, no topological paradox may occur as each pair of neighbor tiles, i.e., tiles having a common border point, share a whole side.

A unit cell of a tessellation is the smallest area polygon such that one can tessellate the whole plane by putting them next to each other in an edge-to-edge manner. In our grid, the same result is obtained by tiling with the green square shown in Fig. 1. Mathematically, we define the green square as $-1 \leq x, y < 1$. In this way, every point $(x, y) \in \mathbb{R}^2$ of the plane is assigned to exactly one unit square: (x, y) is assigned (we also say that it belongs to) to the unit having centre $(2 \cdot \lfloor \frac{x+1}{2} \rfloor, 2 \cdot \lfloor \frac{y+1}{2} \rfloor)$.

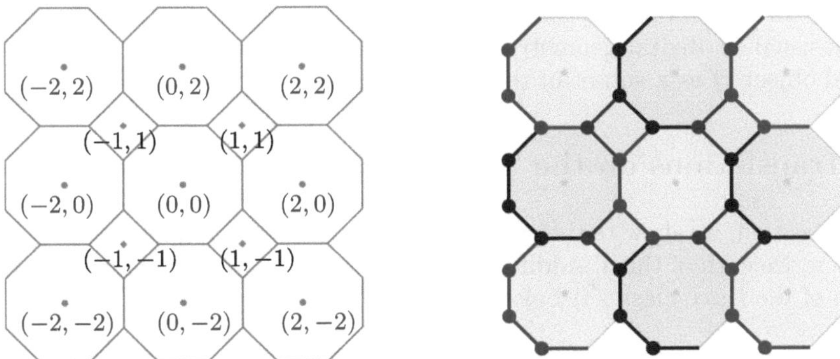

Fig. 2. Left: a segment of the grid $\mathbb{K}$ with coordinates assigned to the (centers of the) tiles. Right: digital cells, showing which sides and corners belong to each cell.

Furthermore, we assign a pixel of the grid to each $(x, y) \in \mathbb{R}^2$, thereby defining the digital cells. Inside any pixel, it is clear where the point belongs, but we also need to deal with points on the edges and corners.

Let b denote the side-length of a square (which is the same as the length of the diagonal direction sides of the octagon) and a the length of the horizontal and vertical sides of the octagon. By our measures, $a = b = 2\sqrt{2} - 2$ as it is shown in Fig. 1 (left).

For characterizing transformations of a digital grid, we define the so-called digital cells as follows.

Let for any octagon with center $(x_0, y_0) \in (2\mathbb{Z})^2$, the region $R_1 \setminus R_2$, where

$$R_1 = \left\{ (x,y) \mid x_0 - 1 \le x < x_0 + 1, y_0 - 1 \le y < y_0 + 1, \right.$$
$$\left. |x - x_0| + |y - y_0| \le \sqrt{2} + \frac{a}{2} \right\}$$
$$R_2 = \left\{ (x,y) \mid x_0 < x < 1, |y - y_0| = \sqrt{2} + \frac{a}{2} - (x - x_0) \right\}$$

be assigned.

Let for any square, with center $(x_0, y_0) \in (2\mathbb{Z} + 1)^2$, the region

$$\left\{ (x,y) \mid x_0 - \left(1 - \frac{a}{2} \right) \le x < x_0 + \left(1 - \frac{a}{2} \right), \right.$$
$$\left. y_0 - \left(1 - \frac{a}{2} \right) < y < y_0 + \left(1 - \frac{a}{2} \right), |x - x_0| + |y - y_0| \le \left(1 - \frac{a}{2} \right) \right\}$$

be assigned.

It is easy to see that in this way, every point of the plane has a uniquely assigned pixel. Figure 2 (right) explains graphically how the points on the sides and the corners of the polygons are assigned to make it uniquely defined.

As usual in digital geometry and digital image processing, we consider a digital object Q as a subset of the grid $\mathbb{K}$.

3   Translations on the Truncated Quadrille Grid

In this section, we show that there are three types of translations on the grid and characterize each of them. Additionally, for any translation vector, we determine which of the three classes it belongs to.

Definition 1. *For any vector $(x, y) \in \mathbb{R}^2$, we define its "integer part" $\lfloor (x,y) \rfloor = (\hat{x}, \hat{y})$ and its "fractional part" $\langle (x,y) \rangle = (\langle x \rangle, \langle y \rangle)$ as follows.*
The vector $\lfloor (x,y) \rfloor = (2 \cdot \lfloor \frac{x+1}{2} \rfloor, 2 \cdot \lfloor \frac{y+1}{2} \rfloor)$ is the integer part, and $\langle (x,y) \rangle = (x,y) - \lfloor (x,y) \rfloor$ is the fractional part.

Observe that $\lfloor (x,y) \rfloor \in (2\mathbb{Z})^2$, i.e., the coordinates $\hat{x}$ and $\hat{y}$ are even integers; and $\langle x \rangle, \langle y \rangle \in [-1, +1) \subset \mathbb{R}$, i.e., they are in the given (left closed, right open) interval.

In this section, we consider translations (of objects/digital polygons/shapes built up by the tiles/pixels) of the grid by a vector $(x, y) \in \mathbb{R}^2$, in general, as follows.

Definition 2. *Let $(x, y) \in \mathbb{R}^2$ be given. Then, consider a finite or infinite digital object $Q \subset \mathbb{K}$. The translation of Q by (x, y) is defined as $Q' = \{(u', v') \in \mathbb{K} \mid$ there is a pixel $(u, v) \in Q$ such that the point $(u + x, v + y) \in \mathbb{R}^2$ belongs to the pixel $(u', v')\}$.*

It is easy to see that by $\lfloor (x,y) \rfloor$, any pixel is translated to the same type of pixel; the integer part of the translation vector keeps the parity of coordinates. Therefore, the important properties of the translation, e.g., if it is bijective and digitally continuous, depend only on the fractional part of the translation vector. We can write this fact formally as follows.

Lemma 1. *For any translation vector (x,y), the integer vector $\lfloor (x,y) \rfloor$ translates the whole grid into itself: each pixel (u,v) of the object is translated to $(u,v) + \lfloor (x,y) \rfloor$. Consequently, the category and the properties of the translation by (x,y) depend only on the fractional vector $\langle (x,y) \rangle$.*

Now, we show how we can decide for any vector $\mathbf{v}(x,y)$ the category of the translation.

Theorem 1. *Let $\mathbf{v}(x,y)$ be the translation vector. Then, there are three possible categories of the translation depending solely on $\langle (x,y) \rangle$. We write the conditions for the cases depending on the sign of $\langle x \rangle$.*

  i) If the sum of the fractional parts of the coordinates fits the given condition, in the case $\langle x \rangle < 0$:

$$- \langle x \rangle + |\langle y \rangle| \leq 2 - \sqrt{2}$$

  and in the case $\langle x \rangle \geq 0$:

$$\langle x \rangle + |\langle y \rangle| < 2 - \sqrt{2},$$

  then the translation is strongly bijective: every pixel is mapped to a pixel with the same shape as the original pixel. Equality is permitted in the first condition due to the definition of digital cells.

  ii) In the case that the fractional part of the translation vector does not fit case i), but still fits the following conditions, then the translation is non-bijective. The condition for horizontal and vertical directions is

$$-1 \leq \langle x \rangle, \langle y \rangle < 1$$

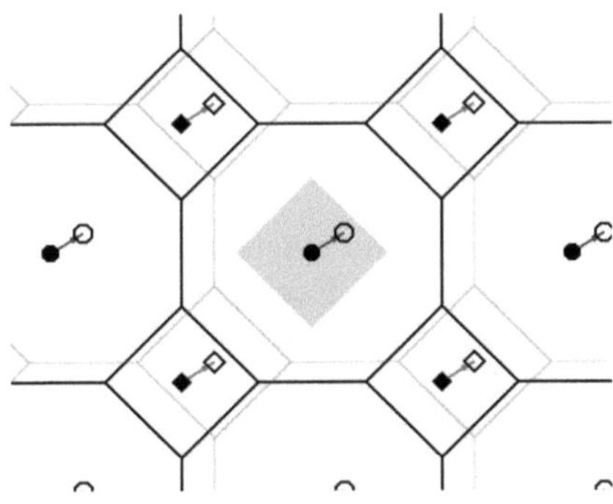

Fig. 3. Graphical explanation for the strongly bijective case.

and the condition diagonally can be expressed as

$$- \langle x \rangle + |\langle y \rangle| \leq \sqrt{2}$$

for values $\langle x \rangle \leq 0$, and

$$\langle x \rangle + |\langle y \rangle| < \sqrt{2}$$

for values $\langle x \rangle > 0$.
Every octagon is mapped to an octagon, but the images of the squares are also octagons.

iii) If the fractional part of the translation vector does not fit the previous cases, i.e., $-1 \leq \langle y \rangle \leq 1$ and for $-1 < \langle x \rangle < 0$

$$- \langle x \rangle + |\langle y \rangle| > \sqrt{2},$$

and for $0 < \langle x \rangle < 1$

$$\langle x \rangle + |\langle y \rangle| \geq \sqrt{2},$$

then the translation is
semi-bijective: every octagon is mapped to a square and vice versa.

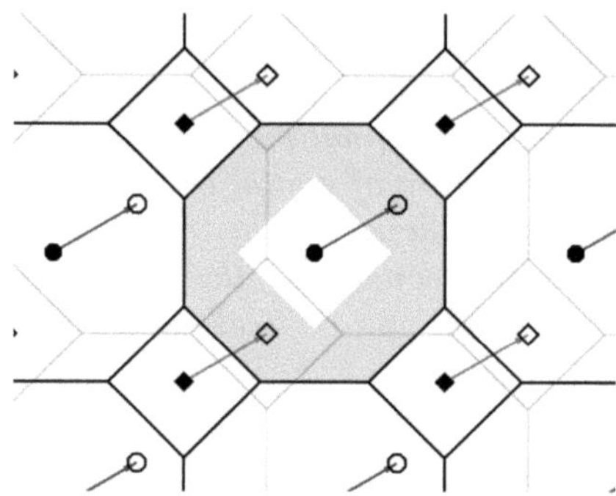

Fig. 4. Graphical explanation for non-bijective translations.

Proof. In case *i)*, as Fig. 3 shows, the fractional part of the translation vector does not change the type of the pixels. Each pixel is mapped to itself by $\langle (x, y) \rangle$. Thus, the shape and size of any translated object are identical to the original object.

In case *ii)*, as Fig. 4 shows, the fractional part of the translation vector changes the type of the square pixels. In fact, both types of pixels are mapped to octagons, and thus, these translations are generally not bijective; the transformation is not one-to-one.

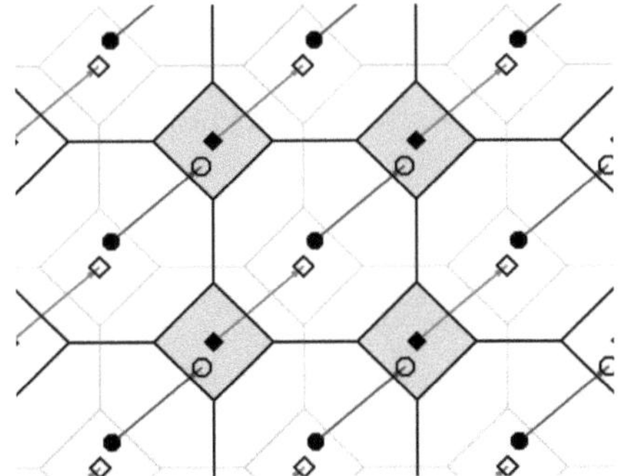

Fig. 5. Graphical explanation for semi-bijective translations.

Finally, in case *iii)*, as Fig. 5 shows, the fractional part of the translation vector changes the type of the pixels to the opposite. These translations are still bijective since the transformation remains one-to-one, but digital continuity fails, and the neighborhood structure changes.   □

Remark 1. No such translations exist where both types of pixels are mapped to squares. This is because the size of the square is smaller than the distance of the midpoints of two neighbor pixels (of any kind).

Now we analyze the three cases shown above.

As shown in Fig. 6, case *i)* is both bijective and digitally continuous. This is the case that one expects when performing a translation: it preserves the shapes, measures (area, perimeter), and connectedness. We refer to these translations as strongly bijective translations.

Case *ii)*, see also Fig. 6, illustrates the non-bijective case where some of the information is lost, as every octagon is mapped to two pixels, both an octagon and a square. These translations are neither one-to-one nor onto. There is nothing to map to the squares. The resulting image will contain many holes (all the squares), and in the octagons, the information (colors) of an octagon and one of its neighbor squares are mixed. Also, the digital continuity [18] fails in these translations, as some neighbor pixels are mapped not to neighbor cells (but to the same pixel). A hole has appeared in the example object shown in our figure. Thus, its topology has also changed, and its area is decreased. Generally, the area could change, e.g., decreasing by losing the square pixels of the translated object and also it may grow by mapping a square to an octagon.

Case *iii)*, as Fig. 6 shows, is in fact, also bijective. Each pixel is mapped into exactly one pixel. Thus, both the one-to-one and onto properties hold. However, the digital continuity can be violated: there are neighbor pixels that are mapped not to neighbor pixels, thus, in this case we refer to weakly bijective translations. As one can see in our example, a connected object becomes disconnected after the translation; thus, the topology of the object has also been changed, as well as its shape and measures, e.g., area.

4   Generalization to Non-Archimedean Cases

Let us make a more general scenario by varying the lengths of the sides of the polygons in the grid.

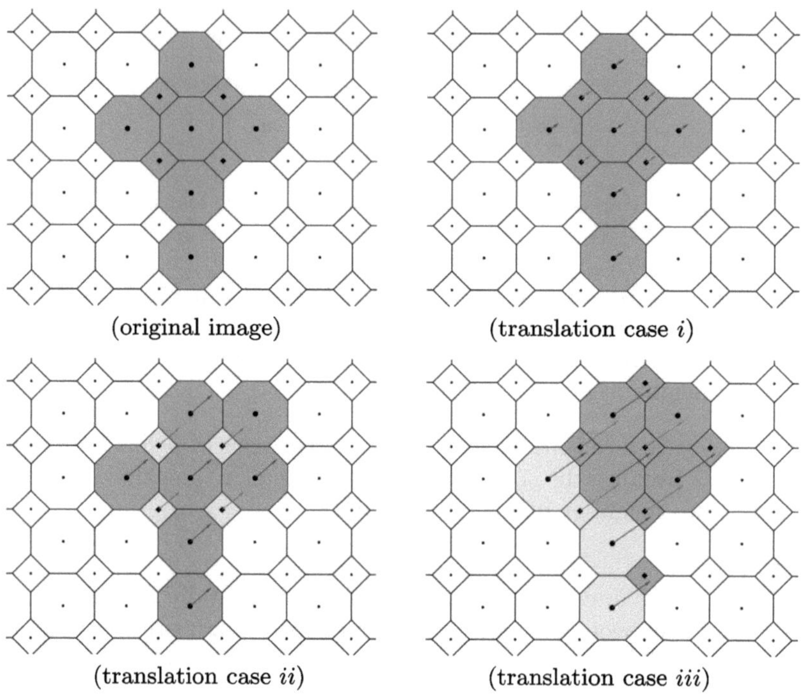

(original image)   (translation case i)

(translation case ii)   (translation case iii)

Fig. 6. Examples for translations of a digital "cross" object shown in the top left; top right: case *i)* translation vector **v** $(0.28, 0.19)$; bottom left: case *ii)* translation vector **v** $(0.62, 0.53)$; and bottom right: case *iii)* with translation vector **v** $(1.08, 0.77)$.

It is clear that the diagonal direction sides of the octagons have lengths exactly the same as the length of the sides of the squares (b). Their length can be varied between $\sqrt{2}$ and 0, while the other four sides of the octagons (the horizontal and vertical ones) can have length a between 0 and 2.

Their relation is described below, as it contains only simple geometric computations; therefore, the proof is omitted.

Lemma 2. *Let* $0 \leq a \leq 2$. *Then,* $b = \sqrt{2}\left(1 - \frac{a}{2}\right)$.
And the other way around, let $0 \leq b \leq \sqrt{2}$. *Then* $a = 2 - \sqrt{2}b$.

Now, let us define the general octagonal-square grids.

Definition 3. *Let* $a \in (0, 2)$ *be the length of the horizontal and vertical sides of the octagons. Let the octagons and squares have alternating centers as follows:*

each point (x, y) with even x and y is a center of an octagon and each point (x, y) with odd x and y is a center of a square. The tessellation/grid obtained in this way with the desired side-lengths a and b is referred to as the a-octagonal-square grid.

Notice that the neighborhood relation of any a-octagonal-square grid is the same as at a truncated square grid, i.e., in our original case. We will come back later to the extremal cases when $a \to 0$ or $a \to 2$.

Theorem 2. *Let $a \in (0, 2)$ and consider the a-octagonal-square grid. Let $\mathbf{v}(x, y)$ be the translation vector. Then, the three possible categories of the translations on the a-octagonal-square grid, based on $\langle (x, y) \rangle$, are as follows. (We list the conditions by the sign of $\langle x \rangle$.)*

i) In the case $\langle x \rangle < 0$, if

$$- \langle x \rangle + |\langle y \rangle| \leq \frac{b}{\sqrt{2}}$$

and in the case $\langle x \rangle \geq 0$, if

$$\langle x \rangle + |\langle y \rangle| < \frac{b}{\sqrt{2}},$$

then the translation is strongly bijective.

ii) If the translation does not fall into the above category but meets the following conditions, it is considered to be non-bijective. The condition for horizontal and vertical directions is

$$- 1 \leq \langle x \rangle, \langle y \rangle < 1,$$

and the condition diagonally can be expressed as

$$- \langle x \rangle + |\langle y \rangle| \leq 1 + \frac{a}{2}$$

for values $\langle x \rangle \leq 0$, and

$$\langle x \rangle + |\langle y \rangle| < 1 + \frac{a}{2}$$

for values $\langle x \rangle > 0$.
Every octagon is mapped to an octagon, but the images of the squares are also octagons.

iii) If the fractional part of the translation vector does not satisfy any of the above conditions, i.e., when it does not belong to the strongly bijective or the non-bijective cases, then for $-1 \leq \langle y \rangle \leq 1$:
and for $-1 < \langle x \rangle < 0$

$$- \langle x \rangle + |\langle y \rangle| > 1 + \frac{a}{2},$$

and for $0 < \langle x \rangle < 1$

$$\langle x \rangle + |\langle y \rangle| \geq 1 + \frac{a}{2},$$

then the translation is semi-bijective: every octagon is mapped to a square and vice versa.

Proof. The proof is similar to the proof of Theorem 1, applying the regions based on the side-lengths a and b. □

Some examples are shown in Fig. 7. As one can see, the category of the translation highly depends on the used grid, i.e., on the parameter a in this case.

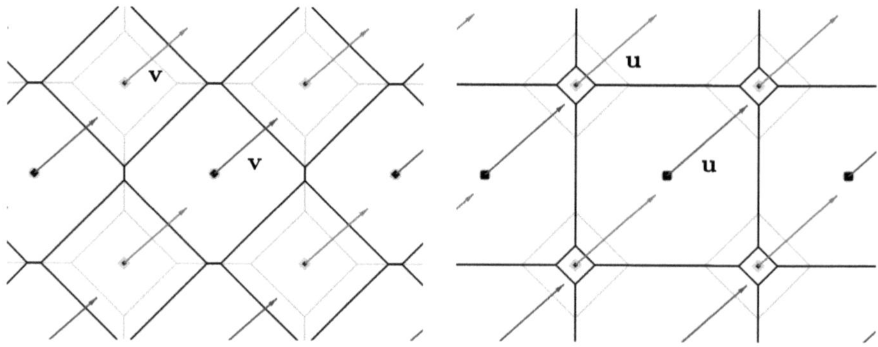

Fig. 7. The gray color grid is the semi-regular grid T(4,8,8). The black color octagonal-square grids are based on various values of a. Left: $a = 0.2$, using the same translation vector **v**, the translation on T(4,8,8) is non-bijective, while it is semi-bijective in the 0.2-octagonal-square grid. Right: $a = 1.7$, using the same translation vector **u**, the translation on T(4,8,8) is semi-bijective, while it is non-bijective on the 1.7-octagonal-square grid.

4.1   The Cases in the Limit

As we can see in the limit, each of the two grids appears to be a square grid; however, this is only apparent as explained below.

Let us start with the case when $a = 2$ (in the limit). Then, in fact, the octagons become squares, and the squares disappear, i.e., become only points and the corners of the large degenerated octagons. In this case, by Theorem 2, cases $i)$ and cases $iii)$ are reduced to very special vectors. Case $i)$ includes only the vectors of $(2\mathbb{Z})^2$, i.e., vectors that transform every octagonal-square grid into itself. This is the strongly bijective case. On the other hand, all vectors from $(2\mathbb{Z} + 1)^2$ swap the squares and octagons; i.e., every polygon is mapped to another type of polygon. These vectors produce the weakly bijective case. In fact, in this limit case, as any two-dimensional object is composed by octagons (of square shape) only, so with these type $iii)$ translations the object disappears (it becomes a set of isolated points). Further, if the translation vector does not fall into the previously described special categories, then it produces a non-bijective translation (case ii). Therefore, most translation vectors in this category and translations by them will cause information loss.

In the case where $b = \sqrt{2}$ (in the limit), the grid appears to be a square grid with the same-sized tiles; however, the neighborhood is still of the Khalimsky type. Translations are strongly bijective if the fractional part $\langle (x, y) \rangle$ of the translation vector is inside (on the boundary w.r.t. the digital cell) of the polygon determined by $|\langle x \rangle| + |\langle y \rangle| \leq 1$. If $\langle (x, y) \rangle$ is outside this region, the translation is weakly bijective. Thus, in this case, the translation is always bijective, but in half of the cases, it is not digitally continuous. This fact is very important since this topological solution, i.e., the Khalimsky topology, can be used in various places with various benefits. However, if one performs translations of images of this grid – even if not carefully looking at it, it promises nice properties – in fact, the digital continuity may fail, i.e., a point with four neighbors is mapped to a point with eight neighbors and vice versa. In this way, a translation of a Jordan curve could be a set of points that do not form a connected curve anymore. An example is shown in Fig. 8. In the top right, a Jordan curve is shown on an octagonal-square grid by red color. Its translated variant is blue. In the top left, on the dual grid representation it is clearly shown that the blue object is not connected. At the bottom, the limit case is shown, where seemingly the translated object is identical to the original. However, in this representation the dark (red and blue) pixels show the 8-adjacent squares (the pixels that originally represent octagons) and the light color pixels (pink and light blue) are the 4-adjacent pixels (representing originally squares). This highlights the fact that even if square pixels are used to represent the objects/image, if the Khalimsky topology is used, we need to be very careful with translations of the image/object because its properties may change.

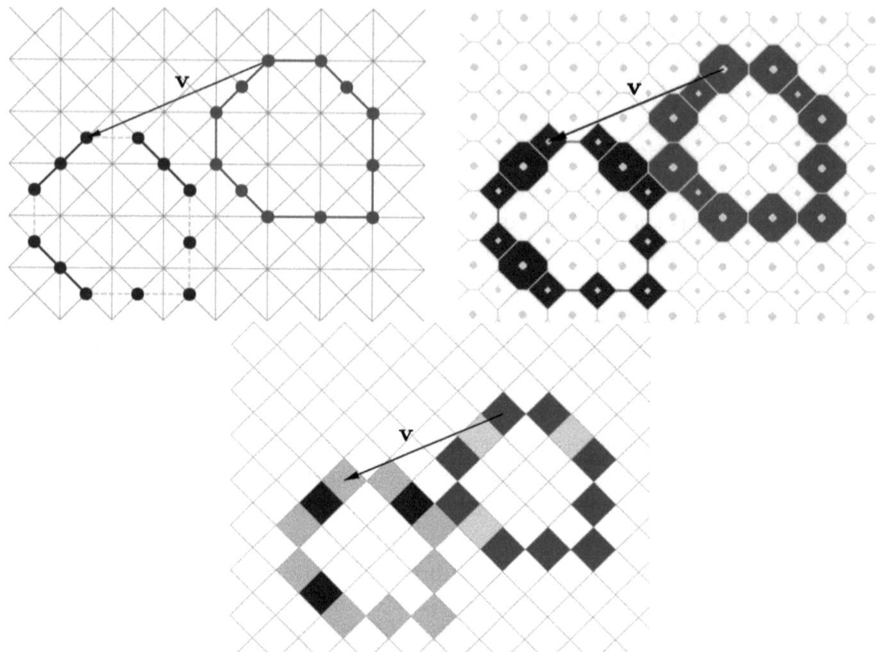

Fig. 8. A Jordan curve (red) translated by vector **v** on the Khalimsky grid and its dual representation. The resulting object (blue) is not connected. (Color figure online)

5   Conclusion

Translation of digital images seems to be the most harmless image transformation, but as we have seen, if the image grid is not rectangular or on the hexagonal grid, or the neighborhood relation used is not the standard one (side neighborhood or closest neighborhood), then we may need to face some unusual properties. Translations are characterized as strongly bijective, weakly bijective, or not bijective on the octagonal-square grids. Integer and fractional parts of the translation vectors were defined, where, roughly speaking, the integer vector gives the quantity of the translation, while the fractional part gives the quality (the category) of the translation. It is concluded that, even in the square grid, if the Khalimsky topology is used, one needs to take care about the vectors of the translations: to have nice properties, the translation vector should map each point to a similar type of point.

References

1. Abuhmaidan, K., Nagy, B.: Bijective, non-bijective and semi-bijective translations on the triangular plane. Mathematics **8**(1), 29 (2020). https://doi.org/10.3390/math8010029
2. Abuhmaidan, K., Nagy, B.: Non-bijective translations on the triangular plane. In: IEEE 16th World Symposium on Applied Machine Intelligence and Informatics (SAMI 2018), Kosice, Slovakia. pp. 183–188 (2018). https://doi.org/10.1109/SAMI.2018.8324836
3. Comic, L., Magillo, P.: Repairing binary images through the 2d diamond grid. In: Combinatorial Image Analysis. IWCIA 2020. Lecture Notes in Computer Science, vol. 12148, pp. 183–198 (2020). https://doi.org/10.1007/978-3-030-51002-2_13
4. Conway, J., Burgiel, H., Goodman-Strauss, C.: The Symmetries of Things. AK Peters (2008)
5. Khalimsky, E., Kopperman, R., Meyer, P.: Boundaries in digital planes. J. Appl. Math. Stoch. Anal. **3**, 27–55 (1990)
6. Khalimsky, E., Kopperman, R., Meyer, P.: Computer graphics and connected topologies on finite ordered sets. Topol. Its Appl. **36**, 1–17 (1990)
7. Kiselman, C.: Elements of Digital Geometry, Mathematical Morphology, and Discrete Optimization. World Scientific, Singapore (2022)
8. Kovács, G., Nagy, B., Stomfai, G., Turgay, N., Vizvári, B.: On chamfer distances on the square and body-centered cubic grids: an operational research approach. Math. Probl. Eng. **2021**(1), 5582034 (2021)
9. Kovács, G., Nagy, B., Vizvári, B.: On Weighted Distances on the Khalimsky Grid. In: Normand, N., Guédon, J., Autrusseau, F. (eds.) DGCI 2016. LNCS, vol. 9647, pp. 372–384. Springer, Cham (2016). https://doi.org/10.1007/978-3-319-32360-2_29
10. Kovács, G., Nagy, B., Vizvári, B.: Weighted distances and digital disks on the Khalimsky grid. J. Math. Imaging Vision **59**(1), 2–22 (2017). https://doi.org/10.1007/s10851-016-0701-5
11. Kovalevsky, V.: Algorithms in Digital Geometry Based on Cellular Topology. In: Klette, R., Žunić, J. (eds.) IWCIA 2004. LNCS, vol. 3322, pp. 366–393. Springer, Heidelberg (2004). https://doi.org/10.1007/978-3-540-30503-3_27
12. Kovalevsky, V.: Image Processing with Cellular Topology. Springer, Singapore (2021). https://doi.org/10.1007/978-981-16-5772-6
13. Magillo, P.: Non-square grids: a new trend in imaging and modeling? Comput. Sci. Rev. **56**, 100695 (2025)
14. Nagy, B.: Non-traditional 2d grids in combinatorial imaging – advances and challenges. In: Combinatorial Image Analysis. IWCIA 2022. Lecture Notes in Computer Science, vol. 13348, pp. 3–27 (2022)
15. Nagy, B.: A Khalimsky-like topology on the triangular grid. In: Discrete Geometry and Mathematical Morphology (DGMM) 2024. Lecture Notes in Computer Science, vol. 14605, pp. 150–162 (2024)
16. Nagy, B.: Cellular Topology on the Triangular Grid. In: Barneva, R.P., Brimkov, V.E., Aggarwal, J.K. (eds.) IWCIA 2012. LNCS, vol. 7655, pp. 143–153. Springer, Heidelberg (2012). https://doi.org/10.1007/978-3-642-34732-0_11

17. Nagy, B., Abuhmaidan, K.: A continuous coordinate system for the plane by triangular symmetry. Symmetry **11**(2), 191 (2019). https://doi.org/10.3390/SYM11020191
18. Rosenfeld, A.: 'Continuous' functions on digital pictures. Pattern Recogn. Lett. **4**(3), 177–184 (1986)
19. Wiederhold, P., Morales, S.: Thinning on Quadratic, Triangular, and Hexagonal Cell Complexes. In: Brimkov, V.E., Barneva, R.P., Hauptman, H.A. (eds.) IWCIA 2008. LNCS, vol. 4958, pp. 13–25. Springer, Heidelberg (2008). https://doi.org/10.1007/978-3-540-78275-9_2

Partial Order and Chain Decomposition of the Set of k-Subsets

Hasmik Sahakyan$^{(\boxtimes)}$ and Levon Aslanyan

Institute for Informatics and Automation Problems of NAS RA, Yerevan, Armenia
{hsahakyan,lasl}@sci.am

Abstract. Let $P(n,k)$ denote the set of all k-tuples with strictly increasing elements from the set $[n] = \{1, 2, \cdots, n\}, 1 \leq k \leq n$. Through a bijective mapping to binary sequences, $P(n,k)$ is associated with the k-th layer L_k of the binary cube B^n. On the other hand, an arbitrary subset $M \subseteq L_k$, with $|M| = r$, $r \leq \binom{n}{k}$ can be viewed as the incidence matrix of a simple k-uniform hypergraph with n vertices and r hyperedges. The hypergraph degree sequence problem is NP-complete for simple k-uniform hypergraphs, starting from $k = 3$. This motivates our study of $P(n,3)$. We investigate properties of the Hasse diagram of the partially ordered set $(P(n,k), \preceq)$, where $\preceq$ is a component-wise partial order. An algorithm is presented that constructs a set of non-intersecting chains covering all elements of $P(n,3)$. The number of these chains equals the width of $(P(n,k), \preceq)$, and they can be used for the identification of monotone functions defined on $P(n,3)$. One motivation for studying these functions is that the degree sequences of the corresponding hypergraphs can be used to generate all degree sequences of uniform hypergraphs. Moreover, monotone Boolean functions defined on B^n play a special role in generating the degree sequences of simple hypergraphs. We consider the case of 3-uniform hypergraphs, the corresponding monotone Boolean functions, and provide complexity estimates for their identification.

Keywords: Partially ordered sets · Chain decomposition · Monotone Boolean functions

1 Introduction

An application-oriented class of partially ordered sets (poset) is considered. Let $(P(n,k), \preceq)$ be a poset, where $P(n,k)$ denotes the set of all k-tuples with strictly increasing elements from the set $[n] = \{1, 2, \cdots, n\}, 1 \leq k \leq n$. The relation $\preceq$ is defined as the component-wise partial order on $P(n,k)$: for $(i_1, \cdots, i_k)$ and $(j_1, \cdots, j_k)$ in $P(n,k)$, we write $(i_1, \cdots, i_k) \preceq (j_1, \cdots, j_k)$ if and only if $i_1 \leq j_1, \cdots, i_k \leq j_k$.

To each $(a_1, \cdots, a_k) \in P(n,k)$, we assign a binary sequence $(\beta_1, \cdots, \beta_n)$, where $\beta_i = 1$ if and only if $i \in \{a_1, \cdots, a_k\}$. This mapping yields the set of vertices in the k-th layer of the binary cube B^n. Any subset M of the k-th layer of

P. Balázs et al. (Eds.): IWCIA 2025, LNCS 15985, pp. 129–149, 2026.
https://doi.org/10.1007/978-3-032-19347-6_9

B^n, with $|M| = r$, $r \leq \binom{n}{k}$ can be interpreted as the incidence matrix of a simple k-uniform hypergraph with n vertices and r hyperedges. The hypergraph degree sequence problem is known to be NP-complete for simple k-uniform hypergraphs, starting from $k = 3$. This serves as the motivation for our investigation of $P(n, 3)$.

Connections between 3-uniform hypergraphs and partially ordered sets on integer triples are also explored in [1, 2]. In [2] the authors establish a connection between 3-uniform hypergraphs with a special class D of degree sequences and a family of ideals of that poset. They further extend the class D to D^{ext}, defined as the degree sequences of the 3-uniform hypergraphs which correspond to the ideals of the poset, and spotlight on its subclass D^{ext-} including only those elements that are unique. The authors define an algorithm that allows a fast reconstruction of some instances of D^{ext-}, they also provide a heuristic to solve the reconstruction problem for the entire class. In [1], the authors study the reconstruction problem according to the number of generators of the ideals, and provide a complete characterization of the degree sequences related to ideals having one or two generators.

We also observe a connection with discrete tomography problems that include an additional constraint: the non-repeatability of rows of the binary matrix to be reconstructed. In [14] , the authors represent these problems within the hypergraph model. In particular, they prove that the set of hypergraphic sequences of simple hypergraphs with n vertices and m hyperedges is nonconvex when viewed as a subset of the n-dimensional $(m+1)$-valued lattice. They also formulate discrete tomography problems with paired projections and establish their connection to the hypergraph degree sequence problem with generalized degrees. In [16] the authors consider different types of paired projections, including the set of all pairs of columns, the set of all pairs with one fixed column, and the set of all pairs of neighboring columns; and investigate the complexity characteristics of the corresponding problems.

In this paper, we investigate properties of the Hasse diagram of the partially ordered set $(P(n, k), \preceq)$. We present an algorithm that constructs a set of non-intersecting chains covering all elements of $P(n, 3)$. The number of these chains is minimal and equals the width of $(P(n, k), \preceq)$, and they can be used for the identification of monotone functions defined on $P(n, 3)$, analogous to Hansel's chains for the identification of monotone Boolean functions (MBF) defined on B^n [11]. One motivation for studying these functions is that the degree sequences of the corresponding hypergraphs can be used to generate all degree sequences of uniform hypergraphs. Moreover, monotone Boolean functions defined on B^n play a special role in generating the degree sequences of simple hypergraphs [13, 17]. The query-based identification of MBFs is known to have a complexity of $\binom{n}{[n/2]} + \binom{n}{[n/2]+1}$ [11]. Several special subclasses of MBFs have been studied, and the complexities of their identification are analyzed – for example, in [3, 10]. In this paper, we focus on a subclass of monotone Boolean functions corresponding to 3-uniform (or $(n-3)$-uniform) hypergraphs and provide complexity estimates for their identification.

The rest of the paper is organized as follows. Section 2 presents the necessary definitions and preliminary results. Section 3 provides formulas for calculating the layer cardinalities of $(P(n,3), \preceq)$ and studies several related properties. Section 4 focuses on partitioning $P(n,3)$ into disjoint chains consisting of sequences of covering pairs. Section 5 examines monotone Boolean functions defined on $P(n,3)$ and their identification using the chain decomposition. Additionally, it considers monotone Boolean functions corresponding to $(n-3)$-uniform hypergraphs and provides complexity estimates for their identification. The paper ends with concluding remarks in Sect. 6.

2   Preliminaries

This section introduces basic concepts related to partially ordered sets (posets), following standard references such as [9,12] , and the Hasse diagram of the poset $(P(n,k), \preceq)$, where $P(n,k)$ is the set of all k-tuples with strictly increasing elements from the set $[n] = \{1, 2, \cdots, n\}$, for some $k, 1 \leq k \leq n$, and $\preceq$ is a component-wise partial order on $P(n,k)$. The necessary definitions from the domains of monotone Boolean functions and hypergraphs are also introduced.

2.1   Partially Ordered Sets

A partially ordered set (or poset) is an ordered pair $(P, \preceq)$, where P is a set and $\preceq$ is a binary relation on P satisfying the following properties: reflexivity, antisymmetry, and transitivity. We write $x \prec y$ when $x \preceq y$ and $x \neq y$.

An element x of P is called minimal if there is no element $y \in P$ such that $y \prec x$. Similarly, x is maximal if there is no element $z \in P$ such that $x \prec z$.

Two elements $x, y \in P$ are comparable if $x \preceq y$ or $y \preceq x$; otherwise, they are incomparable. A chain in a poset $(P, \preceq)$ is a subset of P in which every pair of elements is comparable. An antichain is a subset of P in which no two elements are comparable.

The height of a poset is the cardinality of its largest chain; the width is the cardinality of its largest antichain. We denote the height and width of a poset $(P, \preceq)$ by $h(P)$ and $w(P)$, respectively. In a finite poset, a chain and an antichain can share at most one common element.

Theorem 1. *(Dilworth's Theorem, [12]) Let $(P, \preceq)$ be a finite poset. Then there is a partition of P into $w(P)$ chains.*

Let x and y be distinct elements of P. We say that y covers x if $x \prec y$ but no element z such that $x \prec z \prec y$. The Hasse diagram of a poset $(P, \preceq)$ is a directed graph whose vertex set is P, with edges corresponding to covering pairs. In practice, the Hasse diagram is typically drawn so that if y covers x, the point representing y is placed higher than the point representing x. In this representation, the direction of each edge is implicit, and arrows are usually omitted.

While Dilworth's theorem partitions a poset into chains based on the general partial order (allowing transitive steps), our focus lies on decompositions into chains consisting solely of pairwise covering relations—i.e., chains that appear as directed paths in the Hasse diagram. There is no known general result guaranteeing the existence of such a covering-based chain decomposition of minimal cardinality (equal to the poset's width). In fact, it is easy to construct simple posets that cannot be partitioned into $w(P)$ increasing chains of covering elements. However, such decompositions do exist for several well-known families of posets—for instance, the Boolean cube and the poset of subcubes of the unit cube ordered by inclusion [11].

The poset that we investigate in this work, is a special order of k-element subsets of a finite set. This structure corresponds to the k-th layer of the Boolean cube.

2.2   Monotone Boolean Functions

Let $B^n = \{(x_1, \cdots, x_n)| x_i \in \{0,1\}, i = 1, \cdots, n\}$ denote the set of vertices of the n-dimensional binary cube. We define a coordinate-wise partial order on B^n: for any $\alpha = (\alpha_1, \cdots, \alpha_n)$ and $\beta = (\beta_1, \cdots, \beta_n)$ of B^n, $\alpha \preceq \beta$, if and only if $\alpha_i \leq \beta_i$ for $1 \leq i \leq n$. Let $L_k = \{(\alpha_1, \cdots, \alpha_n) \in B^n| \sum_{i=1}^n \alpha_i = k\}$. We call L_k the k-th layer of B^n.

We define a *partition* of B^n into two $(n-1)$-dimensional sub-cubes based on the values of the binary variables. For an arbitrary x_i, define:

$$B_{x_i=0}^{n-1} = \{(x_1, \cdots, x_n) \in B^n| x_i = 0\}$$

$$B_{x_i=1}^{n-1} = \{(x_1, \cdots, x_n) \in B^n| x_i = 1\}.$$

Any subset $M \subseteq B^n$ is partitioned into: $M_{x_i=1} \subseteq B_{x_i=1}^{n-1}$ and $M_{x_i=0} \subseteq B_{x_i=0}^{n-1}$.

An integer vector $S = (s_1, \cdots, s_n)$ is called the *associated vector of partitions* of the set $M \subseteq B^n$ if $s_i = |M_{x_i=1}|, i = 1, \cdots, n$ ([13]).

As an illustrative example, consider the geometric representation of B^5 shown in Fig. 1, and a subset $M = \{(00111), (01111), (10111), (10011), (11011), (11111)\}$ of B^5, for which the following partitions occur:

$$M_{x_1=0} = \{(00111), (01111)\}, M_{x_1=1} = \{(10111), (10011), (11011), (11111)\},$$

$$M_{x_2=0} = \{(00111), (10111), (10011)\}, M_{x_2=1} = \{(01111), (11011), (11111)\},$$

$$M_{x_3=0} = \{(10011), (11011)\}, M_{x_3=1} = \{(00111), (01111), (10111), (11111)\},$$

$$M_{x_4=0} = \emptyset, M_{x_4=1} = \{(00111), (01111), (10111), (10011), (11011), (11111)\},$$

$$M_{x_5=0} = \emptyset, M_{x_5=1} = \{(00111), (01111), (10111), (10011), (11011), (11111)\}.$$

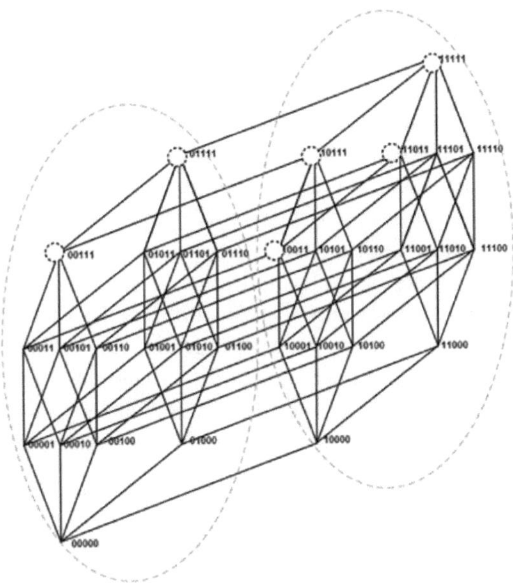

Fig. 1. Geometric representation of B^5 where the left part corresponds to $B^4_{x_1=0}$, and the right part corresponds to $B^4_{x_1=1}$.

Then, the associated vector of partitions for M will be $S = (4, 3, 4, 6, 6)$.

A Boolean function $f : B^n \to \{0, 1\}$ is monotone if, for every pair of vertices $\alpha, \beta \in B^n$, $\alpha \preceq \beta$ implies $f(\alpha) \leq f(\beta)$. The vertices of B^n where f takes the value "1" are called units of the function; vertices, where f takes the value "0" are called zeros.

α^1 is called a *lower unit* of the function if $f(\alpha^1) = 1$, and $f(\alpha) = 0$ for every $\alpha \in B^n$, such that $\alpha \prec \alpha^1$.

α^0 is called an *upper zero* of the function if $f(\alpha^0) = 0$, and $f(\alpha) = 1$ for every $\alpha \in B^n$ such that $\alpha^0 \prec \alpha$.

An example, given in Fig. 1, corresponds to a monotone Boolean function (MBF) in B^5, where the set M is the set of its units (the remaining vertices form the set of zeros), and where (00111) and (10011) represent the lower units.

2.3   Hypergraphs

A *hypergraph* H is a pair (V, E), where $V = \{v_1, v_2, \cdots, v_n\}$ is the vertex set of H, and E, the set of hyperedges, is a collection of non-empty subsets of V [5][1]. The *degree* of a vertex v of H, denoted by $d(v)$, is the number of hyperedges in H that contain v. A hypergraph H is *simple* if it has no repeated hyperedges, and it is *r-uniform* if all hyperedges contain exactly *r*-vertices.

---

[1] In a hypergraph, a hyperedge could contain more copies of the same vertex, in general. In this paper, we use a slightly different definition.

$D(H) = (d(v_1), d(v_2), \cdots, d(v_n))$ is the *degree sequence* of hypergraph H. A sequence $d = (d_1, d_2, \cdots, d_n)$ is *hypergraphic* if there exists a simple hypergraph H with degree sequence d.

The *incidence matrix* of H is an $m \times n$ binary matrix A, where the rows correspond to the hyperedges ($m = |E|$), the columns correspond to the vertices, and $a_{i,j} = 1$ if and only if the vertex v_j belongs to the i-th hyperedge.

Now consider the power set $\mathcal{P}([n])$, partially ordered by inclusion. We identify subsets of $[n]$ with binary sequences of length n, where the i-th entry is 1 if and only if the i-th element of $[n]$ is included in the subset. This establishes a one-to-one correspondence between $\mathcal{P}([n])$ and B^n. In this way, each $M \subseteq B^n$ can be identified with a simple hypergraph H on the vertex set $[n]$, where the hyperedges are non-repetitive and correspond to the elements of M. The degree of i-th vertex is equal to $|M_{x_i=1}|$, and the associated vector of partitions of M corresponds to the degree sequence of H. For example, the subset $M = \{(00111), (01111), (10111), (10011), (11011), (11111)\}$ in Fig. 1 corresponds to the hypergraph $H = ([5], E)$, where
$E = \{\{3,4,5\}, \{2,3,4,5\}, \{1,3,4,5\}, \{1,4,5\}, \{1,2,4,5\}, \{1,2,3,4,5\}\}$, and $D(H) = (4,3,4,6,6)$.

If M is a subset of the k-th layer of B^n, it can be viewed as the incidence matrix of a simple k-uniform hypergraph. The problem of determining whether a sequence is hypergraphic (hypergraph degree sequence problem) is known to be NP-complete for simple k-uniform hypergraphs, starting from $k = 3$ [8].

We also define the notion of steepest degree sequences [6,7,13].

Let $d = (d_1, \cdots, d_n)$ and $d' = (d'_1, \cdots, d'_n)$ be sequences of non-negative integers. d' is an *elementary flattening* of d if there exist indices i and j such that $d_i \geq d_j + 2$, and the components of d' are defined as follows: $d'_i = d_i - 1, d'_j = d_j + 1$, all other components of d' are equal to those of d.

We say that d is *steeper* than d' (or d' is flatter than d), if d' can be obtained from d by elementary flattenings.

A degree sequence $d = (d_1, \cdots, d_n)$ of a uniform hypergraph is called *steepest* [6,7] if and only if every sequence steeper than d is not hypergraphic. For example, $(4,3,4,6,6)$ is steepest sequence.

2.4   The Poset $(P(n,k), \preceq)$

$P(n,k) = \{(i_1, \cdots, i_k) | 1 \leq i_1 < \cdots < i_k \leq n\}$ for some $k, 1 \leq k \leq n$. The number of elements is $|P(n,k)| = \binom{n}{k}$.

We define a component-wise partial order on $P(n,k)$: for two elements $(i_1, \cdots, i_k)$ and $(j_1, \cdots, j_k)$ of $P(n,k)$, $(i_1, \cdots, i_k) \preceq (j_1, \cdots, j_k)$ if and only if $i_1 \leq j_1, \cdots, i_k \leq j_k$. Thus, $(P(n,k), \preceq)$ is a poset, whose minimal element is $(1, 2, \cdots, k)$, and maximal element is $(n-k+1, \cdots, n)$. We define the weight of $(i_1, \cdots, i_k)$ as the sum of its coordinates, $i_1 + \cdots + i_k$.

To each $(a_1, \cdots, a_k) \in P(n,k)$, we assign a binary sequence $(\beta_1, \cdots, \beta_n)$, where $\beta_i = 1$ if and only if $i \in \{a_1, \cdots, a_k\}$. This mapping yields the set of vertices in the k-th layer of the binary cube B^n. For example, the element $(2,3,5)$ of $P(5,3)$ corresponds to the vertex (01101) in B^5.

2.5   Hasse Diagram of $(P(n,k), \preceq)$

The lowest layer of the diagram consists of the unique minimal element $(1, 2, \cdots, k)$. Each successive layer l consists of all elements that cover some element in layer $l - 1$. In each layer, elements are ordered lexicographically[2] with the smallest element positioned on the right. The highest layer contains the unique maximal element $(n - k + 1, \cdots, n)$. Thus, the Hasse diagram consists of $k(n - k) + 1$, layers, numbered from 0 to $k(n - k)$. All vertices of the l layer have weight: $l + (1 + \cdots + k)$.

We define the middle layer(s) as follows:

If $k(n - k)$ is even, then there is a unique middle layer, which is the $\frac{k(n-k)}{2}$ -th layer.

If $k(n - k)$ is odd, then there are two middle layers, which are the $\frac{k(n-k)-1}{2}$ -th and $\frac{k(n-k)+1}{2}$-th layers.

According to the Dilworth's theorem $P(n, k)$ can be partitioned into $w(P(n, k))$ chains. We will prove that $w(P(n, k))$ is achieved by the vertex sets of the layers of $P(n, k)$. Our focus is restricted to the case $k = 3$ in the context of studying 3-uniform hypergraph degree sequences. In this setting, we determine $w(P(n, 3))$, and provide explicit constructions of increasing chains.

3 $(P(n, 3), \preceq)$: Layers, Properties, Quantities

In this section, we provide formulas for calculating the layer cardinalities of $(P(n, 3), \preceq)$ and study several properties that will be used to determine the layer with the largest cardinality and to construct a chain [15].

3.1   Formulas, Properties, Quantities

We enumerate the layers of $P(n, 3)$ starting from 0 to $3(n - 3)$.

Formula. Let L_l denote the l-th $l = 0, 1, \cdots, 3(n - 3)$ layer of $(P(n, 3), \preceq)$, consisting of all elements with weight equal to $l + 6$:

$$L_l = \{(i_1, i_2, i_3) | i_1 + i_2 + i_3 = l + 6, 1 \leq i_1 < i_2 < i_3 \leq n\}$$

We calculate $|L_l|$ by determining the range of feasible values for each coordinate i_j.

1. The minimal feasible value for i_1 is $\max(1, l + 6 - 2n + 1)$, and the maximal value is $\lfloor \frac{l+6}{3} \rfloor - 1$.
2. For a given feasible i_1 the minimal feasible value of i_2 is $\max(i_1 + 1, l + 6 - i_1 - n)$, and the maximal is $\lceil \frac{l+6-i_1}{2} \rceil - 1$.

---

136   H. Sahakyan and L. Aslanyan

3. For given i_1 and i_2, the value of i_3 is uniquely determined.

Thus, the formula for $|L_l|$ can be written as:

$$|L_l| = \sum_{i_1=\max(1,l+6-2n+1)}^{\lfloor \frac{l+6}{3} \rfloor - 1} \left(\left\lceil \frac{l+6-i_1}{2} \right\rceil - \max(i_1+1, l+6-i_1-n) \right)$$

To determine the layer with the greatest cardinality, the given formula is not suitable, and we proceed to study additional properties of the poset.

Symmetry. The middle layer of $(P(n,3), \preceq)$ is $L_{mid} = \frac{3(n-3)}{2}$ for odd n. When n is even, there are two middle layers: the $L_{mid+} = \frac{3(n-3)+1}{2}$-th and the $L_{mid-} = \frac{3(n-3)-1}{2}$ -th layers.

There is a symmetry with respect to the middle layer (layers). Specifically, if the j-th layer (recall that we enumerate the layers starting from 0) contains an element (i_1, i_2, i_3) for some j, then its *opposite* element, defined as: $(n+1-i_3, n+1-i_2, n+1-i_1)$, is located in the $(3(n-3)-j)$-th layer.

As an example, consider $P(9,3)$, illustrated in Fig. 2, and the vertex $(2,3,4)$ from the 3-th layer of $P(9,3)$, its opposite element $(6,7,8)$, is located on the 15-th layer.

We denote by $\hat{P}(n,3)$ the part of $P(n,3)$ above the middle layer(s), and by $\check{P}(n,3)$ the part at or below the middle layer(s) of $P(n,3)$.

Partitioning. Consider partitioning of the structure of $(P(n,3), \preceq)$ into the following three parts, denoted by $P^1(n,3), P^2(n,3)$, and $P^3(n,3)$: $P^1(n,3)$ and $P^3(n,3)$ consist of the first and last $n-3$ layers respectively, and $P^2(n,3)$ consists of the remaining $n-3+1$ layers in the middle part. The cardinalities of the layers increase in $P^1(n,3)$, and decrease in $P^3(n,3)$.

Consider a layer L_i $(0 \leq i \leq n-3-1)$ from the part $P^1(n,3)$. It is straightforward to identify one specific vertex $(1,2,i+3)$ in this layer (lexicographically, the smallest one), which will be used in subsequent analysis. Symmetrically, $P^3(n,3)$ contains the layer opposite to L_i, namely, $L_{3(n-3)-i}$, along with the corresponding vertex $(n-i-2, n-1, n)$, defined as opposite of $(1,2,i+3)$.

Obviously, the middle layer (or middle layers) of $P(n,3)$ belongs to $P^2(n,3)$, and our main focus is on $P^2(n,3)$.

Quantities in $P^1(n,3)$ and $P^3(n,3)$. Elements of i-th layer of $P^1(n,3)$ can be generated starting from the lexicographically smallest element $(1,2,i+3)$. The first part of the elements is generated by simultaneously increasing the second coordinate 2 and decreasing the third coordinate $i+3$, as long as the condition $i_1 < i_2 < i_3$ holds. For example, for $i = 0, 1$, $(1,2,3)$ and $(1,2,4)$ are the only elements, respectively. For $i = 2$, $(1,2,5)$, generates $(1,3,4), \cdots$, for $i = 5$, $(1,2,8)$ generates $(1,3,7), (1,4,6)$. Next, if $3 < i+1$, we continue generating elements, starting from $(2,3,i+1)$, and then continue by increasing 3 and decreasing $i+1$. For instance, for $i = 5$, we begin with $(2,3,6)$, then generate $(2,4,5)$. In general,

L_i consists of the following set of elements:
$\{(i_1, i_2, i_3) | i_1 = 1 + j, i_2 = 2 + j, i_3 = i + 3 - 2j, j = 0, \cdots, \lfloor \frac{1+2+i+3}{3} \rfloor - 2$ (i.e., $j \leq \lfloor \frac{i}{3} \rfloor)\}$, along with additional elements generated by increasing the second coordinate and decreasing the third, as long as this is possible. It follows that the number of such elements increases with i, and therefore $P^1(n, 3)$ has the maximum number of elements on its last $(n - 3 - 1)$-th layer. Completely analogous is the situation for $P^3(n, 3)$. It has the maximum number of elements on its first layer.

Quantities in $P^2(n, 3)$. Now consider the middle zone $P^2(n, 3)$. To simplify the notation, we enumerate the layers in $P^2(n, 3)$ separately, starting from $i = 1$ to $n - 3 + 1$, and use the superscript 2 to indicate this part: $L_1^2, \cdots, L_{n-3+1}^2$.

We construct the vertices of a particular layer L_i^2 within this zone. For simplicity, we will use a, b, c for the first, second and third coordinates, respectively. Let G_{i+1}^a denote the set of elements of L_i^2 with the first coordinate a. Then, $\bigcup_a G_{i+1}^a$ represents the entire set of elements in the layer.

The lexicographically smallest vertex is $(1, i+1, n)$, (hence the notation G_{i+1}^a is used) which serves as the starting point for our construction. For example, in $P^2(9, 3)$, this is the vertex $(1, 2, 9)$ for $i = 1$; $(1, 3, 9)$ for $i = 2$, and so on.

To find the set of all elements with the first coordinate a, we need to determine the smallest possible value for b (denoted as b_{min}) and the largest possible value for c (c_{max} - in this case is uniquely determined from the weight of the layer) for a fixed a. In general, all elements with the first coordinate a can be generated as follows: (a, b_{min}, c_{max}), $(a, b_{min} + 1, c_{max} - 1)$, $(a, b_{min} + 2, c_{max} - 2)$, $\cdots$, $(a, b_{min} + j, c_{max} - j)$, while $b_{min} + j < c_{max} - j$. The number of generated elements is $\lfloor \frac{c_{max} - b_{min} + 1}{2} \rfloor$. This type of operation is referred to as G_1-operation.

Let $a = 1$. Since $(1, i + 1, n)$ is the lexicographically smallest element, therefore, $i + 1$ is exactly b_{min}, and n is c_{max}.

Finding elements with the first coordinate $a + 1$. We need to identify the lexicographically smallest element with the first coordinate $a + 1$. This is $(a + 1, i + 1 - 1, n)$. In general,

$(1, i + 1, n)$ – is the smallest element with the first coordinate 1
$(1 + 1, i + 1 - 1, n)$ – is the smallest element with the first coordinate 2
$(1 + 2, i + 1 - 2, n)$ – is the smallest element with the first coordinate 3
$\cdots$

$(1 + j, i + 1 - j, n)$ – is the smallest element with the first coordinate $1 + j$, while $1 + j < i + 1 - j$.

Termination condition. The process terminates when:

a) for even $i+1$, $j = \frac{i-1}{2}$, and the smallest element is $(1 + \frac{i-1}{2}, i+1 - \frac{i-1}{2}, n)$
(in this case, the difference between the second and first coordinates is 1).
b) for odd $i+1$, $j = \frac{i-2}{2}$, and the smallest element is $(1 + \frac{i-2}{2}, i+1 - \frac{i-2}{2}, n)$
(in this case, the difference between the second and first coordinates is 2).

Applying the G_1-operation. We apply the G_1-operation to each of these elements, and denote the corresponding sets of generated elements as $G1_{i+1}^{1+j}$, where $j = 0, 1, \cdots, \frac{i-1}{2}$ for even $i+1$, and $0, 1, \cdots, \frac{i-2}{2}$ for odd $i+1$.

Let us illustrate this with the example in $P^2(9,3)$ for $i = 4, j = 0, 1$.

$G1_{4+1}^1$ contains $(1, 4+1, 9)$ as its first element, thus $|G1_{4+1}^1| = \lfloor \frac{9-5+1}{2} \rfloor = 2$,
$G1_{4+1}^1 = \{(1,5,9), (1,6,8)\}$.
$G1_{4+1}^2$ contains $(2,4,9)$ as its first element, thus $|G1_{4+1}^2| = \lfloor \frac{9-4+1}{2} \rfloor = 3$,
$G1_{4+1}^2 = \{(2,4,9), (2,5,8), (2,6,7)\}$.

If $i+1$ is odd, one more step is possible after $(1 + \frac{i-2}{2}, i+1 - \frac{i-2}{2}, n)$, this is: $(2 + \frac{i-2}{2}, i+1 - \frac{i-2}{2}, n-1)$, and then we again apply the G_1-operation. The set of elements obtained from this particular step is denoted by G_{i+1}^*.

In our example, G_{4+1}^* contains $(3,4,8)$ as its first element, thus $|G_{4+1}^*| = \lfloor \frac{8-4+1}{2} \rfloor = 2$, $G_{4+1}^* = \{(3,4,8), (3,5,7)\}$.

Let $G1_{i+1}$ denote the union of all these sets:

$$G1_{i+1} = \bigcup_j G1_{i+1}^{1+j} \cup G_{i+1}^*.$$

In our example,

$$G1_{4+1} = \{(1,5,9), (1,6,8), (2,4,9), (2,5,8), (2,6,7), (3,4,8), (3,5,7)\}.$$

Now, a still can be increased until it reaches the greatest possible value for a, which is: $a = \lfloor \frac{1+i+1+n-3}{3} \rfloor$. We continue increasing a. At this stage, increasing a also causes b to increase. We increase both a and b by 1 (this is the smallest b, that is b_{min}) and decrease c by 2 (this is c_{max}). Repeating this operation, which ends at the triple (a, b, c) with $a = \lfloor \frac{1+i+1+n-3}{3} \rfloor$, we obtain a new set of elements (each, followed again by the G_1-operation), which we denote by $G2_{i+1}$.

Continuing from the Last Constructed Element. If we arrive at this point from a) (even $i+1$), then the last element is:

$$\left(1 + \frac{i-1}{2}, i+1 - \frac{i-1}{2}, n\right) = \left(\frac{i+1}{2}, 1 + \frac{i+1}{2}, n\right).$$

We proceed as follows:
compose $\left(\frac{i+1}{2} + j, 1 + \frac{i+1}{2} + j, n - 2j\right)$ for $j = 1, \cdots$, until a reaches $\lfloor \frac{1+i+1+n-3}{3} \rfloor$.
If we arrive at this point from b) (odd $i+1$), then the last element is:

$$\left(2 + \frac{i-2}{2}, i+1 - \frac{i-2}{2}, n-1\right) = \left(1 + \frac{i}{2}, 2 + \frac{i}{2}, n-1\right).$$

We proceed as follows:
compose $(1+\frac{i}{2}+j, 2+\frac{i}{2}+j, n-1-2j)$ for $j = 1, \cdots$, until a reaches $\lfloor \frac{1+i+1+n-3}{3} \rfloor$.

In our example, the last obtained element was $(3, 4, 8)$, and therefore, $G2_{4+1}$ starts with $(4, 5, 6)$, which in this case is the only element.

3.2   Sizes of $G1_{i+1}$ and $G2_{i+1}$

G_{i+1}^{1+j}. $G1_{i+1}^{1+j}$ consists of elements: $(1 + j, i + 1 - j, n)$, therefore, $|G1_{i+1}^{1+j}| = \lfloor \frac{n-i+j}{2} \rfloor$, where $j = 0, 1, \cdots, \frac{i-1}{2}$ for even $i + 1$ and $j = 0, 1, \cdots, \frac{i-2}{2}$ for odd $i + 1$. Thus, we obtain:

for even $i + 1$

$$\left| \bigcup_j G1_{i+1}^{1+j} \right| = \sum_{j=0}^{\frac{i-1}{2}} \lfloor \frac{n-i+j}{2} \rfloor$$

for odd $i + 1$

$$\left| \bigcup_j G1_{i+1}^{1+j} \right| = \sum_{j=0}^{\frac{i-2}{2}} \lfloor \frac{n-i+j}{2} \rfloor.$$

G_{i+1}^*. This set exists only when $i + 1$ is odd. It starts with the element

$$\left(2 + \frac{i-2}{2}, 3 + \frac{i-2}{2}, n - 1 \right) = \left(1 + \frac{i}{2}, 2 + \frac{i}{2}, n - 1 \right),$$

and generates $\lfloor \frac{(n-1-(2+\frac{i}{2})+1)}{2} \rfloor$ elements. Therefore, $\left| G_{i+1}^* \right| = \lfloor \frac{n-2-\frac{i}{2}}{2} \rfloor$.

$G2_{i+1}$. For even $i + 1$, $G2_{i+1}$ consists of elements:

$$\left(\frac{i+1}{2} + j, 1 + \frac{i+1}{2} + j, n - 2j \right),$$

where $j = 1, 2, \cdots$, as long as $\frac{i+1}{2} + j \leq \lfloor \frac{i+n-1}{3} \rfloor$, that is, $j \leq \lfloor \frac{i+n-1}{3} \rfloor - \frac{i+1}{2}$.
For odd $i + 1$, $G2_{i+1}$ consists of elements:

$$\left(1 + \frac{i}{2} + j, 2 + \frac{i}{2} + j, n - 1 - 2j \right),$$

where $j = 1, 2, \cdots$, as long as $\frac{i}{2} + j \leq \lfloor \frac{i+n-1}{3} \rfloor$, that is, $j \leq \lfloor \frac{i+n-1}{3} \rfloor - \frac{i}{2}$.

For each valid j we obtain $\lfloor \frac{n-2j-(1+\frac{i+1}{2}+j)+1}{2} \rfloor$ elements (after applying the $G1$-operation). Therefore,

$$\left| G2_{i+1} \right| = \sum_{j=1}^{\lfloor \frac{i+n-1}{3} \rfloor - \frac{i+1}{2}} \lfloor \frac{n-2j-(\frac{i+1}{2}+j)}{2} \rfloor \text{ for even } i + 1,$$

$$\left| G2_{i+1} \right| = \sum_{j=1}^{\lfloor \frac{i+n-1}{3} \rfloor - \frac{i}{2}} \lfloor \frac{n-2j-(\frac{i}{2}+2+j)}{2} \rfloor \text{ for odd } i + 1.$$

All the above reasoning shows that $G1_{i+1}$ and $G2_{i+1}$ together contain all elements of the i layer, and since they have no elements in common, we can formulate the following statement.

Theorem 2. $G1_{i+1}$ and $G2_{i+1}$ are non-intersecting, and together they cover the i-th layer of $P^2(n,3)$.

Further analysis shows that $|G1_{i+1}| + |G2_{i+1}|$ does not decrease as i increases up to the middle layer of $P^2(n,3)$. Since there is symmetry with respect to the middle layer, this implies that the middle layer (or layers) is among those with the maximum cardinality.

4   Chain Decomposition

In this section, our goal is to partition $P(n,3)$ into disjoint chains consisting of sequences of covering pairs. Each chain must contain exactly one element of the antichain of maximum cardinality, and consequently, will pass through the middle layer/layers. Due to the symmetry of the structure, it is sufficient to construct chains for $\check{P}(n,3)$; the full construction can then be completed by applying symmetry.

Observe that, the antichain of largest cardinality in the middle layer contains the element $(1, \frac{n+1}{2}, n)$ for odd n, and the antichains of largest cardinality in the two middle layers contain $(1, \frac{n}{2}, n)$ and $(1, \frac{n}{2} + 1, n)$ respectively, for even n.

4.1   Algorithm CHAIN–SPLIT (CHS)

The algorithm consists of three steps.

1. Ordering of Elements. We begin by considering the lexicographic order of elements within each layer of $P(n,3)$;

2. Constructing chain fragments in $\check{P}(n,3)$. The first chain starts from the smallest element $(1,2,3)$. Each subsequent chain begins with the smallest (in lexicographic order) unused element of the lowest layer that still contains unused elements, and proceeds upward until it reaches layer L_{mid} for odd n, or layer L_{mid+} for even n. Each chain proceeds upward by increasing the third component of the element until it reaches n or the middle layer. If, at any step, we encounter an element that has been already used in a previous chain, we backtrack by one step and increase the second component by one, then continue increasing the third component. If increasing the second component also leads to an already used element, we backtrack again (by one step) and increase the first component by one, then continue increasing the third, and so on. This process continues until the chain reaches the middle layer or finds a deadlock.

For an element e from L_{mid} (L_{mid+}), we denote by $C(e)$ the chain that reaches this element. The first chain, starting from $(1,2,3)$, reaches $(1, \frac{n+1}{2}, n)$ for odd n, or $(1, \frac{n}{2} + 1, n)$ for even n.

As an example, consider $P(9,3)$. The following lists the elements of L_{mid}: 159, 168, 249, 258, 267, 348, 357, 456.

The chains in $\check{P}(9,3)$ constructed by the algorithm are:

$$C(159) = \{123, 124, 125, 126, 127, 128, 129, 139, 149, 159\},$$
$$C(168) = \{134, 135, 136, 137, 138, 148, 158, 168\},$$
$$C(249) = \{234, 235, 236, 237, 238, 239, 249\},$$
$$C(267) = \{145, 146, 147, 157, 167, 267\},$$
$$C(258) = \{245, 246, 247, 248, 258\},$$
$$C(357) = \{156, 256, 257, 357\},$$
$$C(348) = \{345, 346, 347, 348\},$$
$$C(456) = \{356, 456\}.$$

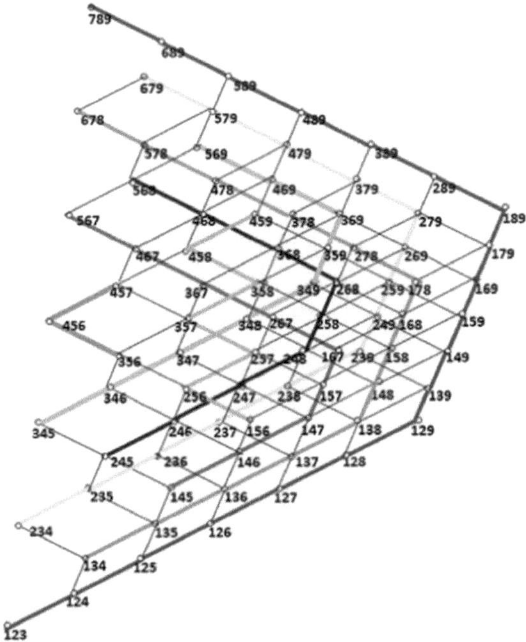

Fig. 2. Chains constructed by CHS algorithm for $P(9,3)$.

3. Extending chains to the $\hat{P}(n,3)$. Complete chains of $P(n,3)$ are constructed by extending the chains of $\check{P}(n,3)$ into the $\hat{P}(n,3)$ region. We use the symmetry property of $P(n,3)$ in the following way.

It is easy to verify that for each element $e = (i_1, i_2, i_3)$ in L_{mid}, its "opposite", i.e., the element $e^{op} = (n+1-i_1, n+1-i_2, n+1-i_3)$ also belongs to L_{mid} for odd n. When n is even, for each element $e = (i_1, i_2, i_3)$ of L_{mid-}, its "opposite" element $e^{op} = (n+1-i_1, n+1-i_2, n+1-i_3)$ belongs to L_{mid+}, and vice versa.

For the example above:

$159^{op} = 159,\ 168^{op} = 249,\ 249^{op} = 168,\ 267^{op} = 348,$
$258^{op} = 258,\ 357^{op} = 357,\ 348^{op} = 267,\ 456^{op} = 456.$

The continuation of a chain $C(e)$ from $\check{P}(n,3)$ into the $\hat{P}(n,3)$ region is represented by the chain $C^{up}(e)$, formed by taking the "opposite" elements of those in $C(e)$, taking in reverse order. For example:

$C^{up}(159) = \{149^{op}, 139^{op}, 129^{op}, 128^{op}, 127^{op}, 126^{op}, 125^{op}, 124^{op}, 123^{op}\} =$
$\{169, 179, 189, 289, 389, 489, 589, 689, 789\}$
$C^{up}(168) = \{178, 278, 378, 478, 578, 678\}$
$C^{up}(249) = \{259, 269, 279, 379, 479, 579, 679\}$
$C^{up}(267) = \{367, 467, 567\}$
$C^{up}(258) = \{268, 368, 468, 568\}$
$C^{up}(357) = \{358, 458, 459\}$
$C^{up}(348) = \{349, 359, 369, 469, 569\}$
$C^{up}(456) = \{457\}.$

As a result, the chains are as follows:

$\{123, 124, 125, 126, 127, 128, 129, 139, 149, 159, 169, 179, 189, 289, 389, 489,$
$589, 689, 789\}$
$\{134, 135, 136, 137, 138, 148, 158, 168, 178, 278, 378, 478, 578, 678\},$
$\{234, 235, 236, 237, 238, 239, 249, 259, 269, 279, 379, 479, 579, 679\},$
$\{145, 146, 147, 157, 167, 267, 367, 467, 567\},$
$\{245, 246, 247, 248, 258, 268, 368, 468, 568\},$
$\{156, 256, 257, 357, 358, 458, 459\},$
$\{345, 346, 347, 348, 349, 359, 369, 469, 569\},$
$\{356, 456, 457\}.$

The entire construction produced by the CHS algorithm is illustrated in Fig. 2 for $P(9,3)$.

About the Complexity. Observe that in constructing chains in $\check{P}(n,3)$ at each element – even if we attempt to continue by checking all three possibilities (first by the third coordinate, in case of fail by the second, and then in case of fail again, by the first) – we perform at most three operations per point. Thus, the total number of operations is at most three times the number of elements of $\check{P}(n,3)$.

Correctness of the algorithm.

1) The algorithm is deadlock-free when growing the chains in $\check{P}(n,3)$.
2) Chains constructed in Step 2 cover all elements of $\check{P}(n,3)$.

The first claim is straightforward.

To prove the second claim, we recursively apply the following Pascal's identity:

$$\binom{n}{r} = \binom{n-1}{r} + \binom{n-1}{r-1}.$$

For $P(n,3)$ we get:

$$\binom{n}{3} = \binom{n-1}{2} + \binom{n-2}{2} + \binom{n-3}{2} + \cdots + \binom{3}{2} + \binom{2}{2}.$$

Thus, $P(n,3)$ can be considered as the union of $P(n-1,2)$, $P(n-2,2)$, $P(n-3,2)$, $\cdots$, $P(3,2)$, $P(2,2)$.

For example, $P(9,3)$ will be partitioned in the following way:

$$
\begin{aligned}
P(8,2) =\{&123, 124, 125, 126, 127, 128, 129,\\
&134, 135, 136, 137, 138, 139,\\
&145, 146, 147, 148, 149,\\
&156, 157, 158, 159,\\
&167, 168, 169,\\
&178, 179,\\
&189\}\\
P(7,2) =\{&234, 235, 236, 237, 238, 239,\\
&245, 246, 247, 248, 249,\\
&256, 257, 258, 259,\\
&267, 268, 269,\\
&278, 279,\\
&289\}\\
P(6,2) =\{&345, 346, 347, 348, 349,\\
&356, 357, 358, 359,\\
&367, 368, 369,\\
&378, 379,\\
&389\}\\
&\cdots\\
P(3,2) = \{&678, 679, 689\}\\
P(2,2) = \{&789\}.
\end{aligned}
$$

The symmetry property for $P(k,2)$ can be formulated as follows:
if the j-th layer contains an element (i_1, i_2), then its opposite element, defined as $(k+1-i_2, k+1-i_1)$, is located in the $(2(k-2)-j)$-th layer.

Since we treat $P(k,2)$ instances as partition blocks for $P(n,3)$, an element (i_1, i_2) in $P(n-1,2)$ becomes $(1, i_1+1, i_2+1)$ in $P(n,3)$. In general, for $P(n-r,2)$, it becomes (r, i_1+r, i_2+r). In this way, while the opposite of (i_1, i_2) in

$P(n-1, 2)$ is $(n - i_2, n - i_1)$, the opposite of $(1, i_1 + 1, i_2 + 1)$ in $P(n, 3)$ is $(n + 1 - (i_2 + 1), n + 1 - (i_1 + 1), n + 1 - 1)$, i.e., $(n - i_2, n - i_1, n)$.

The second claim, when applied to $P(k, 2)$, can be proved easily. Thus, all we need to do is consider the structure of l -th layer of $P(n, 3)$ as a union of $\leq l$ layers of $P(n - i, 2)$, for the corresponding values of i, and then combine the results accordingly.

The chain decomposition algorithm can also be applied for $k > 3$. Figure 3 illustrates this for $P(7, 4)$.

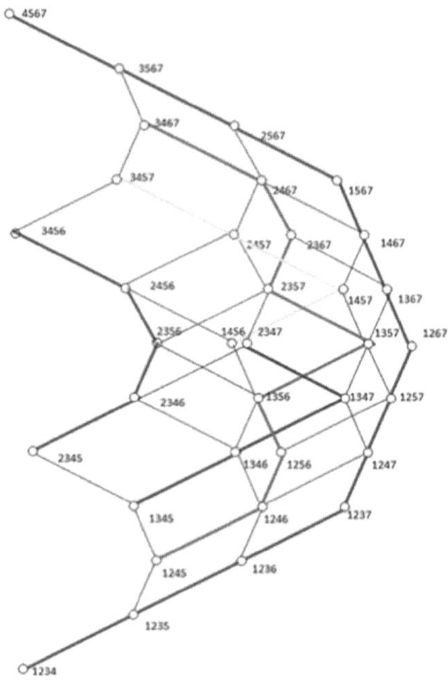

Fig. 3. Chain decomposition for $P(7, 4)$.

5 Monotone Recognition Using Chains

In this section, we consider monotone functions defined on $P(n, 3)$. The chain decompositions described in the previous sections can be used to identify these functions. One motivation for studying these functions is that the degree sequences of the corresponding hypergraphs can be used to generate all possible degree sequences of 3-uniform hypergraphs.

Moreover, monotone Boolean functions defined on B^n play a similar generative role in characterizing hypergraph degree sequences ([13]). In particular, we consider monotone Boolean functions corresponding to 3-uniform (or equivalently, $(n - 3)$-uniform) hypergraphs and provide complexity estimates for their identification.

5.1   Monotone Functions Defined on $P(n,3)$

Let $d = (d_1, \cdots, d_n)$ and $d' = (d'_1, \cdots, d'_n)$ be sequences of non-negative integers. Recall that d is steeper than d' (or d' is flatter than d), if d' can be obtained from d by elementary flattenings, and a degree sequence $d = (d_1, \cdots, d_n)$ of a uniform hypergraph is called steepest if every sequence steeper than d is not hypergraphic.

Theorem 3. *[13] If d is a hypergraphic sequence of a simple hypergraph with n vertices and m hyperedges then every sequence d' that is flatter than d is also hypergraphic.*

The theorem has its analogue for uniform hypergraphs [6]. It follows from Theorem 3 that all degree sequences can be determined by the steepest ones.

It turns out that the incidence matrices of k-uniform hypergraphs whose degree sequences possess the steepest property correspond to monotone functions defined on $P(n,k)$ ([6,7,13]). This motivates our investigation into the query-based recognition of such functions defined on $P(n,3)$.

A function $f : P(n,3) \rightarrow \{0,1\}$ is called monotone if, for every pair of elements $a, b \in P(n,3)$, $a \preceq b$ implies $f(a) \le f(b)$.

It is easy to verify that the following theorem holds.

Theorem 4. *Let $P(k,2)$ be the partition blocks for $P(n,3)$, $k = 2,3, \cdots, n-1$. Let F be a binary function defined on $P(n,3)$, and let f_k be the corresponding parts of the function in $P(k,2)$, such that for every $(i_1, i_2) \in P(k,2)$, $f_k(i_1, i_2) = 1$ if and only if $F(k, i_1 + k, i_2 + k) = 1$. Then, the monotonicity of F implies also the monotonicity of all functions f_k.*

An algorithm for the recognizing F can be described as follows:

1) Partition $P(n,3)$ into blocks $P(k,2)$, $k = 2,3, \cdots, n-1$; and adjust the elements, accordingly, in $P(n,3)$, i.e., an element (i_1, i_2) in $P(n-1,2)$ becomes $(1, i_1 + 1, i_2 + 1)$ in $P(n,3)$, (i_1, i_2) in $P(n-2,2)$ becomes $(2, i_1 + 2, i_2 + 2)$ in $P(n,3)$, and so on.
2) Define a binary function $f_k : P(k,2) \rightarrow \{0,1\}$ for each block, where $f_k((i_1, i_2)) = 1$ if and only if $F((k, i_1 + k, i_2 + k)) = 1$, for every $(i_1, i_2) \in P(k,2)$, and $(k, i_1 + k, i_2 + k)$ is the corresponding element in $P(n,3)$.
3) Recognize each f_k individually using an appropriate algorithm (e.g., using chain decomposition).
4) Combine the recognition results of all f_k to reconstruct F.

5.2   Monotone Boolean Functions on B^n that Take the Value 0 on All Vectors Below the $(n-3)$-Rd Layer

Let $D_m^k(n)$ denote the set of degree sequences of simple k-uniform hypergraphs with n vertices and m hyperedges. Our focus is on $D_m^{n-3}(n)$ (or equivalently, $D_m^3(n)$). Recall that in this case, $m \le \binom{n}{3}$.

Theorem 5. *For every sequence $d \in D_m^{n-3}(n)$, there exists a monotone Boolean function defined on B^n with exactly m units located on $(n-3)$rd layer and/or layers above it in B^n, such that its associated vector of partitions is greater than or equal to d.*

Proof. Let $d \in D_m^{n-3}(n)$, and let M be an m-set on the $(n-3)$-the layer of B^n, whose associated vector of partitions is d. We construct a monotone Boolean function by transforming M into a set M' (by lifting vertices of M upward in the cube as far as possible), where M' represents the set of units of the desired function. The transformation is performed using the algorithm TRANSFORM–MONOTONE, which takes M as input and outputs the required set M' .

Let us first provide an explanation of the stages of the algorithm:

- if all vertices in B^n, which are greater than v, are in M' – this means that v cannot be lifted, as all its upper neighbors are contained in M'. Therefore, v can be considered a unit of a MBF, and we add it to M',
- while there is a vertex $v' \in B^n$ such that $v' > v$ and $v' \notin M'$ – this indicates that there is an upper neighbor v' of v that is not yet in M'. In this case, we can replace v with v' and continue the process.

Algorithm . TRANSFORM–MONOTONE(M)

1: $M' \leftarrow \emptyset$
2: **while** $M \neq \emptyset$ **do**
3:     Take any vertex $v \in M$
4:     $M \leftarrow M \setminus \{v\}$
5:     **if** all vertices in B^n greater (component-wise) than v are in M' **then**
6:         $M' \leftarrow M' \cup \{v\}$
7:     **else**
8:         **while** there exists a vertex $v' \in B^n$ such that $v' > v$ and $v' \notin M'$ **do**
9:             Take any such vertex v'
10:             $v \leftarrow v'$
11:         **end while**
12:         $M' \leftarrow M' \cup \{v\}$
13:     **end if**
14: **end while**
15: **return** M'

After these constructions M' corresponds to a monotone Boolean function, represented by the set of its units.

Notice that during each replacement of a vertex v of M with a greater one, a 0 in some coordinate of v is replaced with 1, thereby increasing that coordinate while keeping all others the same. As a result, the final set M' will have an associated vector d' greater than d. $\qquad \square$

The work of the algorithm is illustrated in the following example.

Let $M = \{(00011), (00110), (01010), (10010), (11000)\}$ consist of elements of the second layer of B^5, and $d = (2, 2, 1, 4, 1)$.

We start with any vertex, let it be (00011). If we first increase the first coordinate, then the second, and then the third, we obtain the following chain: $(00011) < (10011) < (11011) < (11111)$, and thus we put (11111) into M' and delete (00011) from M.

Next, we consider (00110). If we first increase the first coordinate, then the second, we obtain the following chain: $(00110) < (10110) < (11110)$, and the only vertex greater than (11110) is (11111), which is already in M', therefore, we put (11110) into M' and delete (00110) from M.

The next vertex is (01010).
$(01010) < (11010) < (11011)$. The only vertex greater than (11011) is (11111), which is already in M', and thus we put (11011) into M' and delete it from M.

Next is (10010). $(10010) < (11010)$, and there are two vertices greater than (11010) – (11110) and (11011) – both are in M', and thus we put (11010) into M' and delete (01010) from M.

The last is (11000).
$(11000) < (11100) < (11101)$. Such that (11111) is in M', we put (11101) into M', delete (11000) from M, and finish with
$M' = \{(11111), (11110), (11011), (11010), (11101)\}$.
This is the set of units of a monotone Boolean function, and $d' = (5, 5, 3, 4, 3)$.

In this manner, if we characterize $D_m^{n-3}(n)$ in terms of monotone Boolean functions, it is sufficient to consider only those functions whose lower units are located on $(n - 3)$-the and/or higher layers of B^n.

In this regard, we consider a class of monotone Boolean functions f, such that $f(\alpha) = 0$ for all α below the k-th layer of B^n, where $\lfloor \frac{n}{2} \rfloor < k < n$ (the k-th layer itself is not included). We denote this class as M_n^k.

Theorem 6. *([4]) $\varphi(n) = \binom{n}{k} + \binom{n}{k-1}$ for the class of functions M_n^k, where $\varphi(n)$ is the Shannon complexity for query-based recognition.*

Thus, for the functions of interest, the complexity is $\binom{n}{3} + \binom{n}{4}$.

6   Conclusion

We studied the partially ordered set $P(n, 3)$, consisting of all 3-tuples of strictly increasing elements from the set $[n]$, and its interpretation as third layer of the binary cube B^n. Through a bijective mapping to binary sequences, we connected this structure to the representation of simple 3-uniform hypergraphs. We considered subsets of $P(n, 3)$ that serve as the zeros of monotone binary functions defined on $P(n, 3)$. The degree sequences of hypergraphs corresponding to these subsets can be used to generate all degree sequences of 3-uniform hypergraphs.

This motivates the study and recognition of such functions, which can be facilitated through the chain decomposition of $P(n,3)$, developed in this work.

Furthermore, we explored the connection between monotone Boolean functions defined on B^n and the degree sequences of simple 3-uniform hypergraphs. Any subset of this layer can be interpreted as the incidence matrix of a simple 3-uniform hypergraph with n vertices. Symmetrically, any subset of the $(n-3)$rd layer of B^n corresponds to the incidence matrix of a simple $(n-3)$-uniform hypergraph. Any subset M of the $(n-3)$rd layer can be transformed into a set M' by lifting its vertices upward in the cube as far as possible. In doing so, M' becomes the set of units of a MBF that has no units below the $(n-3)$rd layer and whose associated vector of partitions dominates that of M. Thus, the set of all degree sequences of simple $(n-3)$-uniform hypergraphs can be generated by this special subclass of MBFs. Motivated by this observation, we investigated this subclass and provided complexity estimates for its query-based recognition.

Acknowledgments. This research was supported by the Higher Education and Science Committee of MESCS RA (Research project number 25SRNSF-1B022).

References

1. Ascolese, M., Frosini, A., Pergola, E., Rinaldi, S., Vuillon, L.: An algebraic approach to the reconstruction of uniform hypergraphs from their degree sequence. Theoret. Comput. Sci. **1020**, 114872 (2024)
2. Ascolese, M., Frosini, A., Pergola, E., Rinaldi, S., et al.: A heuristic for the p-time reconstruction of unique 3-uniform hypergraphs from their degree sequences. In: CEUR Workshop Proceedings. vol. 3587, pp. 77–91. CEUR-WS (2023)
3. Aslanyan, L., Katona, G., Sahakyan, H.: Shadow minimization boolean function reconstruction. Informatica **35**(1), 1–20 (2024)
4. Aslanyan, L., Sahakyan, H.: Chain split and computations in practical rule mining. Inf. Sci. Comput. Int. Book Ser. **8**, 132–135 (2009)
5. Berge, C.: Hypergraphs: combinatorics of finite sets, vol. 45. Elsevier (1984)
6. Billington, D.: Lattices and degree sequences of uniform hypergraphs. Ars Combin. **21**, 9–19 (1986)
7. Billington, D.: Conditions for degree sequences to be realisable by 3-uniform hypergraphs. J. Comb. Math. Comb. Comput. **3**, 71–91 (1988)
8. Deza, A., Levin, A., Meesum, S.M., Onn, S.: Optimization over degree sequences. SIAM J. Discret. Math. **32**(3), 2067–2079 (2018)
9. Engel, K.: Sperner theory, vol. 65. Cambridge University Press (1997)
10. Gerbner, D., et al.: Identification of a monotone boolean function with k "reasons" as a combinatorial search problem. Discrete Appl. Math. **378**, 703–709 (2026)
11. Hansel, G., et al.: Sur le nombre des fonctions booléennes monotones de n variables. CR Acad. Sci. Paris **262**(20), 1088–1090 (1966)
12. Roman, S.: Lattices and ordered sets. Springer Science & Business Media (2008)
13. Sahakyan, H.: Essential points of the n-cube subset partitioning characterisation. Discret. Appl. Math. **163**, 205–213 (2014)
14. Sahakyan, H.: Some discrete tomography problems in hypergraph model interpretation. Pattern Recognit Image Anal. **34**(1), 116–125 (2024)

15. Sahakyan, H., Aslanyan, L.: Chain split of partial ordered set of k-subsets. New Trends in Information Technologies, pp. 55–65 (2010)
16. Sahakyan, H., Aslanyan, L.: Discrete tomography problems with paired projections and complexity characteristics. In: Olenev, N., Evtushenko, Y., Jaćimović, M., Khachay, M., Malkova, V. (eds) International Conference on Optimization and Applications. pp. 268–280. Springer, Cham (2023). https://doi.org/10.1007/978-3-031-48751-4_20
17. Sahakyan, H., Aslanyan, L., Ryazanov, V.: On the hypercube subset partitioning varieties. In: 2019 Computer Science and Information Technologies (CSIT). pp. 83–88. IEEE (2019)

Similarity of 2D Thinning Algorithms and Their k-Attempt Versions

Gábor Németh$^{(\boxtimes)}$

Department of Image Processing and Computer Graphics, Institute of Informatics,
University of Szeged, Szeged, Hungary
`gnemeth@inf.u-szeged.hu`

Abstract. Thinning algorithms consist of iterative object reduction steps, which are repeated until stability is reached. In each iteration step, a set of points in the outermost layer of the current object is investigated for possible deletion. In the concept of a k-attempt version of a thinning algorithm, any object point is investigated at most k-times ($k = 1, 2, \ldots$). If a point "survives" the deletion attempts k times, then it is considered as an element of the produced centerline and not investigated anymore. This concept yields a faster thinning approach; however, the result may not be the same as we could get without any limitation of deleting attempts. In this work, we compare a set of existing thinning algorithms without any restriction of deletion attempts and their k-attempt versions. Our aim with this is twofold. We would like to investigate that, what the drawback is in the similarity of the k-attempt version of a thinning algorithm regarding the original one. On the other hand, if the k-attempt version of a thinning algorithm produces the same result as the original set of objects, then it is worth investigating the proof in further work.

Keywords: Skeletonization · Thinning algorithms · Quantitative comparison · k-attempt thinning

1 Introduction

Skeleton is a region-based shape descriptor that summarizes the general form of objects [30,31]. An illustrative definition of the skeleton is given using the prairie-fire analogy: the boundary of the object is set on fire and formed by the loci where the fire fronts meet and extinguish each other [5]. Since the mathematical skeleton is continuous, some skeletonization algorithms have been developed, that approximates it for discrete binary objects. On digital binary pictures [16], a frequently used skeletonization method is thinning [30,33], and there are two kinds of skeleton-like shape features in 2D, the centerline and the topological kernel [31]. The centerline is a discrete approximation of the skeleton, and the topological kernel is the minimal set of points that is topologically equivalent to the original object. Note that, in this work, we deal with only thinning algorithms that produce centerlines.

P. Balázs et al. (Eds.): IWCIA 2025, LNCS 15985, pp. 150–167, 2026.
https://doi.org/10.1007/978-3-032-19347-6_10

These features are very useful when the topology or the simplified structure of a shape plays an important role for further processing. It is used for optical character recognition [6,17,18,32], while in computer vision and medical applications (e.g., structural features of blood vessels [30] is still applied as a feature extraction step.

In digital binary pictures, the black points form the objects, while the white ones belong to the background and cavities. A reduction operation alters a set of black points to white ones, which is referred to as deletion. Thinning algorithms consist of iterative reductions, which are repeated until stability is reached. In each iteration step, the set of points in the outermost layer of the current object are investigated for possible deletion. The shape representation is ensured by some geometrical constraints [7,13]. There exist several thinning approaches. In sequential thinning, the reduction operation deletes at most one point at a time, while parallel reductions can remove a set of points simultaneously. On the other hand, there are three different parallel thinning strategies relating to selecting the investigation of points to be deleted: the fully parallel, the subiteration, and the subfield-based approaches [17,21,23,24,24,30]. In fully parallel thinning algorithms, the same parallel reduction operator is applied at every phase of the thinning process [3,4,21,23,34]. An iteration step of subiteration-based algorithms is divided into a number of successive subiterations, where only border points of a certain kind can be deleted [8,10,21,23,29,35]. Subfield-based algorithms partition the digital space into subfields that are alternatively activated, and only some points in the active subfield can be deleted [10,20,21,23,24]. Applying different thinning strategies for the same picture, they may result in various centerlines. The differences in the results may occur as spurious side branches, 1-point thin curve segments in the even-wide object parts, or angular or curvy segments. Note that there exist some solutions in skeletonization that apply multiscale skeletonization to overcome this problem [2].

In conventional thinning algorithms [1,7,13,14,21,23], when a point in the actual object boundary is not deletable in the current iteration step, then it will be investigated in the next iteration step again. Palágyi and Németh introduced the concept of k-attempt thinning and proved this property for some thinning algorithms, i.e., the deletability of any point is sufficient to check at most k-times [24–28]. We do not state that any of the investigated algorithms in this paper holds the k-attempt property; however, it can be seen that the k-attempt version of any algorithm can be faster (because of the limited point checking).

In this paper we investigate two questions:

1. If the k-attempt version of a thinning algorithm does not produce the same centerline as the conventional one (i.e., if each point is investigated at most k times, and then it is considered an element of the final centerline,) what is the difference between the two centerlines, and is it faster than the conventional one?
2. Which conventional algorithm and its k-attempt version produce the same centerlines considering for some $k = 1, 2, 3, 4, 5$ for a set of various test images?

The answer for the first question is interesting when the computation time is more important than the 'accuracy' of the produced centerline. The second one will show us which algorithms can be "*potentially equivalent*" with the k-attempt version and which are not. If an algorithm and its k-attempt version produce the same centerline for a set of test images, then it is worth it to investigate the proven equivalency later, but we bypass those ones, where the results differ for any test image. We have evaluated a quantitative comparison of 17 pairs of parallel 2D thinning algorithms and their k-attempt versions for 160 images.

The rest of this paper is organized as follows: Sect. 2 deals with the concepts and properties of the k-attempt version of any thinning algorithm. Then, the base of the comparison is described in Sect. 3. We summarize our results in Sect. 4. Finally, we round off the paper with some conclusions and remarks.

2 k-Attempt Thinning

In each iteration step of a thinning algorithm, a set of 'deletable' black points is removed. The process is terminated when, in the last iteration step, there is no deletable point remaining in the image.

A drawback of the conventional thinning scheme is that it may consider each point during the successive iteration steps multiple times. In some cases, when the original elongated object has some thick parts (e.g., the lamp) and some thin parts, (e.g., wrought iron part of a lamp) in Fig. 2, the conventional thinning algorithm investigates the already thinned and finished segments during the peeling of the outermost layers of the lamp in iteration steps. This means that checking the border point many times on the wrought iron parts is unnecessary, since there is no point to remove when there are only "finished thin" curve segments on it. To avoid re-checking these kinds of border points, Palágyi and Németh introduced the fixpoints (i.e., safe points in the produced centerlines) in the thinning process [25] and later the concept of k-attempt thinning as well [24,26–28]. In k-attempt thinning, any border point can be investigated at most k-times by a reduction operation, and if a point "survives" these successive deleting attempts, then it becomes an element of the final result. Figure 1 helps to understand the differences between the conventional and the k-attempt version of any thinning algorithm. Let us consider the transition graphs of the points. At the beginning, each point in a binary picture is either black or white. Then, during a traversal scanning of the image, the set of black points are separated into *interior* and *border* points. Interior points are surrounded by a set of black points, while border points are adjacent to at least one white point. When a reduction deletes a point is that adjacent to an interior point, then a point is that interior point becomes a border point. In both of the thinning schemes depicted in Fig. 1, a border point remains border point when it is not 'deletable' or it turns white otherwise. However, in the concept of k-attempt thinning, a counter is assigned to each border point, and when it is not 'deletable', then the counter is increased. After "surviving" k-attempts, then it is considered a firm element of the produced centerline and cannot be deleted anymore, and it is not necessary to visit it again and again in further iteration steps.

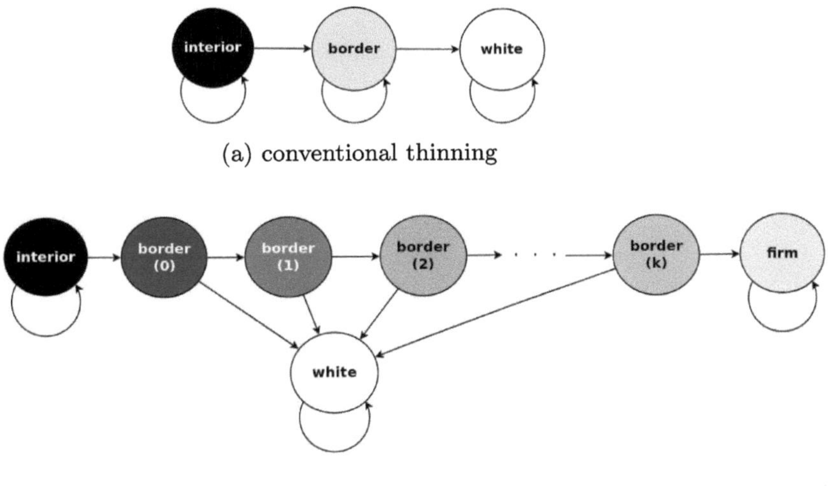

(a) conventional thinning

(b) k-attempt thinning

Fig. 1. Transitions graphs assigned to the conventional and k-attempt thinning schemes.

As shown in Fig. 1, in the conventional thinning scheme, there can be only three states of the points, namely white, (black) interior, and (black) border. In stark contrast, the k-attempt versions of thinning algorithms, we use $k+3$ states. In practical implementation, an element of an image array contains 0 and 1 for white and interior points, respectively, and the border points are denoted by 2 in the conventional thinning scheme. In k-attempt thinning, we count the attempts in image array; hence, the value of the cell becomes $k+1$ (i.e., 2, 3, ..., and so on) when they are border points to be investigated, and $k+2$ after k investigations for possible deletion. Figure 3 illustrates how the object in an image is reduced during thinning steps. Figure 3a shows a crop of an image at iteration 9, where all the states of points can be seen in the highlighted part. Since the points designated to "firm" are not investigated anymore for possible deletion during the thinning process of the k-attempt thinning scheme, this can yield to speed-up the thinning process. Figure 2 shows how the count of border points changes during the iteration steps of conventional thinning and its 1-attempt version.

Since, thanks to the limitation of the attempts for deletion, the produced centerline can be different for the conventional and the k-attempt version of a thinning algorithm. An interesting question is emerging: how different are the extracted centerlines compared to the original results?

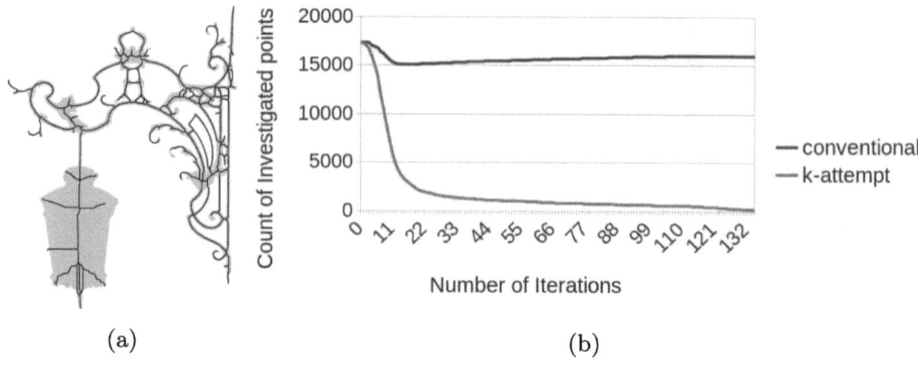

Fig. 2. Number of investigated points in each iteration step. The image of a lamp and its centerline superimposed on it (a). The blue curve and the red curve in the diagram, respectively, show how the counts of border points to be investigated changed during the thinning process (b). (Color figure online)

3   Concept of Comparison of Conventional and k-Attempt Thinning

There are many ways of comparing the results of thinning algorithms [17,18,22]. In this work we investigate whether a k-attempt version of a thinning algorithm could be "potentially equivalent" to the conventional one, and if not, then how wrong it is.

For quantitative comparison of the centerlines, we use the similarity (error) metric introduced by Németh et al. [22]. We have to note that this error metric is designed to compare centerlines and say nothing about how the centerline is related to the original shape of the object. For simplicity, let us denote D_P the Euclidean distance map [9] computed for each white point q from the set of points P, i.e., each white point q has a distance value from the nearest point $p \in P$. A distance map is illustrated in Fig. 4b. For an image I containing the elongated object and its centerline C, we compute the normalized distance maps $\overline{D}$ as follows.

$$\overline{D}_{C,W_I} = \frac{D_C}{D_C + D_{W_I}} \tag{1}$$

where W_I is the set of white points in image I; furthermore, D_{W_I} and D_C are Euclidean distance maps computed for objects from the set of points W_I and centerline C, respectively. A normalized distance map is depicted in Fig. 4e. Let us consider the similarity measure $AA_I(C_1, C_2)$ for two centerlines C_1 and C_2 as the following:

$$AA_I(C_1, C_2) = \frac{\delta(C_2, \overline{D}_{C_1,W_I}) + \delta(C_1, \overline{D}_{C_2,W_I})}{2} \tag{2}$$

$$\delta(P, \overline{D}_{Q,W_I}) = \frac{1}{\#P} \sum_{p \in P} \overline{D}_{Q,W_I}(p) \tag{3}$$

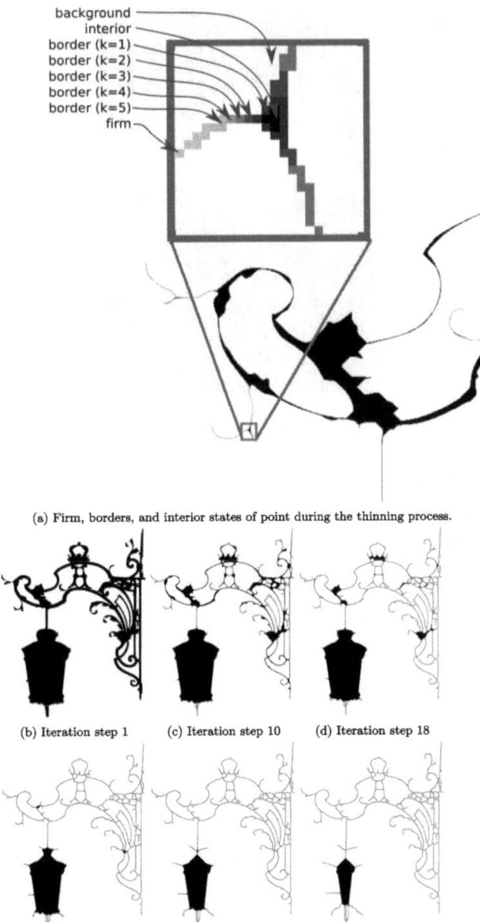

(a) Firm, borders, and interior states of point during the thinning process.

(b) Iteration step 1   (c) Iteration step 10   (d) Iteration step 18

Fig. 3. Snapshots of iteration steps of a 5-attempt thinning. The border points are labeled with different k values according to the attempt of deletion (a). Results in six different iteration steps during the thinning of an image of a lamp.

where $\mathcal{C}_1$ and $\mathcal{C}_2$ are two centerlines, $\delta(P, \overline{D})$ is the average of the normalized distance values located in the elements of P, and AA is the average of the average of normalized distances.

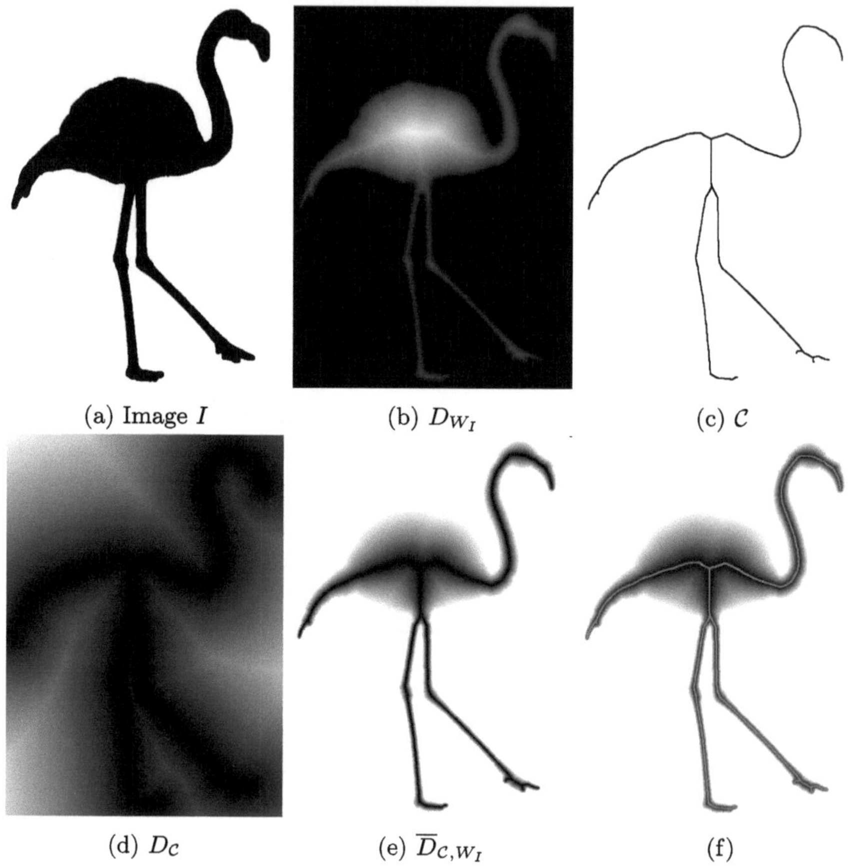

(a) Image I (b) D_{W_I} (c) $\mathcal{C}$

(d) $D_{\mathcal{C}}$ (e) $\overline{D}_{\mathcal{C},W_I}$ (f)

Fig. 4. Building blocks of quantitative evaluation of centerline similarity. (a) The original image. (b) A Euclidean distance map computed for objects in the original image. (c) Centerline of the object. (d) Distance map computer from centerline to background. (e) Normalized distance map computed for the reference image. (f) Centerline superimposed on the normalized distance map.

In Eq. (3), the function $\delta(P, \overline{D}_{Q,W})$ tells the average distance of the set of points P to the set of points Q on the normalized distance map $\overline{D}_{Q,W}$, and $\#P$ indicates the number of elements of the set of points P, while in Eq. (2), the similarity measure is the average of the average distance of two centerlines. Note that similarity measure $AA_I(\cdot,\cdot)$ measures the error, i.e., if $AA_I(\mathcal{C}_1,\mathcal{C}_2) = 0$, then $\mathcal{C}_1$ and $\mathcal{C}_2$ fit perfectly. On the other hand, one of them should contain some extra points.

4   Results

For a fair comparison, each considered algorithm has been implemented in the same way in C++ without any special program library, in a framework where only the deletion rules and thinning strategy differed[1]. The 17 algorithms under comparison were tested on 160 images of "beetles" and "flies" (and their rotated versions), which form a part of *MPEG7 CE Shape-1 Part B* [19] (see Figs. 5 and 6). The algorithms have been tested on a desktop computer (Intel© fpuCore i5-9400 CPU 2.90 GHz × 6; Fedora 34 Cinnamon (x86-64)).

Considering the images, centerline $\mathcal{C}(T)$ is extracted first from the original image I, then the distance map is computed for the objects in I. After that, we extract the centerline from each image of the k-attempt versions of the algorithm T. We use the similarity measure $AA_I(\mathcal{C}(T)), \mathcal{C}(T_k)$ for each image in order to compare the k-attempt versions of algorithm T and the original, conventional one. Our results are summarized in Tables 1 and 2. The algorithms are named according to their names in references [7,13,23]. Note that algorithms SI-2 × 2-1, SI-2 × 2-2, and SI-2 × 2-3 have not been published yet in cited works, but they are evolved from the published 2-subiterational algorithms of [23]. The iteration step in SI-2-NE-SW-1 is divided into two subcycles, and the first deletion direction is NE (i.e., northeast), which is followed by the deletion direction SW (i.e., southwest). Similarly, the iteration step in SI-2-NW-SE-1 is divided into two subcycles, but the sequence of the deletion order is NW (i.e., Northwest) and SE (i.e., Southwest). Combining these two algorithms, we get the algorithm SI-2 × 2-1, where the iteration step is divided into $2 \times 2 = 4$ subiteration steps, and the sequence order of the deletion direction is NE (i.e., northeast), SW (i.e., southwest), NW (i.e., northwest), and SE (i.e., southeast). This combination is already used in [21], but those algorithms use asymmetric reduction operations.

Regarding the execution time, all of the algorithms are very fast. For this set of images, the maximal execution times are under 0.1 s. The k-attempt version of the thinning algorithms can produce high speed-up (>1) if the shape contains long, thin branches and also a thicker part, as it has already been reported in [26–28]. If the shape does not contain long, thin parts, then the attempt at counting takes some extra effort that is not included in the conventional version of thinning algorithms, and it may slow down the execution. Unfortunately, we cannot say yet in advance what will be the advantage or deficit of execution time regarding the shape and an algorithm.

Similarity errors show how different the resulting centerlines of the conventional algorithm and the k-attempt ones are. If the similarity error is zero, then there is no difference between the two centerlines. The differences are also indicated in pixels and percents as well. The difference in percent is the ratio of the difference in pixels and points of the centerline produced by the conventional implementation of the algorithm. Since the numbers cannot really show what the difference in results is, some selected cases are highlighted in Fig. 7. It

---

[1] The implementations are available on the author's GitLab repository: https://infgit. inf.u-szeged.hu/skeleton/k_thinning_algorithm_framework.git.

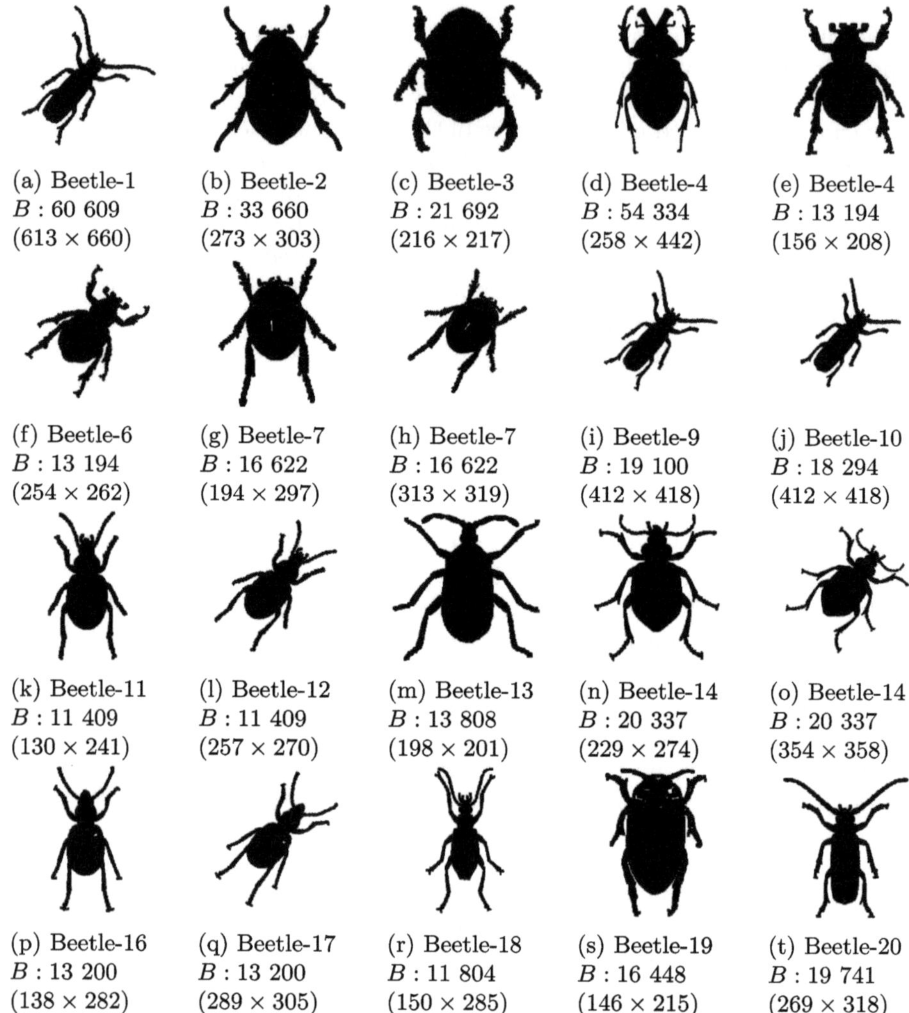

(a) Beetle-1
B : 60 609
(613 × 660)

(b) Beetle-2
B : 33 660
(273 × 303)

(c) Beetle-3
B : 21 692
(216 × 217)

(d) Beetle-4
B : 54 334
(258 × 442)

(e) Beetle-4
B : 13 194
(156 × 208)

(f) Beetle-6
B : 13 194
(254 × 262)

(g) Beetle-7
B : 16 622
(194 × 297)

(h) Beetle-7
B : 16 622
(313 × 319)

(i) Beetle-9
B : 19 100
(412 × 418)

(j) Beetle-10
B : 18 294
(412 × 418)

(k) Beetle-11
B : 11 409
(130 × 241)

(l) Beetle-12
B : 11 409
(257 × 270)

(m) Beetle-13
B : 13 808
(198 × 201)

(n) Beetle-14
B : 20 337
(229 × 274)

(o) Beetle-14
B : 20 337
(354 × 358)

(p) Beetle-16
B : 13 200
(138 × 282)

(q) Beetle-17
B : 13 200
(289 × 305)

(r) Beetle-18
B : 11 804
(150 × 285)

(s) Beetle-19
B : 16 448
(146 × 215)

(t) Beetle-20
B : 19 741
(269 × 318)

Fig. 5. Image of 'Beetles' from *MPEG7 CE Shape-1 Part B* dataset. Numbers in the middle rows are the counts of black points, and ($W \times H$) gives the size of the given test image.

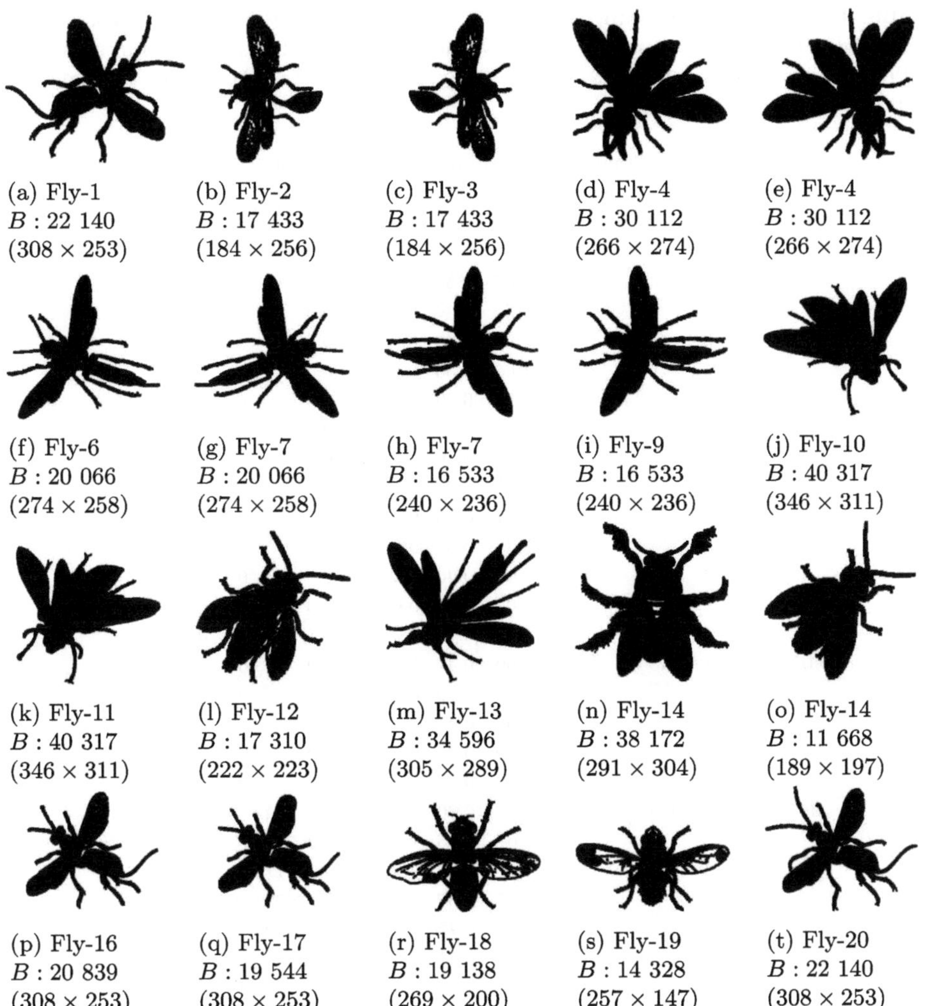

<table>
<tr><td>(a) Fly-1
B : 22 140
(308 × 253)</td><td>(b) Fly-2
B : 17 433
(184 × 256)</td><td>(c) Fly-3
B : 17 433
(184 × 256)</td><td>(d) Fly-4
B : 30 112
(266 × 274)</td><td>(e) Fly-4
B : 30 112
(266 × 274)</td></tr>
<tr><td>(f) Fly-6
B : 20 066
(274 × 258)</td><td>(g) Fly-7
B : 20 066
(274 × 258)</td><td>(h) Fly-7
B : 16 533
(240 × 236)</td><td>(i) Fly-9
B : 16 533
(240 × 236)</td><td>(j) Fly-10
B : 40 317
(346 × 311)</td></tr>
<tr><td>(k) Fly-11
B : 40 317
(346 × 311)</td><td>(l) Fly-12
B : 17 310
(222 × 223)</td><td>(m) Fly-13
B : 34 596
(305 × 289)</td><td>(n) Fly-14
B : 38 172
(291 × 304)</td><td>(o) Fly-14
B : 11 668
(189 × 197)</td></tr>
<tr><td>(p) Fly-16
B : 20 839
(308 × 253)</td><td>(q) Fly-17
B : 19 544
(308 × 253)</td><td>(r) Fly-18
B : 19 138
(269 × 200)</td><td>(s) Fly-19
B : 14 328
(257 × 147)</td><td>(t) Fly-20
B : 22 140
(308 × 253)</td></tr>
</table>

Fig. 6. Image of 'Flies' from *MPEG7 CE Shape-1 Part B* dataset. Numbers in the middle rows are the counts of black points, and ($W \times H$) gives the size of the given test image.

Table 1. Results of k-attempt versions for first eight of the selected 17 parallel thinning algorithms. In the last two columns, the "Max. skel. diff. px." column shows the difference in count of pixels, while the "Max. skel. diff. %." shows the differences in percent.

Algorithm	k	Conv. thinning exec. time (sec.)	Speed-up		Similarity error		Max. skel. diff.	
			Max.	Min.	Min.	Max.	px.	%.
BM99 [3]	1	0.067	6.4	0.28	0.0	0.0	0	0.0
	2		6.67	0.38	0.0	0.0	0	0.0
	3		6.67	0.30	0.0	0.0	0	0.0
	4		6.67	0.36	0.0	0.0	0	0.0
	5		5.67	0.22	0.0	0.0	0	0.0
GH92a [11]	1	0.075	10.67	0.33	0.0	0.0	0	0.0
	2		6.25	0.30	0.0	0.0	0	0.0
	3		8.00	0.30	0.0	0.0	0	0.0
	4		5.5	0.26	0.0	0.0	0	0.0
	5		6.25	0.28	0.0	0.0	0	0.0
GH92b [11]	1	0.065	8.67	0.23	0.0	0.0	0	0.0
	2		5.33	0.38	0.0	0.0	0	0.0
	3		6.00	0.30	0.0	0.0	0	0.0
	4		8.00	0.30	0.0	0.0	0	0.0
	5		6.50	0.35	0.0	0.0	0	0.0
GH92c [11]	1	0.043	5.00	0.41	0.0	0.0	0	0.0
	2		5.00	0.36	0.0	0.0	0	0.0
	3		4.00	0.41	0.0	0.0	0	0.0
	4		4.00	0.21	0.0	0.0	0	0.0
	5		4.00	0.35	0.0	0.0	0	0.0
H87 [15]	1	0.065	4.67	0.30	0.0	0.005742	23	1.83
	2		4.67	0.25	0.0	0.005425	22	1.75
	3		4.67	0.28	0.0	0.005111	21	1.67
	4		4.67	0.37	0.0	0.004799	20	1.59
	5		4.67	0.22	0.0	0.004490	19	1.52
H89 [12]	1	0.063	4.50	0.38	0.0	0.0	0	0.0
	2		4.50	0.41	0.0	0.0	0	0.0
	3		4.50	0.34	0.0	0.0	0	0.0
	4		4.67	0.26	0.0	0.0	0	0.0
	5		4.67	0.36	0.0	0.0	0	0.0
FP-1 [23]	1	0.063	7.25	0.33	0.000347	0.022838	177	10.61
	2		6.14	0.41	0.000202	0.019932	141	8.2
	3		5.00	0.27	0.0	0.017718	119	7.28
	4		6.14	0.35	0.0	0.016144	107	6.55
	5		5.85	0.33	0.0	0.015019	99	6.06
FP-2 [23]	1	0.074	8.50	0.35	0.0	0.002593	23	1.15
	2		9.25	0.29	0.0	0.001772	17	0.85
	3		9.25	0.28	0.0	0.000785	9	0.4
	4		9.25	0.33	0.0	0.000560	7	0.25
	5		9.25	0.15	0.0	0.000355	5	0.17

Table 2. Results of k-attempt versions for the remaining nine of the 17 selected thinning algorithms [23]. In the last two columns, the "Max. skel. diff. px." column shows the difference in count of pixels, while the "Max. skel. diff. %." shows the differences in percent.

Algorithm	k	Conv. thinning exec. time (sec.)	Speed-up		Similarity error		Max. skel. diff.	
			Max.	Min.	Min.	Max.	px.	%.
SI-2-NE-SW-1	1	0.074	4.25	0.51	0.0	0.022789	108	11.04
	2		4.71	0.41	0.0	0.004914	20	1.77
	3		4.12	0.42	0.0	0.004005	17	1.28
	4		4.12	0.39	0.0	0.003712	16	1.21
	5		3.50	0.40	0.0	0.002559	12	0.91
SI-2-NE-SW-2	1	0.074	6.16	0.64	0.0	0.001871	15	1.06
	2		6.16	0.40	0.0	0.000947	4	0.25
	3		4.22	0.34	0.0	0.0	0	0.0
	4		6.16	0.42	0.0	0.0	0	0.0
	5		6.16	0.50	0.0	0.0	0	0.0
SI-2-NE-SW-3	1	0.052	4.28	0.28	0.0	0.001865	15	1.06
	2		3.83	0.42	0.0	0.000944	4	0.25
	3		4.28	0.52	0.0	0.0	0	0.0
	4		4.28	0.26	0.0	0.0	0	0.0
	5		3.75	0.40	0.0	0.0	0	0.0
SI-2-NW-SE-1	1	0.080	6.33	0.50	0.0	0.022789	108	11.04
	2		6.33	0.50	0.0	0.003937	16	1.61
	3		5.42	0.70	0.0	0.002067	11	1.04
	4		5.42	0.34	0.0	0.001726	9	0.72
	5		5.42	0.42	0.0	0.000690	5	0.37
SI-2-NW-SE-2	1	0.082	6.25	0.28	0.0	0.001580	15	1.06
	2		6.25	0.57	0.0	0.000229	1	0.11
	3		5.71	0.45	0.0	0.0	0	0.0
	4		5.00	0.42	0.0	0.0	0	0.0
	5		5.00	0.42	0.0	0.0	0	0.0
SI-2-NW-SE-3	1	0.097	7.00	0.50	0.0	0.001578	15	1.06
	2		6.12	0.57	0.0	0.000228	1	0.11
	3		5.87	0.42	0.0	0.0	0	0.0
	4		5.44	0.32	0.0	0.0	0	0.0
	5		4.37	0.23	0.0	0.0	0	0.0
SI-2 × 2-1	1	0.096	4.41	0.36	0.0	0.000060	1	0.08
	2		4.00	0.33	0.0	0.000187	1	0.12
	3		4.57	0.35	0.0	0.000187	1	0.12
	4		4.17	0.28	0.0	0.0	0	0.0
	5		4.36	0.38	0.0	0.000187	1	0.12
SI-2 × 2-2	1	0.080	5.75	0.42	0.0	0.000065	1	0.07
	2		5.42	0.36	0.0	0.0	0	0.0
	3		4.75	0.32	0.0	0.0	0	0.0
	4		4.75	0.36	0.0	0.0	0	0.0
	5		4.25	0.28	0.0	0.0	0	0.0
SI-2 × 2-3	1	0.043	5.33	0.50	0.0	0.000065	1	0.07
	2		4.57	0.37	0.0	0.0	0	0.0
	3		4.57	0.35	0.0	0.0	0	0.0
	4		4.00	0.25	0.0	0.0	0	0.0
	5		4.00	0.40	0.0	0.0	0	0.0

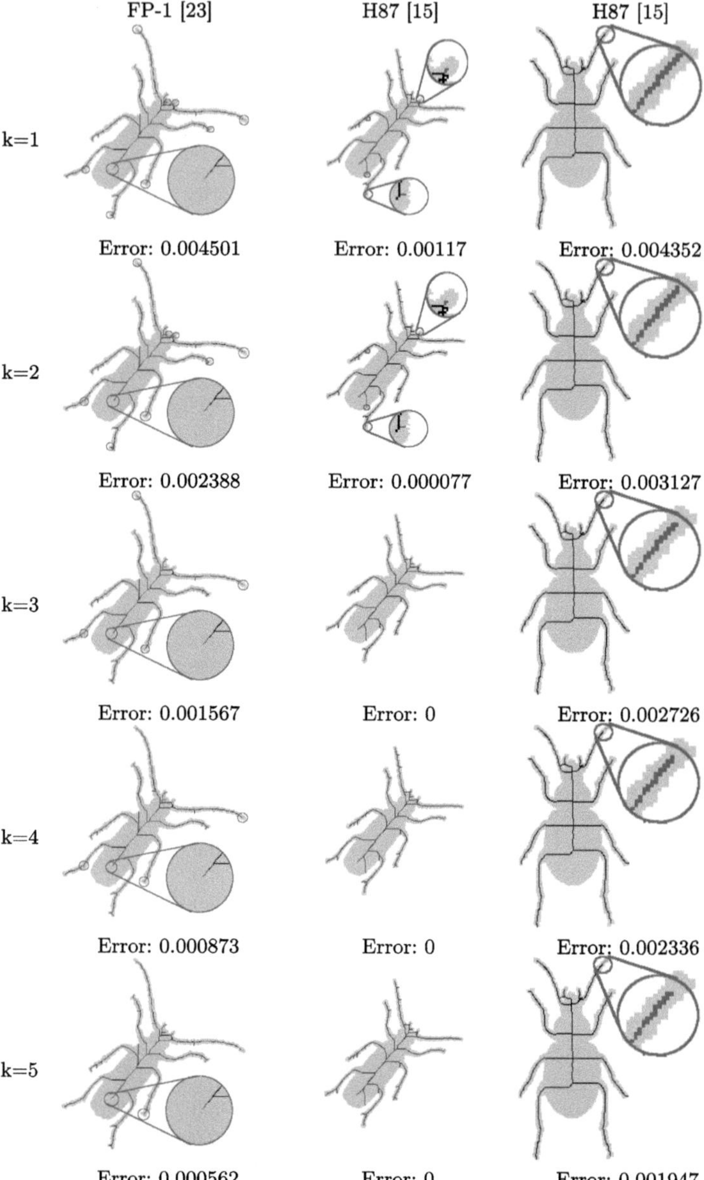

Fig. 7. Illustration of some deviations that produce the k-attempt versions of three selected thinning algorithms. In many cases, for different k, the error is just a few pixels in the centerline. Note that the circles indicate where the produced centerlines differs from the results of the k-attempt versions; however, only some cases are highlighted.

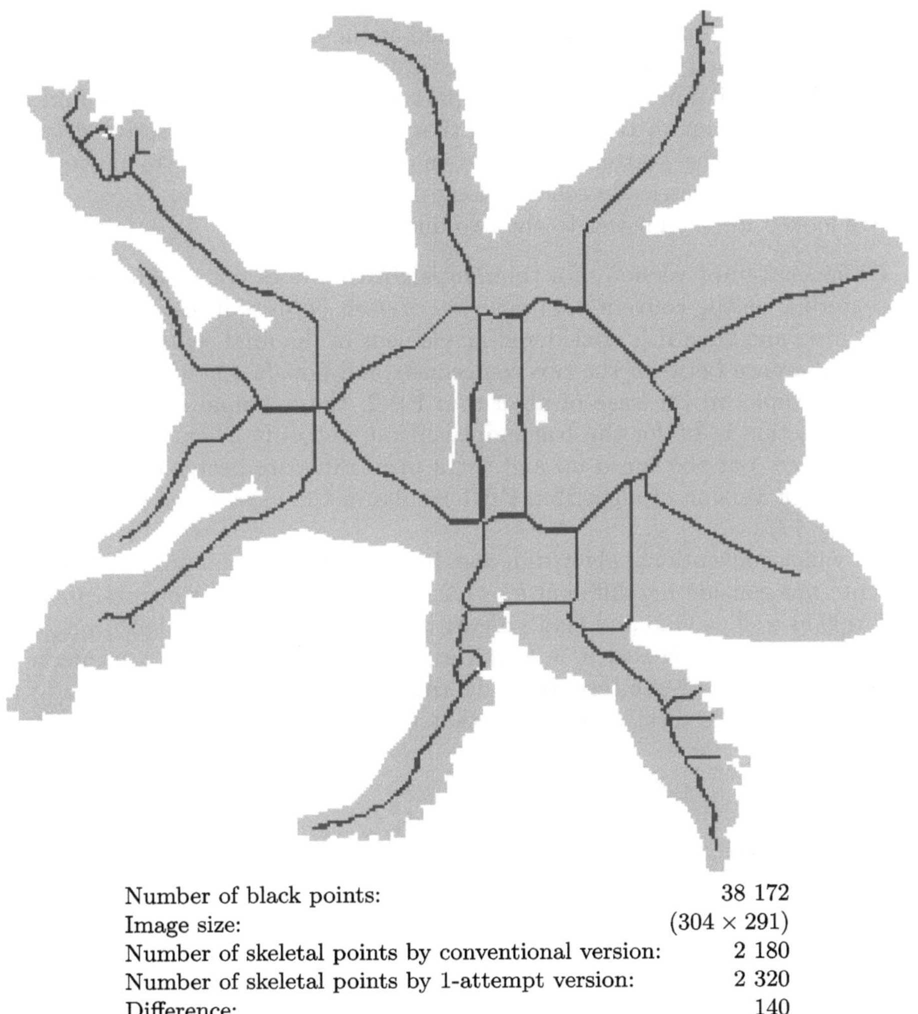

Number of black points:	38 172
Image size:	(304×291)
Number of skeletal points by conventional version:	2 180
Number of skeletal points by 1-attempt version:	2 320
Difference:	140

Fig. 8. An image of a fly contains a long part that is eliminated by the conventional version of the thinning algorithm FP-1 [23] but preserved by the k-attempt version of it. Purple curve segments are those where both kinds of algorithms result in the centerline. The green curve segments are preserved by the 1-attempt version of the FP-1 algorithm. The gray region has been eliminated by both versions of the FP-1 algorithm. (Color figure online)

seems that the difference between the results of the conventional thinning and its k-attempt version appears at the end of the skeletal branches. The reason is that some branches can be shortened only from their ends, and it means that the middle points in the branches are under attack of deletion in each successive iteration step. In the k-thinning version of an algorithm, after k successive attempts, the point cannot be deleted and is not investigated anymore. Hence, the branch reduction from its end is blocked at these fixed points. In an extreme case, a long branch can be pruned drastically by the conventional version of the algorithm if it forms an even-vide part and the algorithm cannot find any geometric constraint in this long part that could stop the successive deletion. For these cases, the k-attempt version may be a good choice to prevent the elimination of geometrically relevant long parts (see Fig. 8).

We looked for the answer to the two questions:

1. If the k-attempt version of a thinning algorithm does not produce the same centerline as the conventional one (i.e., if each point is investigated at most k times, and then it is considered an element of the final centerline), what is the difference between the two centerlines, and how fast is it?
   For example, in the case of algorithm FP-2, the maximal speed-up for the investigation is 8.5 for the 1-attempt version, and only 23 extra points are in the image, but the speed-up and extra pixel ratio are better for $k \geq 2$. The 1-attempt version of algorithm FP-1 results in the most extra points, which is only 177.
2. For which conventional algorithm does its k-attempt version produce the same centerline considering different $k = 1, 2, 3, 4, 5$ for a set of various test images? BM99 as well as Guo and Hall's algorithms [11] and the H89 algorithm, seem to be 1-attempt; hence, it is worth examining their k-attempt property in future work. 2×2-subiterational algorithms SI-2 $\times$ 2-2 SI-2 $\times$ 2-3 seem to be 2-attempt in the best cases. Note that GH92a and GH92b reach very high accelerations.

5   Conclusion

In this work we examined the possibility and benefits of k-attempt versions of 2D thinning algorithms, where after k investigations, the 'survivor' points are considered as elements of the final centerline.

We evaluated the produced centerlines with quantitative comparison. We found that the investigated thinning algorithms produce very similar results; however, the k-attempt version of the algorithm may preserve longer branches of the centerline in comparison to the original one (see Fig. 7). On the other hand, for the investigated test images, the 1-attempt version of a thinning algorithm can be even 10-times faster in comparison to the original versions. We have to note that the speed-up for any algorithm is depends on the images, and it could be higher in special cases when the reduction of the long thinner part is finished in the early iterations and the thicker part can be thinned in further several iteration step (e.g., like a fat balloon with a long strap).

Furthermore, BM99, GH92a, GH92b, GH92c, and H89 algorithms could also be "potentially 1-attempt," which is worth examining deeper in the further works, but SI-2 $\times$ 2-2 and SI-2 $\times$ 2-3 are also promising for $k = 2$.

Inspecting the results of the conventional and k-attempt versions of a thinning algorithm, there may be some even-wide branches that the conventional versions of some thinning algorithms may overshrink, while its k-attempt version preserves them (see Fig. 8). We will investigate this property in future works.

Consequently, it is worth investigating the potentially 1-attempt or 2-attempt version of the algorithms mentioned above, since although these versions may not be equivalent to conventional ones, the difference between the results is negligible, and the k-attempt versions preserve the overshrinking of relevant branches.

Acknowledgments. This research was supported by project no. TKP2021-NVA-09 provided by the Ministry of Culture and Innovation of Hungary from the National Research, Development and Innovation Fund.

References

1. Arcelli, C., Cordella, L., Levialdi, S.: Parallel thinning of binary pictures. Electron. Lett. **11**(7), 148–149 (1975). https://doi.org/10.1049/el:19750113
2. di Baja, G.S., Thiel, E.: A multiresolution shape description algorithm. In: Chetverikov, D., Kropatsch, W.G. (eds.) CAIP 1993. LNCS, vol. 719, pp. 208–215. Springer, Heidelberg (1993). https://doi.org/10.1007/3-540-57233-3_28
3. Bernard, T., Manzanera, A.: Improved low complexity fully parallel thinning algorithm. In: Proceedings 10th International Conference on Image Analysis and Processing. pp. 215–220. IEEE (1999). https://doi.org/10.1109/ICIAP.1999.797597
4. Bertrand, G., Couprie, M.: Two-dimensional parallel thinning algorithms based on critical Kernels. J. Math. Imaging Vision **31**(1), 35–56 (2008). https://doi.org/10.1007/s10851-007-0063-0
5. Blum, H.: A transformation for extracting new descriptors of shape, pp. 362–380 Publication Title: Models for the Perception of Speech and Visual Form Place: Cambridge, MA (1967)
6. Chiu, H.P., Tseng, D.C.: A feature-preserved thinning algorithm for handwritten Chinese characters. Signal Process. **58**(2), 203–214 (1997). https://doi.org/10.1016/S0165-1684(97)00024-8
7. Couprie, M.: Note on Fifteen 2D Parallel Thinning Algorithms. Internal Report, Université de Marne-la-Vallée (2009)
8. Eckhardt, U., Maderlechner, G.: Invariant Thinning. Int. J. Pattern Recognit Artif Intell. **07**(05), 1115–1144 (1993). https://doi.org/10.1142/S021800149300056X
9. González, R.C., Woods, R.E.: Digital image processing, 3rd Edn. Pearson Education (2008)
10. Guo, Z., Hall, R.W.: Parallel thinning with two-subiteration algorithms. Commun. ACM **32**(3), 359–373 (1989). https://doi.org/10.1145/62065.62074
11. Guo, Z., Hall, R.W.: Fast fully parallel thinning algorithms. CVGIP: Image Underst. **55**(3), 317–328 (1992). https://doi.org/10.1016/1049-9660(92)90029-3
12. Hall, R.W.: Fast parallel thinning algorithms: parallel speed and connectivity preservation. Commun. ACM **32**(1), 124–131 (1989). https://doi.org/10.1145/63238.63248
13. Hall, R.W.: Parallel Connectivity-Preserving Thinning Algorithms. In: Machine Intelligence and Pattern Recognition, vol. 19, pp. 145–179. Elsevier (1996). https://doi.org/10.1016/S0923-0459(96)80014-0

14. Hall, R.W., Kong, T., Rosenfeld, A.: Shrinking Binary Images. In: Machine Intelligence and Pattern Recognition, vol. 19, pp. 31–98. Elsevier (1996). https://doi.org/10.1016/S0923-0459(96)80012-7
15. Holt, C.M., Stewart, A., Clint, M., Perrott, R.H.: An improved parallel thinning algorithm. Commun. ACM **30**(2), 156–160 (1987). https://doi.org/10.1145/12527.12531
16. Kong, T., Rosenfeld, A.: Digital topology: introduction and survey. Comput. Vision, Graph. Image Process. **48**(3), 357–393 (1989). https://doi.org/10.1016/0734-189X(89)90147-3
17. Lam, L., Lee, S.W., Suen, C.: Thinning methodologies-a comprehensive survey. IEEE Trans. Pattern Anal. Mach. Intell. **14**(9), 869–885 (1992). https://doi.org/10.1109/34.161346
18. Lam, L., Suen, C.: Evaluation of thinning algorithms from an ocr viewpoint. In: Proceedings of 2nd International Conference on Document Analysis and Recognition (ICDAR '93). pp. 287–290 (1993). https://doi.org/10.1109/ICDAR.1993.395730
19. Latecki, L., Lakamper, R., Eckhardt, T.: Shape descriptors for non-rigid shapes with a single closed contour. In: Proceedings IEEE Conference on Computer Vision and Pattern Recognition. CVPR 2000 (Cat. No.PR00662). vol. 1, pp. 424–429 (2000). https://doi.org/10.1109/CVPR.2000.855850
20. Neusius, C., Olszewski, J., Scheerer, D.: An efficient distributed thinning algorithm. Parallel Comput. **18**(1), 47–55 (1992). https://doi.org/10.1016/0167-8191(92)90110-S
21. Németh, G., Kardos, P., Palágyi, K.: 2D Parallel Thinning and Shrinking Based on Sufficient Conditions for Topology Preservation. Acta Cybernetica **20**(1), 125–144 (2011). https://doi.org/10.14232/actacyb.20.1.2011.10
22. Németh, G., Kovács, G., Fazekas, A., Palágyi, K.: A method for quantitative comparison of 2d skeletons. Acta Polytechnica **13**, 123–142 (2016). https://doi.org/10.12700/APH.13.7.2016.7.7
23. Németh, G., Palágyi, K.: Topology preserving parallel thinning algorithms. Int. J. Imaging Syst. Technol. **21**(1), 37–44 (2011). https://doi.org/10.1002/ima.20272
24. Palágyi, K.: 1-attempt and equivalent thinning on the hexagonal grid. In: Brunetti, S., Frosini, A., Rinaldi, S. (eds.) Discrete Geometry and Mathematical Morphology. pp. 390–401. Springer Nature Switzerland, Cham (2024). https://doi.org/10.1007/978-3-031-57793-2_30
25. Palágyi, K., Németh, G.: Fixpoints of Iterated Reductions with Equivalent Deletion Rules. In: Barneva, R.P., Brimkov, V.E., Tavares, J.M.R.S. (eds.) IWCIA 2018. LNCS, vol. 11255, pp. 17–27. Springer, Cham (2018). https://doi.org/10.1007/978-3-030-05288-1_2
26. Palágyi, K., Németh, G.: k-Attempt Thinning. In: Lukić, T., Barneva, R.P., Brimkov, V.E., Čomić, L., Sladoje, N. (eds.) IWCIA 2020. LNCS, vol. 12148, pp. 258–272. Springer, Cham (2020). https://doi.org/10.1007/978-3-030-51002-2_19
27. Palágyi, K., Németh, G.: 1-Attempt Subfield-Based Parallel Thinning. In: 2021 12th International Symposium on Image and Signal Processing and Analysis (ISPA). pp. 241–246. IEEE, Zagreb, Croatia (2021).https://doi.org/10.1109/ISPA52656.2021.9552163
28. Palágyi, K., Németh, G.: 1-Attempt parallel thinning. J. Comb. Optim. , 1–15 (2021). https://doi.org/10.1007/s10878-021-00744-y
29. Rosenfeld, A.: A characterization of parallel thinning algorithms. Inf. Control **29**(3), 286–291 (1975). https://doi.org/10.1016/S0019-9958(75)90448-9

30. Saha, P.K., Borgefors, G., Sanniti di Baja, G. (eds.): Skeletonization: theory, methods and applications. Computer vision and pattern recognition series, Academic Press, an imprint of Elsevier, London San Diego Cambridge, MA (2017)
31. Siddiqi, K., Pizer, S.M. (eds.): Medial Representations. Mathematics, Algorithms and Applications, Computational Imaging and Vision, vol. 37. Springer Netherlands, Dordrecht (2008). https://doi.org/10.1007/978-1-4020-8658-8
32. Singh, B., Goswami, S., Goyal, P., Mittal, A.: A Robust Thinning Algorithm for Straightening of Curved Text Line. In: Deep, K., Nagar, A., Pant, M., Bansal, J.C. (eds.) Proceedings of the International Conference on Soft Computing for Problem Solving (SocProS 2011) December 20-22, 2011. AISC, vol. 131, pp. 903–910. Springer, New Delhi (2012). https://doi.org/10.1007/978-81-322-0491-6_83
33. Suen, C.Y., Wang, P.S.P.: Thinning Methodologies for Pattern Recognition, Series in Machine Perception and Artificial Intelligence, vol. 8. WORLD SCIENTIFIC (1994). https://doi.org/10.1142/2100
34. Wu, R.Y., Tsai, W.H.: A new one-pass parallel thinning algorithm for binary images. Pattern Recogn. Lett. **13**(10), 715–723 (1992). https://doi.org/10.1016/0167-8655(92)90101-5
35. Zhang, T.Y., Suen, C.Y.: A fast parallel algorithm for thinning digital patterns. Commun. ACM **27**(3), 236–239 (1984). https://doi.org/10.1145/357994.358023

An Order-Independent 2D Parallel 4-Subiteration Thinning Algorithm

Péter Kardos[(✉)]

Department of Image Processing and Computer Graphics, University of Szeged,
Szeged, Hungary
`pkardos@inf.u-szeged.hu`

Abstract. Thinning is a commonly used technique based on iterative object reduction to obtain centerlines of segmented binary objects. Parallel reductions delete multiple points from the picture at the same time. For parallel subiteration-based thinning algorithms, an iteration step is composed of some subiterations, each corresponding to a specific deletion direction. The results produced by such algorithms usually depend on the order of these directions. This work presents the very first order-independent 4-subiteration thinning algorithm, which generates the same centerlines for arbitrary permutations of the four deletion directions and preserves topology for all possible objects.

Keywords: Skeleton · Subiteration-based thinning · Topology preservation · Order Independence

1 Introduction

Skeletonization is a process to produce a reduced-dimension medial representation of binary digital objects [16]. *Thinning* is an easy-to-implement technique based on iterative object reduction to generate centerlines [4,12], which can be accelerated on GPU [17]. Although it is considered an old approach, it still plays a keyrole in some recent applications (see for example [13,18]).

The *reductions* applied by thinning algorithms change only black points to white points, which is referred to as the deletion of black points. Sequential reductions remove only one point at a time, while parallel reductions can delete a set of points simultaneously. We can make a distinction among three major parallel thinning strategies [4]. *Fully parallel* algorithms perform the same parallel reduction in each iteration step. *Subiteration-based* (or *directional*) algorithms decompose an iteration step into some successive parallel reductions according to different deletion directions, and a subset of border points associated with the actual direction are deleted by a parallel reduction. *Subfield-based* algorithms are based on the partition of the digital space into some subfields which are alternatively activated, and at a given iteration step successive parallel reductions assigned to these subfields are carried out, deleting only some black points in the active subfield.

An important requirement for such operations is to preserve the topology of objects [12]. Without this property, some artifacts such as disconnected parts and unwanted loops may be introduced in the centerline during thinning, which limits its usefulness for some purposes. Couprie analyzed fifteen frequently referred 2D parallel thinning algorithms, five of which proved to be not topology-preserving [3].

Topology preservation is difficult to ensure for fully parallel algorithms, in comparison to the sequential and to the other two types of parallel algorithms. On the other hand, most sequential algorithms suffer from the drawback that they can result various skeletons for different point-visiting orders, while subiteration-based and subfield-based algorithms are not invariant under the permutation of deletion directions or under the permutation of subfields to activate. Earlier some order-independent sequential thinning algorithms were already proposed [5, 6, 10] (i.e., algorithms that produce identical skeletons for any sequential visiting orders), however, according to our best knowledge, no study was focused on the order independence of directional and subfield-based algorithms.

This work presents the first order-independent parallel subiteration-based thinning algorithm in regards to the order of deletion directions. The rest of this paper is organized as follows. Section 2 deals with some basic notions and results of digital topology. In Sect. 3, the new algorithm is introduced. In Sects. 4 and 5, we prove the topological correctness and order independence of the algorithm. Finally, we round off the paper with some concluding remarks.

2 Basic Notions and Results

In this work, we utilize the core principles of digital topology as presented by Kong and Rosenfeld [12].

The elements of the 2D digital space $\mathbb{Z}^2$ are referred to as *points*, which can be represented by the elements of the square grid due to its duality to $\mathbb{Z}^2$. Two points are considered 4-*adjacent* if they share an edge and they are 8-*adjacent* if they share an edge or a vertex, see Fig. 1. A point p is j-adjacent to a non-empty set of points X if there is a point $q \in X$ such that p and q are j-adjacent ($j = 4, 8$). Let us denote by $N_j(p)$ the set of points being j-adjacent to a point p in $\mathbb{Z}^2$, and let $N_j^*(p) = N_j(p) \setminus \{p\}$.

A sequence of distinct points $\langle p_0, p_1, \ldots, p_m \rangle$ is called a *j-path* from p_0 to p_m in a non-empty set of points $X \subseteq \mathbb{Z}^2$ if each point of the sequence is in X and p_i is j-adjacent to p_{i-1} for each $i = 1, 2, \ldots, m$. Two points are said to be *j-connected* in a set X if there is a j-path in X between them. A set of points X is *j-connected* in the set of points $Y \supseteq X$ if any two points in X are j-connected in Y. A *j-component* of a set of points X is a maximal (with respect to inclusion) j-connected subset of X.

An $(8, 4)$ *binary digital picture* is a quadruple $(\mathbb{Z}^2, 8, 4, B)$ [12], where set $\mathbb{Z}^2$ contains all points of the considered grid, $B \subseteq \mathbb{Z}^2$ denotes the set of *black points*, and each point in $\mathbb{Z}^2 \setminus B$ is said to be a *white point*. A *black component* or *object* is an 8-component of q, while a *white component* is a 4-component of $\mathbb{Z}^2 \setminus B$. Let us denote by $C(p)$ the number of 8-components in picture $(\mathbb{Z}^2, 8, 4, B \cap N_8^*(p))$.

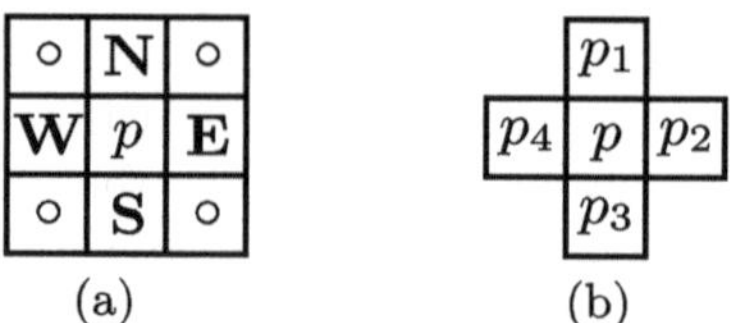

Fig. 1. The studied two adjacency relations on the square grid (a). Points being 4-adjacent to the central point p are marked 'N', 'E', 'S', and 'W', while points that are 8-adjacent but not 4-adjacent to p are represented by 'o'. Indexing scheme for the elements in $N_4^*(p)$ (b).

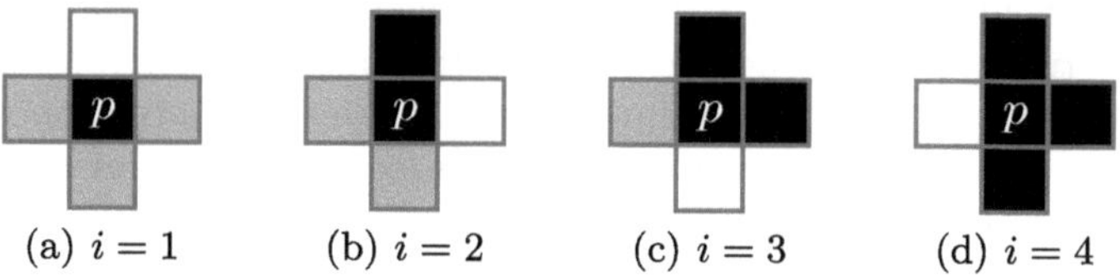

(a) $i = 1$ (b) $i = 2$ (c) $i = 3$ (d) $i = 4$

Fig. 2. The four types of strict i-border points ($i = 1, 2, 3, 4$). Points depicted in gray can be either black or white points.

A black point is said to be a *border point* if it is 4-adjacent to at least one white point. A border point is called an i-border point if p_i is white in Fig. 1. Alternatively, we can also refer to 1-, 2-, 3-, and 4-border points as **N**-, **E**-, **S**-, and **W**-border points, respectively (see Fig. 1(a)). Furthermore, a border point p is a *strict i-border point* ($i \in 1, 2, 3, 4$) if i is the lowest index such that $p_i \in \mathbb{Z}^2 \setminus B$ (see Fig. 2). Note that, while an i-border point may also be a j-border point at the same time for $i \neq j$, this cannot be stated for strict i- and j-border points. In other words, the latter, stricter definition classifies border points into four disjunct groups. We introduce the notation $\beta(p)$ to refer to the type of a strict border point p: $\beta(p) = i$ means that p is a strict i-border point.

A border point p is called *isolated* if all points in $N_8^*(p)$ are white (i.e., $\{p\}$ is a singleton object). An object is considered to be *small* if it is composed of mutually 8-adjacent points but it is not constituted by an isolated border point (see Fig. 3).

The *lexicographical order relation* $\prec$ between two distinct points $p = (p_x, p_y)$ and $q = (q_x, q_y)$ is defined as follows:

$$p \prec q \quad \Leftrightarrow \quad p_y < q_y \vee (p_y = q_y \wedge p_x < q_x).$$

Let Q be a finite set of points in $\mathbb{Z}^2$. Then, point $p \in Q$ is the *smallest element* of Q if for any $q \in Q \setminus \{p\}$, $p \prec q$. The smallest elements of the possible small objects are indicated in Fig. 3.

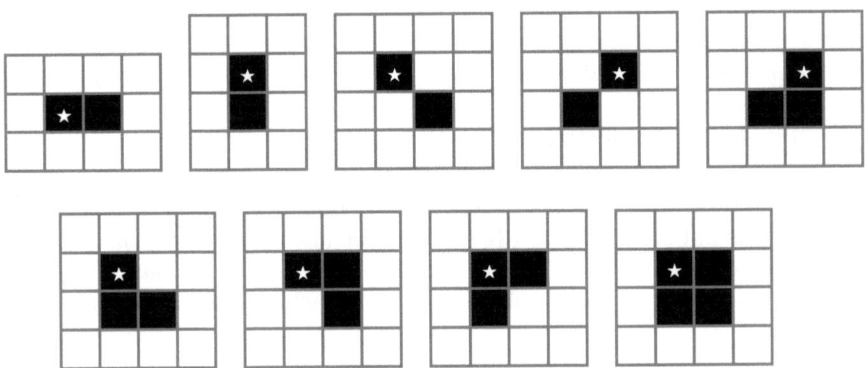

Fig. 3. The nine possible configurations of the considered small objects. Their smallest elements (with respect to relation $\prec$) of them are marked stars.

A black point is said to be *simple* for a set of black points (or in a picture) if its deletion is a topology-preserving reduction [11,12]. The following theorem gives a characterization of simple points.

Theorem 1. [4] *Black point p is simple if and only if p is a border point and* $C(p) = 1$.

Figure 4 shows some examples for simple and non-simple points.

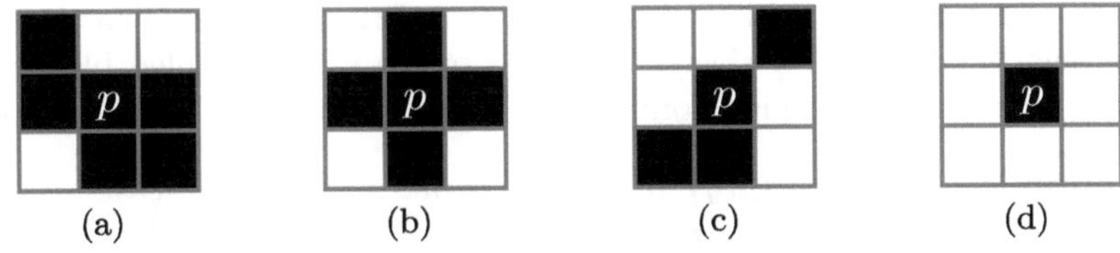

(a)   (b)   (c)   (d)

Fig. 4. Examples for simple and non-simple points in (8,4) pictures. Point p is only simple in case (a), since in configuration (b), p is not a border point, while in (c) and (d), $C(p) = 2$ and $C(p) = 0$, respectively.

A 2D reduction is *topology-preserving* if each object in the input picture contains exactly one object in the output picture, and each white component in the output picture contains exactly one white component in the input picture [12].

The following sufficient condition can be used for (8,4) pictures to verify if a reduction preserves topology or not [7].

Theorem 2. [7] *A reduction* $\mathcal{R}$ *is topology-preserving if all of the following conditions hold for any picture* $(\mathbb{Z}^2, 8, 4, B)$.

1. *Only simple points are deleted by* $\mathcal{R}$.
2. *For any two 4-adjacent points,* $p, q \in B$ *that are deleted by* $\mathcal{R}$*,* p *is simple for* $B \setminus \{q\}$.
3. $\mathcal{R}$ *never deletes completely any small object.*

Let us suppose that two 4-adjacent simple points p, q do not form a small object, and p is not simple for $B \setminus \{q\}$, which means, that the set $\{p, q\}$ does not fulfill Condition 2 of Theorem 2. We call such a set a *decision pair*. Proposition 1 states an important property of decision pairs, which we will rely on later in this paper.

Proposition 1. [10] *If* $\{p, q\}$ *is a decision pair, then* $N_8(p, q)$ *matches at least one of the configurations in Fig. 5 or their rotations by* $90°$*,* $180°$*, or* $270°$.

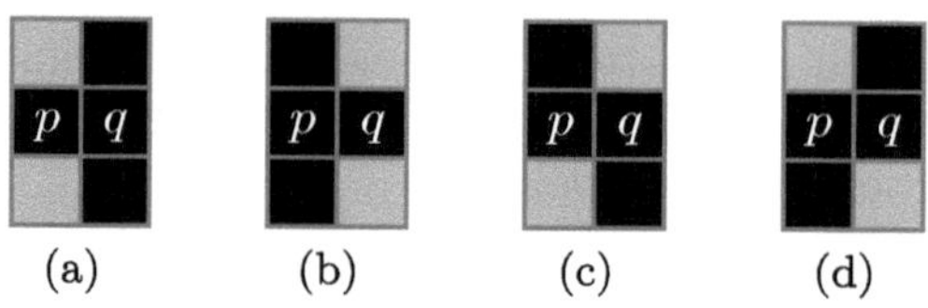

(a)   (b)   (c)   (d)

Fig. 5. Possible configurations of the horizontal decision pair $\{p, q\}$. The points in positions depicted gray can be either white or black.

If both p and q correspond to simple points in Fig. 5, then it is easy to verify that the deletion of p and q always disconnects the two unlabeled black points in $N_8(p, q)$, hence the following converse of Proposition 1 can be also stated.

Proposition 2. *If the set* $\{p, q\}$ *of simple points matches at least one of the configurations in Fig. 5 or their rotations by* $90°$*,* $180°$*, or* $270°$*, then* $\{p, q\}$ *is a decision pair.*

We say that a simple point p is *unbound* if it is neither an element of a decision pair nor of a small object, else it is *bound*. Figure 6 helps the reader to compare non-simple, bound, and unbound points.

Another concept of a *P-simple set* introduced by Bertrand serves as a basis of a further method for verifying topology preservation:

Definition 1. [1] *Let* $Q \subset B$ *be a set of black points in a picture. A set* Q *is called a P-simple set if any point* $q \in Q$ *is simple for* $B \setminus R$ *for any* $R \subseteq Q \setminus \{q\}$. *Furthermore, the elements in a P-simple set are called P-simple points.*

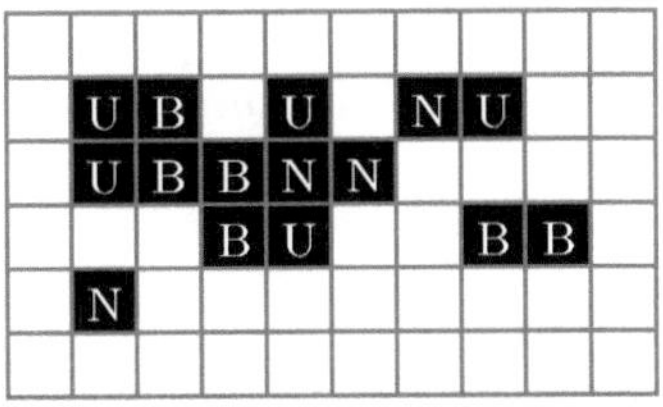

Fig. 6. Examples for non-simple, bound, and unbound points in a picture labelled with letters 'N', 'B', and 'U', respectively.

Bertrand suggested the following sufficient condition for topology preservation.

Theorem 3. [1] *A reduction that deletes a P-simple set is topology-preserving.*

Later, Kardos and Palágyi proved the equivalence between the conditions in Theorems 2 and 3.

Theorem 4. [8] *A reduction satisfies all conditions of Theorem 2 if and only if it deletes only P-simple points.*

Furthemore, the above authors proposed the concept of *super-deletable* points as well [9], which relies on the following two definitions.

Definition 2. [9] *Let p be a simple strict m-border point in picture $(\mathbb{Z}^2, 8, 4, B)$ $(m \in \{1, 2, 3, 4\})$, such that it is not an element of a small object. Point p is m-deletable if for any $1 \leq n < m$ and any simple n-border point $q \in N_4^*(p)$, p is simple for $B \setminus \{q\}$ or q is not n-deletable.*

Definition 3. [9] *An object point is* super-deletable *if one of the following conditions hold:*

1. *It is m-deletable $(m \in \{1, 2, 3, 4\})$.*
2. *It is an element of a small object but not its smallest element.*

In this paper, we will also make use of the following relationship.

Theorem 5. [9] *A reduction is topology-preserving if it deletes only super-deletable points.*

Besides topological correctness, another requirement for thinning is to preserve geometrical information about the shape of objects. A possible way to meet this expectation is to determine and retain the endpoints of line segments. Bertrand and Couprie proposed another strategy, which is based on the accumulation of so-called isthmuses [2]. A *2D isthmus* is a non-simple border point. Here we extend this notion to the non-deletable points in decision pairs (as those points also become non-simple after the deletion of the other elements of those pairs):

Definition 4. *A non-isolated m-border point p is an* isthmus *in an (8,4) picture if it is neither an m-deletable point nor an element of a small object (m ∈ {1, 2, 3, 4}).*

3   Proposed Algorithm

In this section, we present our subiterational thinning algorithm, which is based on the deletion of super-deletable points and on the accumulation of isthmuses (see Sect. 2). The choice of the former strategy is motivated by the fact that similarly to super-deletability, "directional" deletability depends on the type of border points. Besides that, isthmus preservation as a geometric condition can be easier combined with the former condition for topology preservation than with the retaining of endpoints. However, there is an important difference between existing isthmus-based directional algorithms and the new method: while the former ones filter isthmuses from all actual border points simultaneously (see for example [14]), the latter one restricts this operation to border points only in a given direction at a time. In this regard, the new algorithm better aligns with the directional nature of the subiteration-based approach then existing similar ones.

For a better understanding of the exact steps of the algorithm to be introduced shortly, some problematic cases must be taken into account from the perspective of order independence, if we want to apply the mentioned strategies. For this purpose, let us consider the binary picture in Fig. 7(a) and the parallel reductions $\mathcal{R}_\mathbf{N}$ and $\mathcal{R}_\mathbf{E}$ that delete the super-deletable $\mathbf{N}$- and $\mathbf{E}$-border points in the actual picture respectively. Let us perform these two reductions succesively in both possible orders $\langle \mathbf{N}, \mathbf{E} \rangle$ and $\langle \mathbf{E}, \mathbf{N} \rangle$, which is demonstrated in Figs. 7(b)-(c) and (d)-(e), respectively. During these steps we detect and collect isthmus points (represented by stars in the mentioned figures), which must be kept in the output skeleton. The following three issues can be observed:

(i) The interior point above point p depicted '▲' in Fig. 7 becomes a strict 1-border point after first performing $\mathcal{R}_\mathbf{N}$, and the latter point then forms a decision pair with the 2-border point p, hence p is considered as an isthmus in the second subiteration. However, in the opposite order $\langle \mathbf{E}, \mathbf{N} \rangle$, p is deleted, since it is visited before it's upper neighbor becomes a border point.

(ii) While the non-simple point q in the input picture remains non-simple in the order $\langle \mathbf{N}, \mathbf{E} \rangle$, in case $\langle \mathbf{E}, \mathbf{N} \rangle$ it becomes a simple point. Note that in the local neihgborhood of q, there are two components, but the component composed of the points marked '•' is completely removed in Fig. 7(d), before q is visited.

(iii) Finally, the strict 2-border point r is super-deletable in the input picture, since it constitutes a decision pair with a strict 4-border point marked '♠'. However the latter one becomes a strict 1-border point after the deletion of the point depicted '♣' in Fig. 7(b) (before r is visited), hence for the order $\langle \mathbf{N}, \mathbf{E} \rangle$, r is no more super-deletable in Fig. 7(b).

For handling the first problem mentioned in (i) in the above listing, the interior points at the beginning of an iteration step must be added to a constraint set together with isthmus points. Hence these points will not be taken into consideration during the whole iteration. The issues in (ii) and (iii) can be solved if the deletion of the critical super-deletable points and the detection of isthmuses are always simultaneously carried out. For this aim, we may delay the deletion of some earlier visited super-deletable points for a later subiteration.

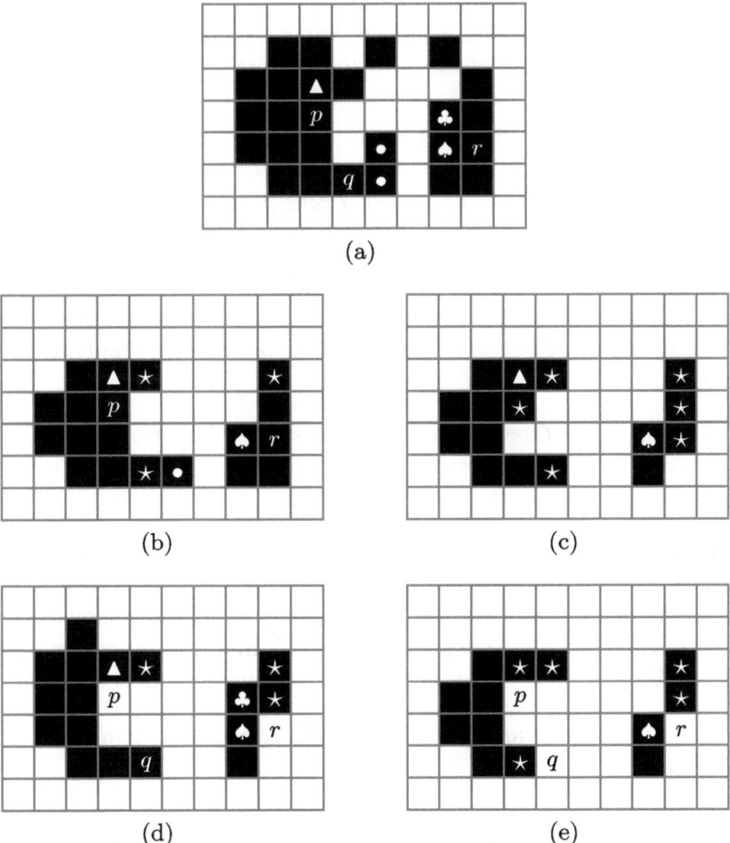

Fig. 7. Examples for demonstrating the critical situations regarding order-independence. The original picture (a), the outputs of succesive reductions $\mathcal{R}_\mathbf{N}$ (b) and $\mathcal{R}_\mathbf{E}$ (c), and the outputs of succesive reductions $\mathcal{R}_\mathbf{E}$ (d) and $\mathcal{R}_\mathbf{N}$ (e). Detected isthmus points are represented by stars.

In order to formulate the algorithm according to the above ideas, here we extend the definitions of m-deletable points, super-deletable points, isthmuses, unbound and bound points (see Definitons 5- 8), and we define what we mean by "simultaneously visitable points" (see Definition 9).

Definition 5. *Let p be a simple strict m-border point in picture $(\mathbb{Z}^2, 8, 4, B)$ $(m \in \{1, 2, 3, 4\})$ such that p is not an element of a small object, and let us introduce a constraint set of points $C \subset B$. Point p is m-C-deletable if $p \notin C$ and for any $1 \leq n < m$ and any simple n-border point $q \in N_4^*(p)$, p is simple for $B \setminus \{q\}$ or q is not n-deletable.*

Definition 6. *An object point p is super-C-deletable for a constraint set C if one of the following conditions hold:*

1. *Point p is m-C-deletable $(m \in \{1, 2, 3, 4\})$.*
2. *If p is an element of a small object that does not contain any point in C, then p is not the smallest element of that object.*

Definition 7. *Let p be a non-isolated strict m-border point in a picture $(\mathbb{Z}^2, 8, 4, B)$ $(m \in \{1, 2, 3, 4\})$ such that p is not an element of a small object, and let $C \subset B$ a constraint set of points. Point $p \notin C$ is called a C-isthmus if it is not m-C-deletable.*

Definition 8. *Let p be a simple point in picture $(\mathbb{Z}^2, 8, 4, B)$ $(m \in \{1, 2, 3, 4\})$ and let $C \subset B$ a constraint set of points. Point p is called C-unbound if any decision pair or small object containing p also contains an element of C, else it is C-bound.*

Note that the case $C = \emptyset$ in Definitions 6-8 corresponds to the original notions of m-deletable points, super-deletable points, isthmuses, unbound and bound points.

Definition 9. *Let $\mathcal{P} = (\mathbb{Z}^2, 8, 4, B)$, $C \subset B$ be a constraint set, and let $p, q \in B$. We denote by $\Gamma(p, q)$ the set of directions $i \in \{\mathbf{N}, \mathbf{E}, \mathbf{S}, \mathbf{W}\}$ such that p and q are both i-border points, and we say that p and q are simultaneously visitable if $\Gamma(p, q) \neq \emptyset$. Furthermore, p is called a critical C-unbound point from a direction $j \in \{\mathbf{N}, \mathbf{E}, \mathbf{S}, \mathbf{W}\}$ if it is C-unbound and the following conditions hold:*

- *p is also a j-border and super-C-deletable point in Y, and*
- *$j \notin \Gamma(p, q)$.*

Figure 8 gives illustrative examples on the notions of Definition 9.

Definition 10. *A black point p is a critical corner point if it matches the template in Fig. 9.*

Now we can proceed to discuss our parallel 4-subiteration thinning algorithm **K-4SI** which is formulated in Algorithm 1.

At the beginning of an iteration (see the **repeat** loop), interior points in the actual picture are collected in set I. An iteration step is composed of four subiterations according to the deletion directions **N**, **E**, **S**, and **W** (see the **for** loop). The order of these directions is not prescribed. During subiteration i, detected isthmuses are collected into set K. By using the constraint set $\mathcal{C}$ (which

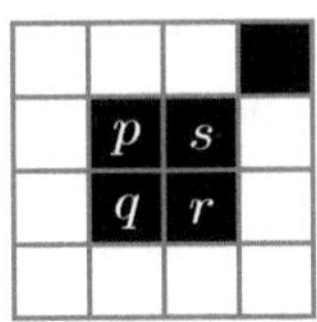

Fig. 8. Example for the notions in Definition 9. In this picture, p, q, and r all unbound points (or equivalently, C-unbound points for $C = \emptyset$). $\Gamma(p, s) = \{\mathbf{N}\}$, $\Gamma(q, s) = \emptyset$, and $\Gamma(r, s) = \{\mathbf{E}\}$. Hence non-simple point s is simultaneously visitable with p and r but not with q. Furthermore, p is critical from direction $\mathbf{W}$, while r is critical from direction $\mathbf{S}$.

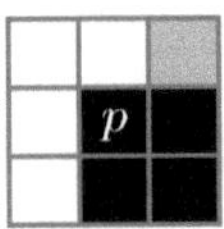

Fig. 9. The configuration of critical corner points. The position colored gray represents a point that can be either black or white.

Algorithm 1. Thinning algorithm **K-4SI**

Input: picture $(\mathbb{Z}^2, 8, 4, X)$ and an arbitrary permutation $\langle d_1, d_2, d_3, d_4 \rangle$ of directions $\{\mathbf{N}, \mathbf{E}, \mathbf{S}, \mathbf{W}\}$
Output: picture $(\mathbb{Z}^2, 8, 4, Y)$
1: $\mathcal{C} = \emptyset$
2: $K = \emptyset$
3: $Y \leftarrow X$
4: **repeat**
5:     $I \leftarrow \{p \mid p$ is an interior point in $Y\}$
6:     **for** $i \leftarrow \{1, 2, 3, 4\}$ **do**
7:         $K \leftarrow K \cup \{p \mid p \in Y \setminus I, p$ is both a d_i-border point and a $\mathcal{C}$-isthmus in $Y\}$
8:         $\mathcal{C} \leftarrow I \cup K$
9:         $D(i) \leftarrow \{p \mid p$ is a super-$\mathcal{C}$-deletable d_i-border point in $Y\}$
10:       $Q \leftarrow \{p \mid p$ is a critical $\mathcal{C}$-unbound point from direction i in $Y\}$
11:       **if** $i \neq \mathbf{W}$ **then**
12:         $Q \leftarrow Q \cup \{p \mid p$ is a critical corner point in $Y \setminus \mathcal{C}\}$
13:       **end if**
14:       $D(i) \leftarrow D(i) \setminus Q$
15:       $Y \leftarrow Y \setminus D(i)$
16:     **end for**
17: **until** $D(d_1) \cup D(d_2) \cup D(d_3) \cup D(d_4) = \emptyset$

---

is initialized to $\emptyset$) we exclude from deletion the points that were interior at the beginning of the given iteration and the previously detected isthmuses being i-border points. When i is the actual direction, only super-$\mathcal{C}$-deletable i-border points can be deleted. Furthermore, here we prevent the deletion of $\mathcal{C}$-unbound points being critical from direction i, and if the actual direction differs from $\mathbf{W}$, then critical corner points may neither be removed. As these two types of points are actually deletable, they must not be added to set K. Instead, we collect those to a separate set $\dot{Q}$, which is simply subtracted from set $D(i)$ of the actual deletable points. This process is repeated until stability is reached (i.e., no point is deleted within the last four subiterations).

In experiments the proposed algorithm **K-4SI** was tested on objects of different shapes, and the obtained results were compared with the centerlines of two existing 4-subiteration algorithms by Rosenfeld and Palágyi et. al [14,15]. Figures 10,11,12 and 13 present four illustrative examples. The fused images in Fig. 10 combine the results for orders of deletion $\langle \mathbf{N}, \mathbf{E}, \mathbf{S}, \mathbf{W} \rangle$ and $\langle \mathbf{W}, \mathbf{S}, \mathbf{E}, \mathbf{N} \rangle$. The three numbers in parentheses are the counts of points in the centerlines for orders $\langle \mathbf{N}, \mathbf{E}, \mathbf{S}, \mathbf{W} \rangle$, $\langle \mathbf{W}, \mathbf{S}, \mathbf{E}, \mathbf{N} \rangle$, and the count of common skeletal points, respectively. Figures 11, 12 and 13 compare only the results for the same order $\langle \mathbf{N}, \mathbf{E}, \mathbf{S}, \mathbf{W} \rangle$, hence there is only one number in parentheses (i.e., the count of skeletal points corresponding to the former permutation).

4   Topology Preservation

To validate the topological correctness of algortihm **K-4SI**, first we prove the following statement.

Lemma 1. *Let $\mathcal{P} = \{\mathbb{Z}^2, 8, 4, Y\}$ an arbitrary picture, and let $C \subset Y$ be a constraint set. Then, the deletion of super-C-deletable points is a topology-preserving reduction.*

Proof. We use induction on the elements of C.

If $C = \emptyset$, then super-C-deletable points coincide with super-deletable points, whose deletion preserve the topology by Theorem 5.

Let us assume that for set C, the statement in the theorem holds. Let $C' = C \cup \{q\}$, and let $\mathcal{R}$ be a reduction that deletes super-C'-deletable points. By Definition 5 and 6, $\mathcal{R}$ deletes only simple points, which corresponds to Condition 1 of Theorem 2.

If q is not super-C-deletable in Y or it is an unbound super-C-deletable point in Y, then, by Definition 5, this obviously does not influence the deletability of the other elements in C. Similarly, if q is a super-C-deletable element of a small object $\mathcal{O}$, then $\mathcal{O}$ is not fully deleted since $q \in C'$. Hence, $\mathcal{R}$ fulfills condition 3 of Theorem 2.

Next, we investigate the case when q is super-C-deletable in Y such that it is not an element of a small object. Then, q must be an element of one or more decision pairs in $Y \setminus C$. Let us suppose that point r in the decision pair $\{q, r\}$ is not super-C-deletable in Y but it is super-C'-deletable in Y.

If r is also an element of another decision pair $\{r, s\}$, where s is super-C-deletable in Y, then by carefully studying the recursive condition in Definition 5, we can observe that $\beta(s) > \beta(r) > \beta(q)$ and s is not super-C'-deletable in Y.

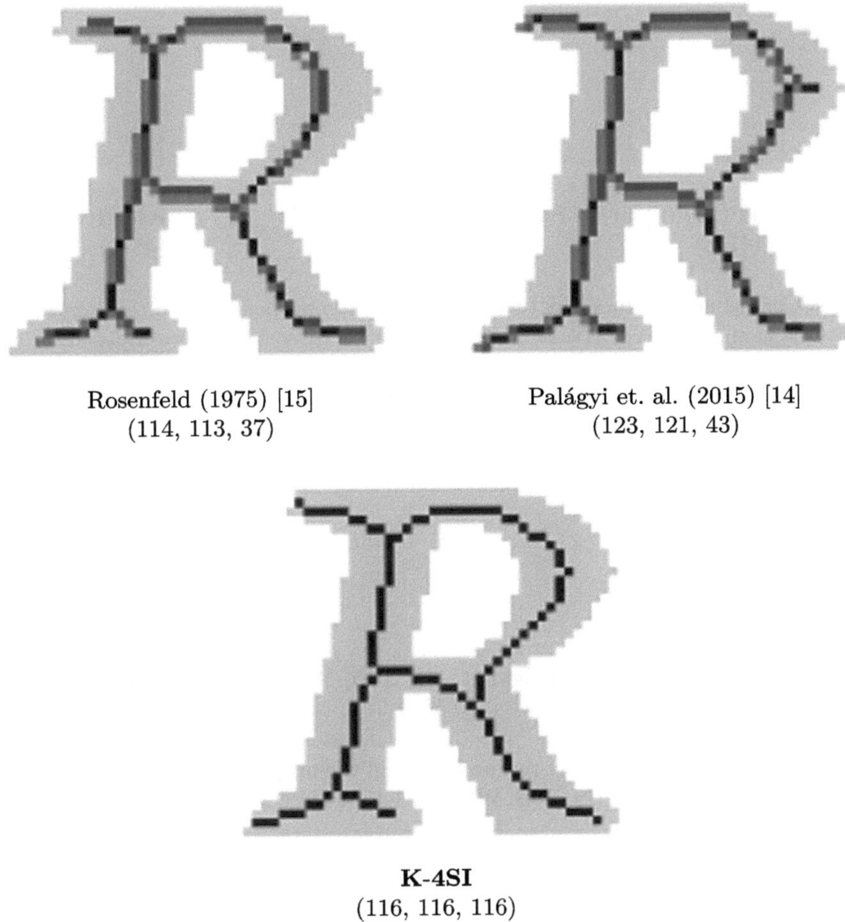

Rosenfeld (1975) [15]
(114, 113, 37)

Palágyi et. al. (2015) [14]
(123, 121, 43)

K-4SI
(116, 116, 116)

Fig. 10. Centerlines produced by two existing and the proposed 4-subiteration thinning algorithms for two different orders of deletion directions superimposed on a 51×46 image of a letter. Red and green pixels show the points contained only in the centerline corresponding to order $\langle \mathbf{N}, \mathbf{E}, \mathbf{S}, \mathbf{W} \rangle$ and $\langle \mathbf{W}, \mathbf{S}, \mathbf{E}, \mathbf{N} \rangle$, respectively, while blue pixels depict the common points of these two centerlines.

In addition, if s is also an element of a further decision pair $\{s, t\}$, where t is not super-C-deletable in Y, then we can conclude, that $\beta(t) > \beta(s) > \beta(r) > \beta(q)$, and t is super-C'-deletable in Y. Note, that this can only happen if $\beta(q) = 1$, $\beta(r) = 2$, $\beta(s) = 3$, and $\beta(t) = 4$, and since there are only four different types of border points, there is not any longer chain of such decision pairs starting from q.

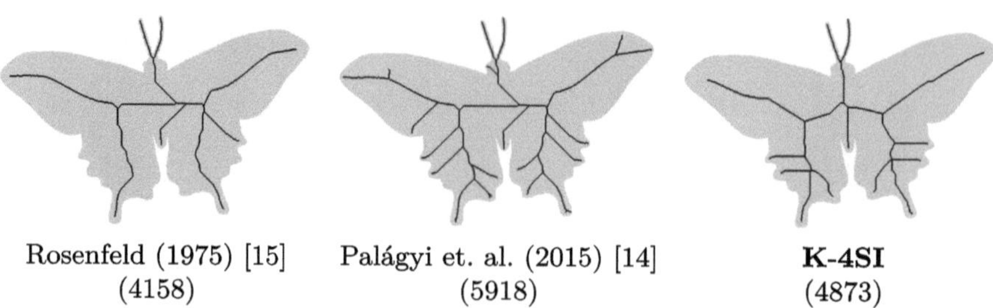

Rosenfeld (1975) [15] Palágyi et. al. (2015) [14] **K-4SI**
(4158)   (5918)   (4873)

Fig. 11. Centerlines produced by two existing and the proposed 4-subiteration thinning algorithms using the order $\langle \mathbf{N}, \mathbf{E}, \mathbf{S}, \mathbf{W} \rangle$ of deletion directions superimposed on a 652 $\times$ 446 image of a butterfly.

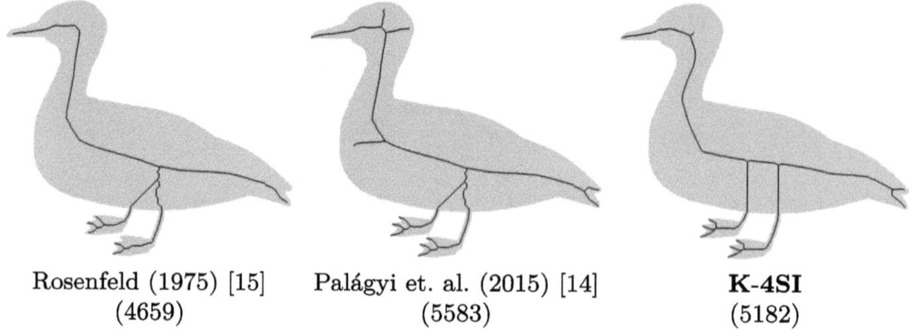

Rosenfeld (1975) [15] Palágyi et. al. (2015) [14] **K-4SI**
(4659)   (5583)   (5182)

Fig. 12. Centerlines produced by two existing and the proposed 4-subiteration thinning algorithms using the order $\langle \mathbf{N}, \mathbf{E}, \mathbf{S}, \mathbf{W} \rangle$ of deletion directions superimposed on a 760 $\times$ 678 image of a duck.

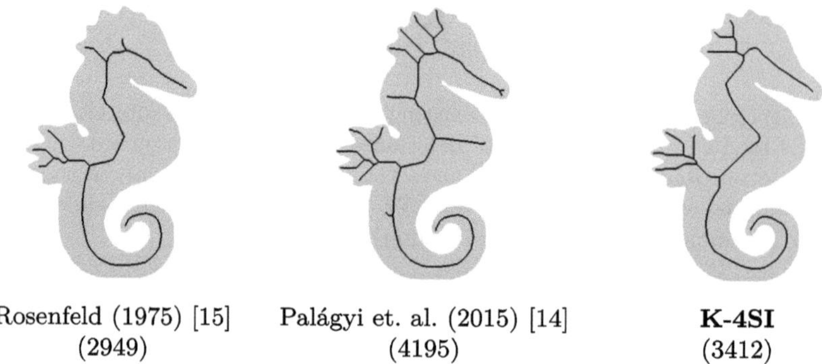

Rosenfeld (1975) [15] Palágyi et. al. (2015) [14] **K-4SI**
(2949)   (4195)   (3412)

Fig. 13. Centerlines produced by two existing and the proposed 4-subiteration thinning algorithms using the order $\langle \mathbf{N}, \mathbf{E}, \mathbf{S}, \mathbf{W} \rangle$ of deletion directions superimposed on a 530 $\times$ 530 image of a seahorse.

If we apply similar train of thoughts for any other possible chain of decision pairs starting from q, we can observe that reduction $\mathcal{R}$ does not delete both elements of any decision pairs, which means that $\mathcal{R}$ also satisfies Condition 2 of Theorem 2.

Consequently, $\mathcal{R}$ preserves topology by Theorem 2, hence the induction is complete. $\qquad\square$

Theorem 6. *Algorithm* **K-4SI** *is topology-preserving.*

Proof. For a given constraint set C, algorithm **K-4SI** deletes only super-C-deletable points in any subiteration step. Therefore, by Lemma 1, each subiteration step of the algorithm performs a topology-preserving reduction. Consequently, each iteration step preserves the topology as well, and thus, so does the whole algorithm. $\qquad\square$

5 Order Independence

In this section we show that algorithm **K-4SI** is order-independent (i.e., invariant under the order of deletion directions). In preparation for this, below we establish some key properties of the algorithm. Lemma 2 states that the deletability of a point in a decision pair does not change during a subiteration. Then, relying on this relationship, Lemma 3 points out that non-deletable points remain non-deletable during a subiteration. Furthermore, Lemma 4 states similar property for deletable points.

Lemma 2. *Let $\{p, q\}$ be a decision pair in picture $(\mathbb{Z}^2, 8, 4, Y)$, where $\beta(p) < \beta(q)$, and let S be the set of deleted points in a given subiteration i of an iteration of Algorithm* **K-4SI** *($i \in \{1, 2, 3, 4\}$). If $\{p, q\}$ remains a decision pair in $Y \setminus S$, then $\beta(p) < \beta(q)$ still holds in $Y \setminus S$, or point q is added to the constraint set C in subiteration i.*

Proof. Let β_0 and β_1 denote the value of $\beta(q)$ (i.e., the type of strict border direction between 1 and 4) at the beginning and at the end of the given subiteration, respectively. The inequality in the lemma can be only reversed if $\beta_1 - \beta_0 \geq 2$. This implies that $\beta_0 \geq 3$. In addition, $\beta(p) \neq 1$ both in sets Y and $Y \setminus S$, or else $\beta(q)$ could not exceed the value of $\beta(p)$, and $\beta(p) \neq 4$, or else $\beta(p) < \beta(q)$ could not be fulfilled. Figure 14 depicts the possible arrangements of sets $N_8(p)$ and $N_8(q)$ in picture $\mathcal{P} = (\mathbb{Z}^2, 8, 4, Y)$.

In Fig. 14(a), $t \notin Y$, or else q would be an interior point, which is not possible. This together with Proposition 1 implies $s \in Y$, and as p can neither be an interior point, $r \notin Y$. This means that $\beta(p) = 4$ and $\beta(q) = 3$ in Y, which contradicts the condition of the lemma.

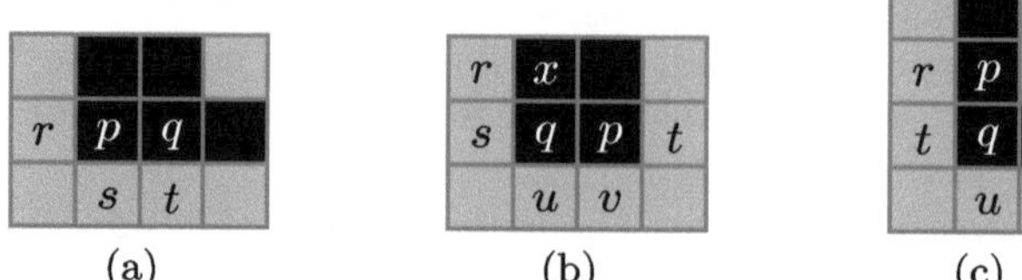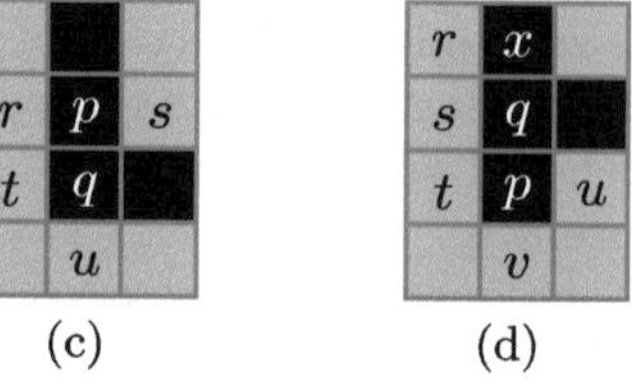

(a)   (b)   (c)   (d)

Fig. 14. The possible configurations for the decision pair $\{p, q\}$ mentioned in Lemma 2. Each element depicted in gray matches either a black or a white point.

Let us consider Fig. 14(b). By careful examination of the possible colors of points r, s, t, u, and v, we can conclude that $\beta(q) = 3$ and $\beta(p) = 2$, or $\beta(q) = 4$ and $\beta(p) = 3$. In either situation, $\beta_1 > \beta_0$ may only occur if $x \in S$ (i.e., black point x is deleted in the given subiteration). This requires $\{r, s\} \cap Y = \emptyset$, or else set $\{q, x\}$ would be a decision pair in Y by Prop. 2, and q would be a non-simple point in $Y \setminus S$, which would again contradict the assumption of the lemma on q. This means, that x may be either a strict 4-border point or a strict 1-border point. In the former case x and q are simultaneously visited in the given subiteration, which means, that q is added to the set K of $\mathcal{C}$-isthmuses, thus $q \in \mathcal{C}$ from this subiteration onwards. In the latter case, by observing the possible colors of the upper (north) neighbors of r and x, we can conlude that x is a critical corner point (see Fig. 9). Hence, the algorithm deletes x only if $i = \mathbf{W}$. This again concludes $q \in \mathcal{C}$ from subiteration i onwards.

In Fig. 14(c), $s \notin Y$, or else p could be only a strict 4-border point in Y, which is not possible due to our initial assumptions. Hence, q is a strict 2-border point, however, $\beta(p) > \beta(q)$ could only occur in $Y \setminus S$, if p would be deleted, which again contradicts the condition of the lemma on $\{p, q\}$.

Finally, let us investigate case (d) in Fig. 14. Similarly to case (b) we can derive that $\beta(p)$ may only exceed the value of $\beta(q)$ in $Y \setminus S$ if black point x is a strict 4-border point or a critical corner point in Y, which is deleted in subiteration $i = W$, when q is added to the set $\mathcal{C}$.

Therefore, we can conclude that the statement of the lemma holds. $\qquad\square$

Lemma 3. *Let us suppose that during subiteration i of an iteration of algorithm* **K-4SI**, *the set of black points Y in the actual picture is reduced to $Y \setminus D(i)$ $(i \in \{1, 2, 3, 4\})$. If point p is not super-$\mathcal{C}$-deletable in Y, then it is still not super-$\mathcal{C}$-deletable in $Y \setminus D(i)$.*

Proof. If non-deletable point p is in the constraint set $\mathcal{C}$ at the beginning of subiteration i, then the lemma trivially holds. Hereafter, we will assume that $p \notin \mathcal{C}$.

If p is an element of a small object $\mathcal{O}$, then by Condition 2 of Definition 6, it must be the smallest element of $\mathcal{O}$, which evidently remains the smallest in $Y \subset S$, hence the lemma is satisfied in this case.

Otherwise, p must violate Condition 1 of Definition 6, which means that it is not an i-$\mathcal{C}$-deletable point. First, let us suppose that p is a non-simple point in Y. Then, by Theorem 1, $C(p) > 1$. It is sufficient to prove that each object in $N_8^*(p) \cap Y$ contains a point x such that x is not super-$\mathcal{C}$-deletable, or it is a $\mathcal{C}$-unbound point x and $\Gamma(p, x) \neq \emptyset$ (i.e., p and x are simultaneously visitable). The latter property ensures that p is added to the set of $\mathcal{C}$-isthmuses K before $C(p)$ decreases to 1 (i.e., before p becomes a simple point).

We make a distinction among three possible cases shown in Fig. 15(a)-(c):

- Let us suppose that there exists an object $\mathcal{O}$ in $Y \in N_8^*(p)$ such that $q, r \in \mathcal{O}$, and q and r are 8-adjacent but not 4-adjacent, see Fig. 15(a). Then, it can be easily verified that both points in $N_4^*(p) \setminus \{q, r\}$ must be white in the actual picture, or else p would be a simple point or an interior point. From here, we distinguish three subcases:
    * If $y \in Y$, then we can observe that $\Gamma(q, r) = \emptyset$, hence either $q \in Y \setminus D(i)$ or $r \in Y \setminus D(i)$.
    * Let $y \notin Y$ and $x \notin Y$. If $u \in Y$, then q is non-simple. If $u \notin Y$ then it is easy to verify that either q is non-simple again (for $t \in Y$ or $v \in Y$), or it is a $\mathcal{C}$-unbound point such that $\mathbf{W} \in \Gamma(p, q)$.
    * If $y \notin Y$ and $x \in Y$, then either q or w is non-simple, or $\{q, w\}$ is a decision pair (by Prop 2 for $t \in Y$ or $u \in Y$), or x is a $\mathcal{C}$-unbound point (if $w \notin Y$) where $\mathbf{W} \in \Gamma(p, x)$.
- Let us consider configuration (b). If set $\{q, r\}$ is a decision pair or at least on of points q and r is non-simple, then $q \in Y \setminus D(i)$ or $r \in Y \setminus D(i)$. Otherwise, $s \in Y$ must hold by Prop. 2. Now, if $t \in Y$, then r is non-simple, and if $t \notin Y$, then r is a $\mathcal{C}$-unbound point such that $\mathbf{E} \in \Gamma(p, r)$.
- Finally, let us examine situation (c). If $r \in Y$ or $s \in Y$, then q is non-simple. Otherwise, q is a $\mathcal{C}$-unbound point and $\{\mathbf{N}, \mathbf{E}\} \subseteq \Gamma(p, q)$.

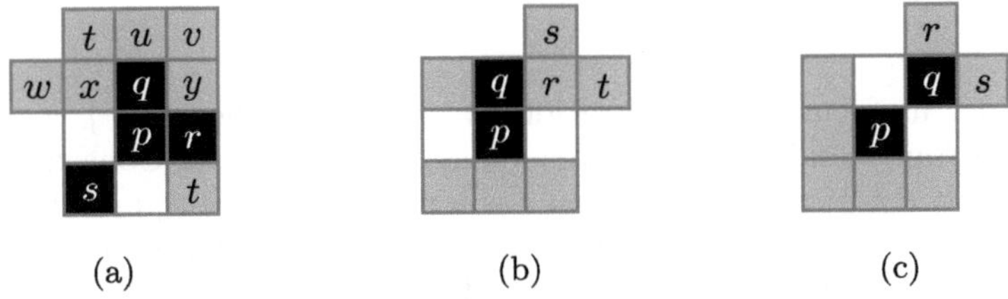

(a)   (b)   (c)

Fig. 15. The possible base configurations for non-simple points. Each element depicted in gray matches either a black or a white point.

It can be readily seen that any other possible configurations of $N_8(p)$ are rotated or reflected versions of the above ones, where the proof can be similarly carried out.

Until now we proved the lemma for non-simple points. Next, let us assume that p is simple but not super-$\mathcal{C}$-deletable in Y. Then there must be a super-$\mathcal{C}$-deletable point $q \in N_4(p) \cap Y$ such that $\{p, q\}$ is a decision pair in Y. Let us assume that each of the decision pairs containing p remain in $Y \setminus D(i)$. Since

Lemma 2 ensures that the deletability of the points in decision pairs does not change during a subiteration, we can state that p is still not super-$\mathcal{C}$-deletable in $Y \setminus D(i)$.

Hereafter we examine the situation when $\{p, q\}$ is not a decision pair anymore in $Y \setminus D(i)$. Since the deletability of q cannot change by Lemma 2, $q \notin Y \setminus D(i)$ (i.e., q is deleted in subiteration i of the actual iteration). Without loss of generality, we can assume that q is an **N**-border point and p is its south neighbor. Depending on the colors of the west and east neighbors of p, we can make a distinction among four cases, three of which are depicted in Fig. 16. Since in this proof we do not rely on the border type of black points, the two cases for the west and east neighbors having different colors can be similarly examined. (Fig. 16(b) depicts one of those cases.) Furthermore, the proof can also be similarly established for the rotated and reflexed versions of the mentioned configurations. It is obvious that if only q is deleted in the depicted cases, then points r and s are elements of two different objects in $Y \cap N_8^*(p)$. Hence, it is sufficient to show that at least one point in each of these objects remains in $Y \setminus D(i)$.

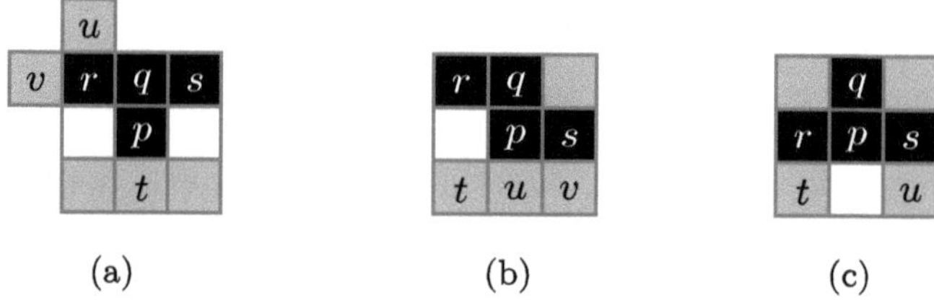

(a)   (b)   (c)

Fig. 16. The possible base configurations for decision pairs. Each element depicted in gray matches either a black or a white point.

- First, let us focus on configuration (a). If $u \in Y$, then $\{q, r\}$ is a decision pair by Prop. 2 and thus, r cannot be super-$\mathcal{C}$-deletable. Consequently, $r \in Y \setminus D(i)$. If $u \notin Y$ and $v \in Y$, then r is a non-simple point, and if $u \notin Y$ and $v \notin Y$, then $\Gamma(q, r) = \mathbf{W}$, hence r is a critical $\mathcal{C}$-unbound point. This implies $r \notin Y \setminus D(i)$. Due to the symmetry of $N_8(p)$ in Fig. 16(a), the statement $s \notin Y \setminus D(i)$ can be proved similarly as for r.
- Now, let us examine case (b). We can derive $r \notin Y \setminus D(i)$ in the same way as in the previous case. Furthemore, if $t \in Y$, then $Gamma(q, s) = \emptyset$, hence q or s remains in $Y \setminus D(i)$. Let us suppose that $t \notin Y$. Here we investigate three subcases depending on the possible colors of t and u:
  - ⋆ Let $\{t, u\} \subset Y$. The deletion of q implies that either $d_i = \mathbf{N}$ or $d_i = \mathbf{E}$. Since t is not an **E**-border point and u is not an **N**-border point, at least one of the points t and u must remain in $Y \setminus D(i)$.
  - ⋆ If $t \in Y$ and $u \notin Y$, then p must be non-simple, however, this is a contradiction with the initial assumption. (Elements of a decision pair must be both simple points.)

⋆ If $u \notin Y$, then $v \in Y$ implies that $\mathbf{W} \in \Gamma(p, v)$, while $v \notin Y$ implies that $\mathbf{S} \in \Gamma(p, s)$. From here, it can be similarly to the previous cases verified that either v or s is $\mathcal{C}$-unbound point or a non-simple point, or they are elements of a decision pair in $N_8(p)$.

- Finally, let us investigate configuration (c). If $t \in Y$, then the arrangement of the elements in $N_8(t)$ is the rotated version of $N_8(r)$ of Fig. 16(a). Thus, we can similarly derive $t \in Y \setminus D_i$ as in case (a) for point r. If $t \notin Y$, then $\mathbf{S} \in \Gamma(p, r)$. Thus, r is a critical $\mathcal{C}$-unbound point, which means that $r \in Y \setminus D(i)$. Due to the symmetry of configuration (c), we can similarly show that $s \in Y \setminus D(i)$ or $u \in Y \setminus D(i)$.

Consequently, the lemma holds. $\qquad\qquad\qquad\qquad\qquad\qquad\qquad\qquad$ □

The following propositions are straightforward consequences of Definition 1, Theorem 4, and Theorem 6.

Proposition 3. *In each iteration step of algorithm* **K-4SI***, the set of deleted points is a P-simple set.*

Proposition 4. *A subset of a P-simple set is also P-simple.*

Lemma 4. *Let us suppose that during subiteration i of an iteration of algorithm* **K-4SI***, the set of black points Y in the actual picture is reduced to $Y \setminus D(i)$ ($i \in \{1, 2, 3, 4\}$). If point p is super-$\mathcal{C}$-deletable in Y but it is not deleted in subiteration i, then p is also super-$\mathcal{C}$-deletable in $Y \setminus D(i)$.*

Proof. If p is an element of a small object, then it is not the smallest element in that object, and obviously, this property does not change during a subiteration. Furthermore, it can be easily seen, that if we extend any small object $\mathcal{O}$ to a non-small object, then at least one of the points in $\mathcal{O}$ in the extended object must be an interior point or a $\mathcal{C}$-isthmus point. This implies that if during a subiteration, a non-small object is reduced to a small object $\mathcal{O}$, then the algorithm will add at least one element of $\mathcal{O}$ to the constraint set $\mathcal{C}$, thus any other elements in $\mathcal{O}$ can be then deleted by Condition 2 of Definition 6.

Hereafter, we assume that p is neither in Y, nor in $Y \setminus S$ an element of a small object. Let U denote the set of super-$\mathcal{C}$-deletable points in picture $(\mathbb{Z}^2, 8, 4, Y)$. As $p \in U$, p must be simple in Y. By Proposition 3, U is a P-simple set in Y, and by Proposition 4, set $S \cup \{p\} \subseteq U$ is also P-simple in Y. Thus by Definition 1, p is simple in $Y \setminus S$ as well.

Furthermore, if p is $\mathcal{C}$-unbound in Y, then, by combining Proposition 2, Lemmas 2-3, and Definition 8, we can verify that p remains $\mathcal{C}$-unbound in $Y \setminus S$.

Now, let us suppose that p is $\mathcal{C}$-bound in Y, which means that it constitutes a decision pair $\{p, q\}$ with a simple point $q \in Y \setminus$. If $q \in \mathcal{C}$, then q may not be deleted in the actual subiteration i, hence it does not affect the deletability of p. Let us consider that $q \notin \mathcal{C}$. Then, by Definition 5, it is easy to verify that one of the following two cases may occur:

- $\beta(p) < \beta(q)$, or

- $\beta(p) > \beta(q)$, and there is an m-$\mathcal{C}$-deletable point $r \in Y$ such that $\{q, r\}$ is a decision pair in $Y \setminus \mathcal{C}$ and $\beta(q) > \beta(r)$.

In both cases, Lemma 2 implies that the above inequalities also hold in $Y \setminus S$, or the algorithm adds point q to the set K of detected $\mathcal{C}$-isthmuses in subiteration i, which means that q is not super-$\mathcal{C}$-deletable in $Y \setminus S$. Hence, again, q does not influence the deletablity of p.

This completes the proof of the lemma, since we showed for all possible situations that p remains super-$\mathcal{C}$-deletable in $Y \setminus S$. $\qquad\square$

Finally, we have arrived to the main theorem of the paper.

Theorem 7. *Algorithm* **K-4SI** *is order-independent.*

Proof. By Lemmas 3 and 4, the deletability of a point does not change in a subiteration of the algorithm. Consequently, the set $D(d_1) \cup D(d_2) \cup D(d_3) \cup D(d_4)$ of deleted points at the end of an iteration is independent of the chosen permutation of directions **N**, **E**, **S**, and **W**, which concludes the order independence of algorithm **K-4SI**. $\qquad\square$

6   Conclusions

This work introduces the very first order-independent parallel subiteration-based thinning algorithm which produces centerlines of 2D binary pictures. The proposed method combines the strategies of deleting super-deletable points and accumulating isthmus points. In addition, the algorithm is proved to be topology-preserving as well.

Acknowledgments. This research was supported by project no. TKP2021-NVA-09 provided by the Ministry of Culture and Innovation of Hungary from the National Research, Development and Innovation Fund.

References

1. Bertrand, G.: On P-simple points. Compte Rendu de l'Académie des Sciences de Paris, Série Math. **I**(321), 1077–1084 (1995)
2. Bertrand, G., Couprie, M.: Transformations topologiques discretes. In: Coeurjolly, D., Montanvert, A., Chassery, J. (eds.) Géométrie discrète et images numériques, pp. 187–209. Hermes Science Publications (2007)
3. Couprie, M.: Note on fifteen 2d parallel thinning algorithms. Tech. Rep. IMG2006-01, Université de Marne-la-Vallée (2006)
4. Hall, R.W.: Parallel connectivity-preserving thinning algorithms. In: Kong, T.Y., Rosenfeld, A. (eds.) Topological Algorithms for Digital Image Processing, pp. 145–179. Elsevier Science B. V. (1996). https://doi.org/10.1016/S0923-0459(96)80014-0

5. Iwanowski, M., Soille, P.: Order independence in binary 2d homotopic thinning. In: Kuba, A., Nyúl, L., Palágyi, K. (eds.) DGCI 2006, LNCS 4245. pp. 592–604. Springer (2006). https://doi.org/10.1007/11907350_50

6. Kardos, P., Palágyi, K.: Isthmus-based order-independent sequential thinning. In: Proceeding 9th International Conference on Signal Processing, Pattern Recognition and Applications, SPPRA 2012. pp. 28–34. ACTA Press (2012). https://doi.org/10.2316/p.2012.778-025

7. Kardos, P., Palágyi, K.: On topology preservation in triangular, square, and hexagonal grids. In: Proc. 8th International Symposium on Image and Signal Processing, ISPA 2013, pp. 782–787 (2013). https://doi.org/10.1109/ispa.2013.6703844

8. Kardos, P., Palágyi, K.: A single-step 2d thinning scheme with deletion of P-simple points. In: Progress in Pattern Recognition, Image Analysis, Computer Vision, and Applications, CIARP 2017, LNCS, vol. 10657, pp. 475–482. Springer (2017). https://doi.org/10.1007/978-3-319-75193-1_57

9. Kardos, P., Palágyi, K.: Maximal P-simple sets on (8,4) pictures. In: Proceeding 6th International Symposium on Computational Modeling of Objects Presented in Images: Fundamentals, Methods, and Applications, CompIMAGE'18, pp. 23–32. Springer (2019). https://doi.org/10.1007/978-3-030-20805-9_3

10. Kardos, P., Németh, G., Palágyi, K.: An order-independent sequential thinning algorithm. In: IWCIA 2009, LNCS 5852, pp. 162–175. Springer-Verlag (2009). https://doi.org/10.1007/978-3-642-10210-3_13

11. Kong, T.Y.: On topology preservation in 2-d and 3-d thinning. Int. J. Pattern Recognit Artif Intell. **9**, 813–844 (1995). https://doi.org/10.1142/S0218001495000341

12. Kong, T.Y., Rosenfeld, A.: Digital topology: Introduction and survey. Comput. Vision Graph. Image Process. **48**, 357–393 (1989). https://doi.org/10.1016/0734-189X(89)90147-3

13. Matejek, B., et al.: Scalable biologically-aware skeleton generation for connectomic volumes. IEEE Trans. Med. Imaging **41**(9), 2360–2370 (2022). https://doi.org/10.1109/TMI.2022.3164343

14. Palágyi, K., Németh, G., Kardos, P.: Topology-preserving equivalent parallel and sequential 4-subiteration 2d thinning algorithms. In: Proc. 9th International Symposium on Image and Signal Processing and Analysis, IEEE/EURASIP, ISPA 2015 (2015). https://doi.org/10.1109/ISPA.2015.7306077

15. Rosenfeld, A.: A characterization of parallel thinning algorithms. Inf. Control **29**(3), 286–291 (1975). https://doi.org/10.1016/s0019-9958(75)90448-9

16. Saha, P.K., Borgefors, G., Sanniti di Baja, G.: Skeletonization: Theory, Methods and Applications. Academic Press (2017). https://doi.org/10.1016/B978-0-08-101291-8.00017-1

17. Wagner, M.G.: Real-time thinning algorithms for 2D and 3D images using GPU processors. J. Real-Time Image Proc. **17**(5), 1255–1266 (2019). https://doi.org/10.1007/s11554-019-00886-7

18. Wang, R., Guo, X., Liu, J., Wan, Y.: concrete crack skeleton analysis: a machine vision approach to feature extraction. Adv. Civil Eng. **2024**(1), 6942295 (2024). https://doi.org/10.1155/2024/6942295

A Discrete Bijective Reflection with Near-Equivalence to Shear Rotation Based Reflection

Gaëlle Largeteau-Skapin[1]([envelope]), Lidija Čomić[2], Rita Zrour[1], and Eric Andres[1]

[1] University of Poitiers, Laboratory XLIM, ASALI, UMR CNRS 7252, BP 30179, 86962 Futuroscope Chasseneuil, France
`{gaelle.largeteau.skapin,rita.zrour,eric.andres}@univ-poitiers.fr`
[2] Faculty of Technical Sciences, University of Novi Sad, Novi Sad, Serbia
`comic@uns.ac.rs`

Abstract. We introduce an improved version of the discrete bijective reflection proposed in [Andres 2019] and we show that this new reflection is almost always identical to the reflection based on the quasi shear rotation for integer centers. The difference between the two reflections is restricted to a discrete set of angles of the form $\arctan(k + \frac{1}{2})$, where $k \in \mathbb{N}$. The *quasi shear rotation* (QSR) based reflection yields among the lowest errors compared to the continuous reflection. Our *improved pivot reflection* (IPR) provides the same error in most cases and is easier to compute. Our new approach therefore offers a promising direction for the computation of nD rotations.

Keywords: Discrete geometry · Reflection · Bijectivity · Pivot

1 Introduction

Rotations and reflections are fundamental operations in geometry, with applications in physics, computer science, and engineering. While these transformations are well understood in the continuous setting, their discrete counterparts remain the subject of active research, particularly in the field of discrete geometry. The challenge lies in defining discrete analogues that are both bijective and computationally efficient, while faithfully approximating their continuous versions. Reflection is one of the most fundamental linear transformations [6]. Nevertheless, it has received relatively little attention in the discrete setting, despite its central role in the construction of n-dimensional rotations [3,4,13].

Digitized rotations have many applications such as template matching [5], object tracking [14], light wave simulation [7,9,15], etc. The accurate and efficient computation of rotations in discrete spaces remains a fundamental problem in areas such as discrete geometry, computer graphics, and image processing. While continuous models provide elegant theoretical tools, their direct application to discrete settings often leads to inaccuracies and loss of bijectivity. In

P. Balázs et al. (Eds.): IWCIA 2025, LNCS 15985, pp. 188–199, 2026.
https://doi.org/10.1007/978-3-032-19347-6_12

recent years, bijective transformations have emerged as a powerful alternative, preserving essential properties such as invertibility and structural consistency. The theoretical discussion on the subject of digitized rigid motions can be found in [11].

In this context, the discrete bijective reflection introduced in [2] offered a promising approach, bridging the gap between discrete and continuous rotational models. The idea of this paper was to divide discrete space into parallel discrete lines orthogonal to the reflection line and to chose a pivot point for each of these lines. The orthogonal lines are mapped on themselves this way which is an interesting property. Let us call this reflection the *Pivot based Bijective Reflection (PBR)*. Another way of defining a discrete bijective reflection with a reflection line passing through $(0,0)$ is to use a discrete bijective rotation with center $(0,0)$ and a trivial reflection such as one with reflection line $y = 0$: a rotation can be defined by two reflections, and therefore a reflection can be defined by a rotation and a (in this case trivial) reflection since reflections are involutions. There are two known bijective rotations: the first one is based on Pythagorean triplets [8]. Although it provides the smallest errors compared to the continuous rotations, it is not defined for all angles. The second one is the quasi-shear rotation [1] that is defined for all angles. We noticed that the reflection based on the quasi-shear rotation, named *quasi shear reflection* (QSR) had lower error measures than PBR and preserved the discrete orthogonal lines like a PBR. The question was therefore: is the QSR a PBR with another choice of pivot points?

We present an improved version of this digitized bijective reflection with integer center (meaning that the reflection line passes through an integer point), which closely approximates the bijective shear reflection for almost all angles. We demonstrate that the new formulation coincides almost everywhere with the QSR, with discrepancies confined to a discrete set of angles of the form $\arctan(k + \frac{1}{2})$, where $k \in \mathbb{N}$. Owing to the low approximation error of the QSR and the computational simplicity of the proposed method, this approach offers a promising direction for discrete modeling of rotations in arbitrary dimensions. The problem with a high dimensional approach completely based on shears is that the number of shears needed increases quadratically with the dimension, while the number of reflections grows linearly. The proposed method not only preserves bijectivity but also opens the door to significantly simpler computations, making it well-suited for high-dimensional applications [13].

The rest of the paper is organized as follows. In Subsect. 2.1, we briefly review the theoretical background. In Subsect. 2.2, we present the reflection based on the *quasi shear discrete rotation*. In Sect. 3, we present the pivot based discrete reflection and introduce the new way of finding the pivot point. In Sect. 4, we provide a detailed comparison between the *improved pivot reflection* and the *quasi shear reflection* highlighting both the similarities and the discrete discrepancies. In Sect. 5, we present numerical experiments and error analysis. Finally, in Sect. 6 we conclude the paper and outline possible directions for future research.

2   Preliminaries

2.1   Basic Notation

Let $\{i, j\}$ denote the canonical basis of the two-dimensional Euclidean space $\mathbb{R}^2$. The discrete grid $\mathbb{Z}^2$ is defined as the set of all points in $\mathbb{R}^2$ with integer coordinates. We refer to elements of $\mathbb{R}^2$ as *Euclidean points*, and to those of $\mathbb{Z}^2$ as *discrete points*.

Two discrete points $p = (p_x, p_y)$ and $q = (q_x, q_y)$ are said to be 8-*connected* if $|p_x - q_x| \leq 1$ and $|p_y - q_y| \leq 1$.

For a real value $x \in \mathbb{R}$, we denote by $\lfloor x \rfloor$ the greatest integer less than or equal to x, and by $\lceil x \rceil$ the smallest integer greater than or equal to x. We use $\lfloor x + 0.5 \rfloor$ to obtain the closest integer to the real value x.

A *naive discrete straight line* (DSL) is the set of all integer points (x, y) satisfying the inequality:

$$-\frac{\omega}{2} \leq ax - by < \frac{\omega}{2},$$

where a and b are integers such that $\frac{a}{b}$ represents the slope, and $\omega = \max(|a|, |b|)$ defines the *arithmetical thickness* [12]. Naive DSLs are 8-connected and minimal: removing any point from the set breaks the 8-connectivity [12].

We denote by $\mathcal{R}_\theta : \mathbb{R}^2 \to \mathbb{R}^2$ the reflection of an Euclidean point with respect to the line passing through $(0,0) \in \mathbb{R}^2$ with direction vector $v = (\sin\theta, \cos\theta)$. The corresponding discrete operator, which maps discrete points to discrete points, is denoted $Refl_\theta : \mathbb{Z}^2 \to \mathbb{Z}^2$.

Let us suppose, w.l.o.g for what follows, that the center of the rotation is the integer point $(0,0)$. A continuous rotation of center $(0,0)$ and angle θ can be defined as the composition of two reflections:

$$\mathcal{R}ot_\theta = \mathcal{R}_{\alpha+\theta/2} \circ \mathcal{R}_\alpha,$$

for any arbitrary angle α.

2.2   The Quasi Shear Reflection

By reversing the previous formulation, it is easy to see that it is possible to define a Reflection using a Rotation:

$$\mathcal{R}_\alpha = \mathcal{R}_0 \circ \mathcal{R}ot_{2\alpha}$$

The *quasi shear rotation* proposed in Algorithm 1 is similar to the one proposed in [1], although in a much simpler form since we are only considering an integer center. The idea is to simply *round* the horizontal and vertical shifts in the shear based rotation introduced by A.W. Paeth [10].

Definition 1. *The* quasi shear reflection *(QSR) of angle α is defined using the* quasi shear rotation *(see Algorithm 2):*

$$Refl_\alpha = Refl_0 \circ Rot_{2\alpha}$$

Algorithm 1: QUASISHEARROTATION

---

    input : Discrete point (x, y), angle α
    output: Discrete point (x_s, y_s)
1 **begin**
2     $x_1 = x - \lfloor 0.5 + y \tan(\alpha/2) \rfloor$;
3     $y_s = y + \lfloor 0.5 + x_1 \sin \alpha \rfloor$;
4     $x_s = x_1 - \lfloor 0.5 + y_s \tan(\alpha/2) \rfloor$;

Algorithm 2: QUASISHEARREFLECTION

---

    input : Discrete point (x, y), angle α
    output: Discrete point (x_S, y_S)
1 **begin**
2     $pt2 = (x, -y)$; (* Refl(0) *)
3   QUASISHEARROTATION($pt2$, 2α) (* Rotation of angle 2α *)

---

3   Improved Pivot Reflection

3.1   Recall: Pivot Based Bijective Reflection [2]

In [2] a bijective reflection algorithm in two dimensions has been proposed along with an associated rotation. The reflection line is defined by a straight line passing through $(0, 0)$ (w.l.o.g. since we here consider reflection and rotation centered on integer points). The reflection line is digitized and the 2D space is paved by discrete straight lines perpendicular to the reflection line. For each perpendicular line, integer points are reflected by central symmetry with respect to the reflection line. Two consecutive discrete reflections are combined to define a discrete bijective rotation about arbitrary center.

The idea behind a completely discrete framework is the following [2]: Firstly, the reflection line is digitized. The *discrete reflection straight line* (DRSL) is defined by:

$$L = \left\{ (x, y) \in \mathbb{Z}^2 : -\frac{\max\left(|a|, |b|\right)}{2} \le ax - by < \frac{\max\left(|a|, |b|\right)}{2} \right\}.$$

The DRSL is defined as a naive line. Secondly, discrete space is partitioned (with parameter k) with DSLs that are perpendicular to the DRSL. The *perpendicular discrete straight lines* (PDSL) are defined by:

$$P(k) : \left\{ (x, y) \in \mathbb{Z}^2 : \frac{(2k - 1)\max\left(|a|, |b|\right)}{2} \le bx + ay < \frac{(2k + 1)\max\left(|a|, |b|\right)}{2} \right\}.$$

The perpendicular discrete straight lines are defined as naive lines as well.

For angles θ verifying $0 \le \theta \le \pi/4$, we can order all the points on a given PDSL with respect to a *pivot point* that lies on or is close to the DRSL. Because the lines are naive, and for $0 \le \theta \le \pi/4$, there is one point (x, y) for each y and k. The value of the abscissa x is given by: $computeX(y, k, a, b) := \lceil (2k-1)/2 - (a/b)y \rceil$. For angles between $\pi/4$ and $\pi/2$, the formulas are similar with the computation of the ordinate based on $computeY(x, k, a, b) := \lceil ((2k-1)a - 2bx)/(2a) \rceil$ (see [2] for details).

Let $pivot(x_p, y_p)$ be the symmetry center, the symmetric image of a point $P(x, y)$ is given by : $(computeX(2y_p - y, k, a, b), 2y_p - y)$.

The use of $computeX$ (resp. $computeY$) ensures that the symmetric image of a point remains in the same discrete orthogonal line as the starting point.

In the original paper [2], the choice of the pivot point was such that the reflection transformation was not performing so well as the Shear based reflection for integer centers.

3.2   New Pivot Choice Using the Quasi Shear Reflection

We introduce the *improved pivot reflection*, which relies on the pivot-based framework, but differs from the original approach in [2] (see Sect. 3.1) by locally incorporating the *quasi shear reflection* in the pivot computation. Starting from the same initial point P, we compute its image P_s using the *quasi shear reflection*. The pivot is then defined by the average x_p or y_p of x- or y-coordinates of the two points, depending on the rotation angle.

This modification enables the pre-computation of a lookup table containing all pivot positions for each of the orthogonal line to the reflection line, significantly accelerating the reflection process by avoiding per-point shear computations across the image.

Given the order k of an orthogonal line and the angle α of the reflection, Algorithm 3 computes the pivot point for the considered orthogonal line. This algorithm also uses $computeX$ and $computeY$ to ensure orthogonal line stability.

Algorithm 3: COMPUTEIMPROVEDPIVOT

input : DLS order k, angle α
output: Discrete point (x_p, y_p)

1 **begin**
2 $\quad$ $a = \sin(\alpha)$; $b = \cos(\alpha)$;
3 $\quad$ (* computing the reference point P *)
4 $\quad$ $P = $ **if** abs$(\alpha) \leq \pi/4$ **then**
5 $\quad\quad$ $(computeX(\lceil abk \rceil, k, a, b), \lceil abk \rceil)$
6 $\quad$ **else**
7 $\quad\quad$ $(\lceil abk \rceil, computeY(\lceil abk \rceil, k, a, b))$
8 $\quad$ (* computing the image of P through the Quasi Shear Reflection *)
9 $\quad$ $P_s = $ QUASISHEARREFLECTION$[P, \alpha]$;
10 $\quad$ (* result is the midpoint of P and P_s *)
11 $\quad$ $(x_P, y_P) = $ Midpoint(P, P_s)

The *improved pivot reflection* (see Algorithm 4) of a point (x, y) is computed by first determining the index $k = ComputeK(\{x, y\}, a, b, \{xo, yo\}) = (b(x - xo) + a(y - yo))/(Max[Abs[a], Abs[b]])$ of the orthogonal line it belongs to, and then using the corresponding pivot point (x_P, y_P) for that line. Depending on the angle, we either use $computeX$ with y_P, or $computeY$ with x_P, to compute the final reflected point. The list of pivot points for an image are stored in a LookUp Table (see LUT in Algorithm 4).

Algorithm 4: IMPROVEDPIVOTREFLECTION

> **input** : Discrete point (x, y), LUT , angle α , k_{min}
> **output:** Discrete point (x_r, y_r)

1 **begin**
2   k = computeK$((x, y), \sin(\alpha), \cos(\alpha))$;
3     $(x_P, y_P) = LUT(k + Abs(k_{min}) + 1)$;
4   **if** abs$(\alpha) \leq \pi/4$ **then**
5   (x_r, y_r)=$(computeX(2y_P - y, k, a, b), 2y_P - y)$
6   **else**
7   (x_r, y_r)=$(2x_P - x, computeY(2x_P - x, k, a, b))$

---

4   Comparison Between the Quasi Shear Reflection and the Improved Pivot Reflection

In this section, we compare both types of reflections that are seemingly based on two different approaches, although the IPR uses the QSR to compute the pivot point. While comparing the two reflection transforms, we noticed that, in most cases, the *quasi shear reflection* preserves the orthogonal discrete reflection lines when the center of the associated rotation has integer coordinates. Thus the following proposition:

Proposition 1. *For integer centers, the* improved pivot reflection *is identical to the* quasi shear reflection *except for angles* α *such that* $tan(\alpha) = k + 0.5$, $\forall k \in \mathbb{N}$.

Proof. We consider w.l.o.g. angles $\alpha \in [0, \pi/4]$ and therefore use $computeX$ and express X through Y.

From the definition of the *improved pivot reflection* of angle α with center $(0, 0)$, we obtain the following equation for the image (X_P, Y_P) of any discrete point $(x, y), x \in \mathbb{Z}, y \in \mathbb{Z}$:

$$Y_P = -y + (A + \lfloor 0.5 - A \rfloor + \lfloor 0.5 + (\lfloor 0.5 + \lceil -0.5 + B - A \tan(\alpha) \rceil \rfloor$$
$$- \lfloor 0.5 + \lfloor 0.5 - A \rfloor \tan(\alpha) \rfloor) \sin(2\alpha)])$$
$$X_P = -\lfloor 0.5 - B + Y_P \tan(\alpha) \rfloor$$

With
$$A = \lceil \cos(\alpha) \lfloor 0.5 + x + y \tan(\alpha) \rfloor \sin(\alpha) \rceil$$
$$B = \lfloor 0.5 + x + y \tan(\alpha) \rfloor$$

From the definition of the *quasi shear reflection* with same angle α and same center $(0,0)$, we obtain the following point:

$$Y_S = -y + \lfloor 0.5 + (x - \lfloor 0.5 - y \tan(\alpha) \rfloor) \sin(2\alpha) \rfloor$$
$$X_S = x - \lfloor 0.5 - y \tan(\alpha) \rfloor - \lfloor 0.5 + Y_S \tan(\alpha) \rfloor$$

We use the following equalities in the proof:

$$\lfloor x + 0.5 \rfloor = x, \ \forall x \in \mathbb{Z} \tag{1}$$

$$- \lfloor 0.5 - x \rfloor = \lfloor 0.5 + x \rfloor, \ \forall x \in \mathbb{R}, x \neq k \pm 0.5, \ \forall k \in \mathbb{N} \tag{2}$$

We can transform ceiling into floor and vice versa via the following formulas:

$$\lfloor -x \rfloor = -\lceil x \rceil \tag{3}$$

$$\lceil -x \rceil = -\lfloor x \rfloor \tag{4}$$

Since we have expressed X_P and X_S using Y_P and Y_S, we will first prove the equality for Y_S and then for X_S. We have the expression for both Y_S and Y_P: $Y_S = -y + \lfloor 0.5 + (x - \lfloor 0.5 - y \tan(\alpha) \rfloor) \sin(2\alpha) \rfloor$ and $Y_P = -y + (A + \lfloor 0.5 - A \rfloor + \lfloor 0.5 + (\lfloor 0.5 + \lceil -0.5 + B - A \tan(\alpha) \rceil \rfloor - \lfloor 0.5 + \lfloor 0.5 - A \rfloor \tan(\alpha) \rfloor) \sin(2\alpha) \rfloor)$.

From Eq. 1, we can simplify $Y_P = -y + (A - A + \lfloor 0.5 + (\lfloor 0.5 + \lceil -0.5 + B - A \tan(\alpha) \rceil \rfloor - \lfloor 0.5 - A \tan(\alpha) \rfloor) \sin(2\alpha) \rfloor)$

Thus: $Y_P = -y + (\lfloor 0.5 + (\lfloor 0.5 + \lceil -0.5 + B - A \tan(\alpha) \rceil \rfloor - \lfloor 0.5 - A \tan(\alpha) \rfloor) \sin(2\alpha) \rfloor)$

We have to verify : $(x - \lfloor 0.5 - y \tan(\alpha) \rfloor) = (\lfloor 0.5 + \lceil -0.5 + B - A \tan(\alpha) \rceil \rfloor - \lfloor 0.5 - A \tan(\alpha) \rfloor)$

We can simplify $\lfloor 0.5 + \lceil -0.5 + B - A \tan(\alpha) \rceil \rfloor = \lceil -0.5 + B - A \tan(\alpha) \rceil$
This leads to : $(\lceil -0.5 + B - A \tan(\alpha) \rceil - \lfloor 0.5 - A \tan(\alpha) \rfloor)$

B is an integer and can be extracted from the ceiling function:
$(x - \lfloor 0.5 - y \tan(\alpha) \rfloor) = (B + \lceil -0.5 - A \tan(\alpha) \rceil - \lfloor 0.5 - A \tan(\alpha) \rfloor)$

Since $\lceil -(0.5 + x) \rceil = - \lfloor (0.5 + x) \rfloor$, we obtain:
$(B + \lceil -0.5 - A \tan(\alpha) \rceil - \lfloor 0.5 - A \tan(\alpha) \rfloor)$
$\quad = (B - \lfloor 0.5 + A \tan(\alpha) \rfloor - \lfloor 0.5 - A \tan(\alpha) \rfloor)$

In all cases where $A\tan(\alpha)$ is not a half integer, we have: $\lfloor 0.5 + A\tan(\alpha)\rfloor + \lfloor 0.5 - A\tan(\alpha)\rfloor = 0$ which leads to the result on the Y coordinates.

We want now to prove that $X_P = X_S$:

$X_S = x - \lfloor 0.5 - y\tan(\alpha)\rfloor - \lfloor 0.5 + Y_S\tan(\alpha)\rfloor$

$X_P = - \lfloor 0.5 - B + Y_P\tan(\alpha)\rfloor$

We replace B by its expression to obtain:

$X_P = - \lfloor 0.5 - \lfloor 0.5 + x + y\tan(\alpha)\rfloor + Y_P\tan(\alpha)\rfloor$

Since x and $\lfloor 0.5 + y\tan(\alpha)\rfloor$ are integers we can pull them out of the floor function as follows: $X_P = x + \lfloor 0.5 + y\tan(\alpha)\rfloor - \lfloor 0.5 + Y_P\tan(\alpha)\rfloor$

We therefore have to have both $-\lfloor 0.5 - y\tan(\alpha)\rfloor = +\lfloor 0.5 + y\tan(\alpha)\rfloor$ and $-\lfloor 0.5 + Y_S\tan(\alpha)\rfloor = -\lfloor 0.5 + Y_P\tan(\alpha)\rfloor$.

From Eq. 2 and since y is integer, the first equality is true except for any α having a half integer tangent and we already proved that we also have $Y_P = Y_S$ in that case.

We therefore have $(X_P, Y_P) = (X_S, Y_S)$ except for α such that $tan(\alpha) = k \pm 0.5 \;\forall k \in \mathbb{N}$. $\qquad\square$

In conclusion, the image of a point obtained by the *quasi shear reflection* and its image obtained by our *improved pivot reflection* are the same except for angles α such that $\tan(\alpha) = k + \frac{1}{2}$ with $k \in \mathbb{N}$.

5   Results and Images

Figure 1 illustrates two reflections for a randomly chosen angle $\alpha = 0.681236$ and a center at $(0,0)$. The naïve reflection line is shown in black. Each corresponding orthogonal line is color-coded for clarity. The result of our *improved pivot reflection* is displayed on the left side of the figure, while the *quasi shear reflection* result appears on the right. As proven, both results are identical.

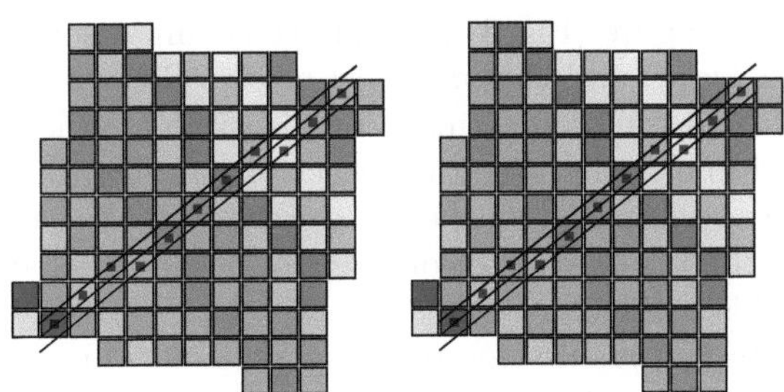

Fig. 1. For a random angle $\alpha = 0.681236$ and center $(0,0)$: the *improved pivot Reflection* (left) and the *quasi shear reflection* (right) are identical.

On the left of Fig. 2, we present the original image with the orthogonal straight lines (in color) and the naive reflection line (in black) for angle $\alpha = arctan(0.5)$ and center $(0,0)$. The result of our *improved pivot reflection* is displayed in the middle part of the figure. The *quasi shear reflection* result is displayed on the right side of the figure. We observe that the *improved pivot reflection* preserves the orthogonal lines (as expected since the algorithm is designed to do so) whereas the *quasi shear reflection* alters their structure.

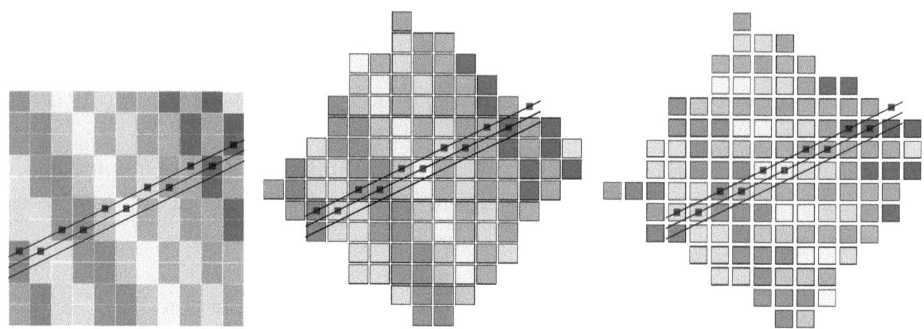

Fig. 2. For angle $\alpha = arctan(0.5)$ and center $(0,0)$: the original image (left), *improved pivot reflection* (middle) and the *quasi shear reflection* (right).

In the following we compare the quality of the approximation for the digitized continuous reflection (which is not bijective) with some discrete bijective versions: the digitized *real rounded*, the *quasi shear reflection*, the *pivot reflection* of [2] and our *improved pivot reflection*. The image size is 301. The error is measured as the maximum and mean distances between the obtained discrete result and the ideal Euclidean result for each point in the image.

In Figs. 3 and 4 we can see the error (on the y axis) obtained while varying the angle α (x axis) between 0 and $\pi/4$. The top curves represent the max error, the lower ones correspond to the mean error. We started from angle 0 to $\pi/4$ with steps 0.01. We added the result for the specific angle $arctan(0.5)$ to obtain the error curves. The first figure (Fig. 3) presents the comparison between the *quasi shear reflection* (in green) and the *improved pivot reflection* (in dashed purple). The difference between the transforms is only visible for the angle $arctan(0.5)$ (marked by the red line), otherwise the reflections are identical and the dashed purple IPR line is superimposed on the QSR green line. We can see, on the right part of Fig. 3 that the mean error of our version is better than the *quasi shear reflection* one for this particular case.

Figure 4 presents the results for other reflections. On the left we can see the results all together, on the right we can see each reflection separately. The *real rounded reflection* (in red) has both the best max and mean curves but it is not bijective and creates holes in the image. The *quasi shear reflection* (in green) has the second best results. The *pivot reflection* [2] (in blue) is simpler to compute and preserves the orthogonal lines, but it does not have a very good

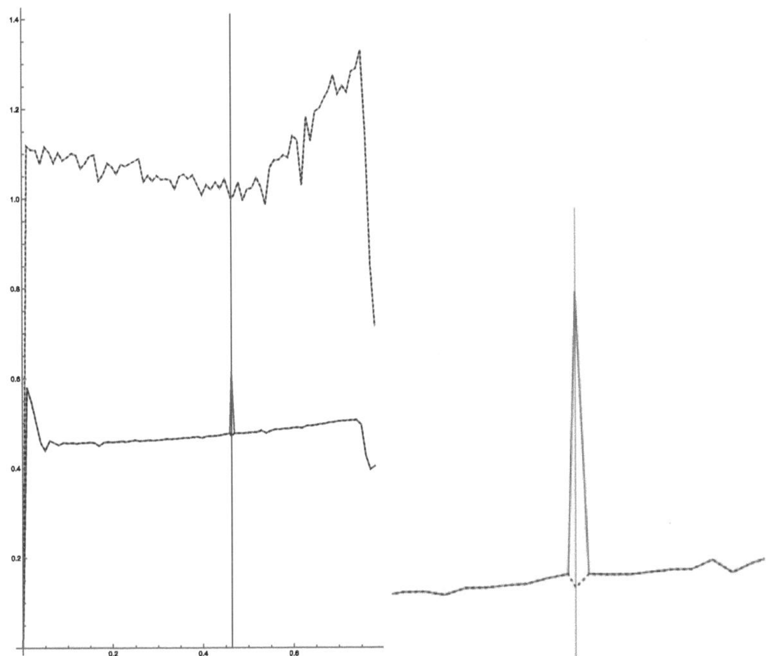

Fig. 3. Comparison between the *quasi shear reflection* (in green) and our *improved pivot reflection* (dashed purple). Max error (top), mean error (bottom). The red line mark the angle *arctan*(0.5). On the right, we have zoomed in on the mean error for angle *arctan*(0.5). (Color figure online)

error measure. And finally, our algorithm *Improved Pivot Reflection* (in dashed purple), which is almost equivalent to the shear reflection except for the very particular case $\alpha = arctan(0.5)$ where its mean error on the overall image is the best among all the bijective reflections. Our solution also ensures the preservation of the orthogonal lines.

For the particular angle $arctan(0.5)$, we obtain the following numerical error measures:

Reflection	Max error	Mean error
Real Rounded	0.447214	0.357767
Pivot [2]	1.34164	0.573477
Quasi Shear Based	1.	0.59415
Improved Pivot	1.	**0.490541**

Figure 5 presents the result for the IPR and the QSR algorithms on a 100×100 pixels image for the angle $arctan(0.5)$.

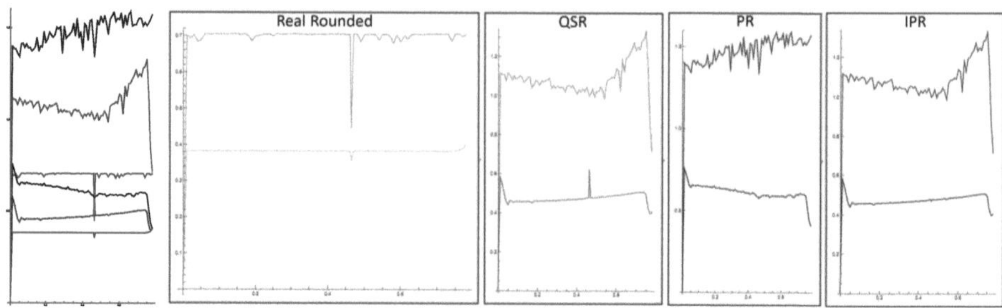

Fig. 4. Comparison between several versions of discrete reflections: the Real Rounded Reflection (red), the Quasi Shear Reflection (green), the Pivot Reflection [2] (blue) and our Improved Pivot Reflection (purple). Max error (top), mean error (bottom). (Color figure online)

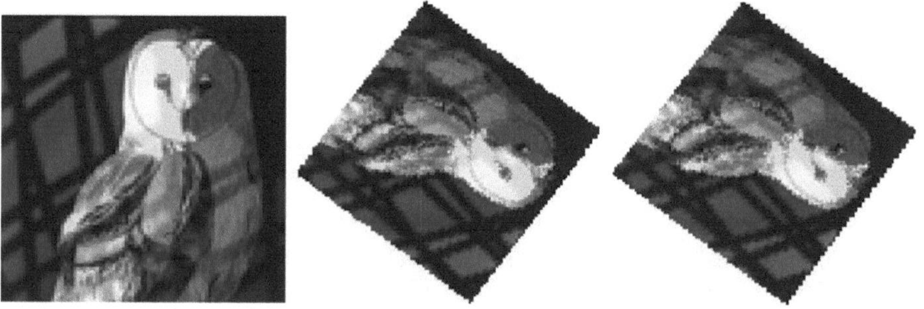

Fig. 5. Left: the original image. Middle: the IPR reflection $arctan(0.5)$. Right: the QSR reflection $arctan(0.5)$.

6   Conclusion and Future Work

In this paper, we have introduced an improved version of the discrete bijective reflection and proved its near-equivalence to the *quasi shear reflection*, when considering a reflection line passing through an integer point. This equivalence holds for all angles except a discrete set of the form $\arctan(k + \frac{1}{2})$, where $k \in \mathbb{N}$.

Thanks to its simpler formulation and lower computational cost, the proposed *improved pivot reflection* offers a practical and efficient alternative to the *quasi shear reflection* in the discrete setting. Despite its simplicity, it preserves bijectivity and approximates the continuous reflection in a way that preserves the orthogonal lines. These results reinforce the idea that structurally simple discrete operators can still provide high-fidelity approximations of their continuous counterparts. The proposed reflection thus constitutes a promising building block for defining and implementing higher-dimensional discrete rotations that are inherently defined by reflections [13].

Further research will investigate generalizations to higher dimensions and further applications with discrete geometric transformations are also interesting research topics.

Acknowledgement. This research has been partially supported by the Science Fund of the Republic of Serbia through PRIZMA program, GRANT No 7632 (MaMIPU).

References

1. Andres, E.: The Quasi-Shear rotation. In: Miguet, S., Montanvert, A., Ubéda, S. (eds.) DGCI 1996. LNCS, vol. 1176, pp. 307–314. Springer, Heidelberg (1996). https://doi.org/10.1007/3-540-62005-2_26
2. Andres, E., Dutt, M., Biswas, A., Largeteau-Skapin, G., Zrour, R.: Digital Two-Dimensional Bijective Reflection and Associated Rotation. In: Couprie, M., Cousty, J., Kenmochi, Y., Mustafa, N. (eds.) DGCI 2019. LNCS, vol. 11414, pp. 3–14. Springer, Cham (2019). https://doi.org/10.1007/978-3-030-14085-4_1
3. Breuils, S., Coeurjolly, D., Lachaud, J.O.: Construction of Fast and Accurate 2D Bijective Rigid Transformation, pp. 80–92 (2024). https://doi.org/10.1007/978-3-031-57793-2_7
4. Breuils, S., Kenmochi, Y., Sugimoto, A.: Visiting Bijective Digitized Reflections and Rotations Using Geometric Algebra, pp. 242–254 (05 2021). https://doi.org/10.1007/978-3-030-76657-3_17
5. Fredriksson, K., Mäkinen, V., Navarro, G.: Rotation and lighting invariant template matching. Inf. Comput. **205**(7), 1096–1113 (2007)
6. Goodman, R.: Alice through looking glass after looking glass: the mathematics of mirrors and kaleidoscopes. Am. Math. Monthly **111**(4), 281–298 (2004). http://www.jstor.org/stable/4145238
7. Heckbert, P.S., Hanrahan, P.: Beam tracing polygonal objects. ACM SIGGRAPH Comput. Graph. **18**(3), 119–127 (1984)
8. Jacob, M., Andres, E.: On discrete rotations. In: Discrete Geometry for Computer Imagery. p. 161 (1995)
9. Mora, F., Ruillet, G., Andres, E., Vauzelle, R.: Pedagogic discrete visualization of electromagnetic waves. Eurographics poster session, Granada, Spain (2003)
10. Paeth, A.: A fast algorithm for general raster rotation. In: Proc. Graphics Interface 86. pp. 77–81. Vancouver (Canada) (1986)
11. Pluta, K., Romon, P., Kenmochi, Y., Passat, N.: Bijective digitized rigid motions on subsets of the plane. J. Math. Imaging Vision **59**(1), 84–105 (2017)
12. Reveillès, J.: Calcul en Nombres Entiers et Algorithmique. Ph.D. thesis, Université Louis Pasteur, Strasbourg, France (1991)
13. Richard, A., Fuchs, L., Largeteau-Skapin, G., Andres, E.: Decomposition of nd-rotations: classification, properties and algorithm. Graph. Models **73**(6), 346–353 (2011)
14. Yilmaz, A., Javed, O., Shah, M.: Object tracking: A survey. ACM Comput. Surv. **38**(4) (2006)
15. Zrour, R., Chatelier, P., Feschet, F., Malgouyres, R.: Parallelization of a Discrete Radiosity Method. In: Nagel, W.E., Walter, W.V., Lehner, W. (eds.) Euro-Par 2006. LNCS, vol. 4128, pp. 740–750. Springer, Heidelberg (2006). https://doi.org/10.1007/11823285_77

Rotational and Rotation-Free Multidirectional Q-Concavity Shape Descriptors

Péter Balázs[1]($\boxtimes$) , Péter Bodnár[1] , and Sara Brunetti[2]

[1] Department of Image Processing and Computer Graphics, University of Szeged, Árpád tér 2, 6720 Szeged, Hungary
{pbalazs,bodnaar}@inf.u-szeged.hu
[2] Dipartimento di Ingegneria dell'Informazione e Scienze Matematiche, Via Roma, 56, 53100 Siena, Italy
sara.brunetti@unisi.it

Abstract. In this paper we extend a scalar Q-convexity shape descriptor in two ways towards rotational and rotation-free multidirectional descriptors. Since the original descriptor is defined with respect to the horizontal and vertical directions, we increase its ability to represent a shape by repeatedly rotating the shape and computing the descriptor, in the first extension. However, since rotation is generally not a safe transformation in the 2D digital space, as a second approach, we develop an alternative that works directly with an arbitrary pair of orthogonal lattice directions: the shape remains fixed and a certain number of pairs of directions are employed. We provide the algorithmic details of this approach and its computational complexity. In case of both approaches, the Q-concavity values measured along different directions can be gathered into a vector multidirectional descriptor (signature). We conducted experiments on the MPEG-7 dataset with different sets of lattice directions. We conclude that even though the rotational signature is faster to compute, the rotation-free signature can ensure better classification accuracy. Depending on whether computational time or accuracy is more important in the given application, one of the two methods can be preferred to the other, and hence both of them seem to provide a viable option.

Keywords: Shape descriptors · Classification · Q-convexity · Rotation

1 Introduction

Shape descriptors are widely used for image analysis and classification. Among these, we focus on numerical shape characteristics that quantify how closely the considered object resembles a given shape. Examples of such classical descriptors include moments, orientation, elongation, and circularity [9,15]. This permits us to decide whether certain shape characteristics have suitable discriminative

P. Balázs et al. (Eds.): IWCIA 2025, LNCS 15985, pp. 200–213, 2026.
https://doi.org/10.1007/978-3-032-19347-6_13

potential that make them appropriate for the intended task. In the literature, there are several descriptors that measure the convexity of a shape (see, for example, [5,11,13,14,16,17]). In this context, we studied a less restrictive notion of convexity, defined first in [7] for the reconstruction problem in Discrete Tomography, and then used for shape representations [1–3,6].

In this paper, we extend the scalar Q-convexity shape descriptor proposed in [6] in two ways to develop multidirectional descriptors. Whereas the original descriptor is defined with respect to the horizontal and vertical directions, here we obtain a multidirectional descriptor by repeatedly rotating the shape and computing the descriptor, in the first extension, gaining a so-called rotational multidimensional descriptor. However, since rotation is generally not a safe transformation in the 2D digital space, in the second extension we generalize the descriptor to any arbitrary pair of orthogonal lattice directions, and we utilize a certain number of pairs of directions to compute the so-called rotation-free descriptor. We provide the algorithmic details of this approach as well as its computational complexity analysis.

We conducted experiments on the MPEG-7 dataset with different sets of lattice directions to compare the behavior of the two descriptors. The results show that the multidirectional rotational descriptor can be computed faster than the rotation-free descriptor, but the latter outperforms the former with respect to the classification accuracy in all the experiments.

We conclude that depending on the given application, whether computational time or accuracy is critical, one of the two methods can be preferred to the other, and hence both of them seem to provide a viable option.

The paper is structured as follows. In Sect. 2 we give the preliminaries. In Sect. 3 we propose a rotational and a rotation-free Q-concavity vector descriptor. The algorithmic details and complexity analysis of the latter approach are given in Sect. 4. In Sect. 5 we design and carry out experiments and report on their results. Finally, Sect. 6 is for the conclusion.

2 Preliminaries

Any binary image $\mathcal{F}$ is a binary matrix of m rows and n columns, and can be represented by a set of black (foreground) and white (background) pixels (unit squares) in the $m \times n$ bounding box $\mathcal{B}$ (see Fig. 1 left). Equivalently, black pixels can be regarded as points of $\mathbb{Z}^2$, thus any binary image can be viewed as a finite lattice set F (bounded by an $m \times n$ rectangle $\mathcal{B} \subset \mathbb{Z}^2$) (see Fig. 1 middle). Throughout the paper, we will use both representations: the set notation is more suitable to describe geometrical properties, and images are illustrated as sets of black and white pixels. We assume a right-handed Cartesian coordinate system rotated by 90 degrees clockwise, therefore the order in both representations is the same. Furthermore, even as in the matrix representation (see Fig. 1 right), the origin of the lattice set is in the top left corner, and pixels in the image have equivalent non-negative coordinates in both representations.

We now introduce the main definitions concerning Q-convexity [7]. Throughout the paper, let us consider any two orthogonal lattice direction vectors

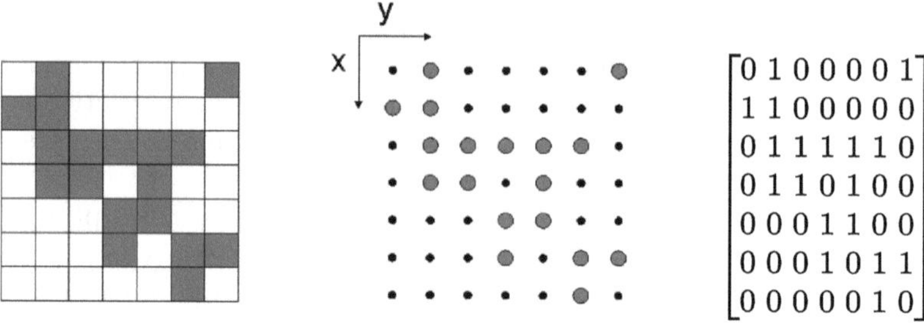

Fig. 1. A binary image represented by black and white pixels (left), as a set of lattice points (middle) and as a binary matrix (right).

$\mathbf{r} = (r_1, r_2)$ and $\mathbf{s} = (s_1, s_2)$ such that either $r_1, s_1 > 0$ with $gcd(r_1, r_2) = 1$ and $gcd(s_1, s_2) = 1$, or $(r_1, r_2) = (1, 0)$ and $(s_1, s_2) = (0, 1)$.

Let us denote the coordinates of any point $M \in \mathcal{B}$ by (M_r, M_s) in the system defined by $\mathbf{r}$, $\mathbf{s}$, where $M_\mathbf{r} = \frac{(M-O)\cdot\mathbf{r}}{||\mathbf{r}||^2}$, and $M_\mathbf{s} = \frac{(M-O)\cdot\mathbf{s}}{||\mathbf{s}||^2}$ and O is the origin. Then, point M and the directions determine the following four quadrants:

$$Z_0{}^{\mathbf{r},\mathbf{s}}(M) = \{N \in \mathcal{B} : 0 \leq N_\mathbf{r} \leq M_\mathbf{r},\ 0 \leq N_\mathbf{s} \leq M_\mathbf{s}\} ,$$
$$Z_1{}^{\mathbf{r},\mathbf{s}}(M) = \{N \in \mathcal{B} : 0 \leq N_\mathbf{r} \leq M_\mathbf{r},\ M_s \leq N_\mathbf{s} < n\} ,$$
$$Z_2{}^{\mathbf{r},\mathbf{s}}(M) = \{N \in \mathcal{B} : M_\mathbf{r} \leq N_\mathbf{r} < m,\ M_s \leq N_\mathbf{s} < n\} ,$$
$$Z_3{}^{\mathbf{r},\mathbf{s}}(M) = \{N \in \mathcal{B} : M_\mathbf{r} \leq N_\mathbf{r} < m,\ 0 \leq N_\mathbf{s} \leq M_\mathbf{s}\} .$$

To simplify notation, we do not specify the directions when either it is clear from the context or the claim holds in general. By the definition of quadrants, if $N \in Z_k(M)$, then $Z_k(N) \subset Z_k(M)$. An illustration of the four zones determined by a point and the specified orthogonal lattice directions is shown in Fig. 2.

Definition 1. *Let* $\mathbf{r} = (r_1, r_2)$ *and* $\mathbf{s} = (s_1, s_2)$ *be two lattice directions. A lattice set F (or equivalently, the binary image $\mathcal{F}$) is $Q(\mathbf{r}, \mathbf{s})$-convex if $Z_k{}^{\mathbf{r},\mathbf{s}}(M) \cap F \neq \emptyset$ for all $k = 0, \ldots, 3$ implies $M \in F$. Otherwise we say it is non-$Q(\mathbf{r}, \mathbf{s})$-convex.*

As noticed before, we discard to specify the directions when either it is clear from the context or the claim holds in general. Thus, for simplicity, in all these cases we will write 'Q-convex'. By definition, if F is Q-convex and $M \in \mathcal{B} \setminus F$, a quadrant $Z_k(M)$ exists such that $Z_k(M) \cap F = \emptyset$. Note that a Q-convex image is also convex along the prescribed directions. Recall that a lattice set E is digitally convex if the intersection of its convex hull $\mathcal{CH}(E) \cap \mathbb{Z}^2 = E$, where $\mathcal{CH}(E)$ denotes the convex hull of E. Thus, on the other hand, for an arbitrary Q-convex lattice set F, $\mathcal{CH}(F) \cap \mathbb{Z}^2 = F$ does not necessarily hold, i.e., F (or equivalently, $\mathcal{F}$) is generally not digitally convex, and so Q-convexity is weaker than digital convexity ([8]). Figure 2 illustrates the definition in the case of directions $\mathbf{r} = (1, -1)$ and $\mathbf{s} = (1, 1)$. If $M \notin F$ and all the quadrants contain points of F, then F is not Q-convex and actually M is a "Q-concave" point.

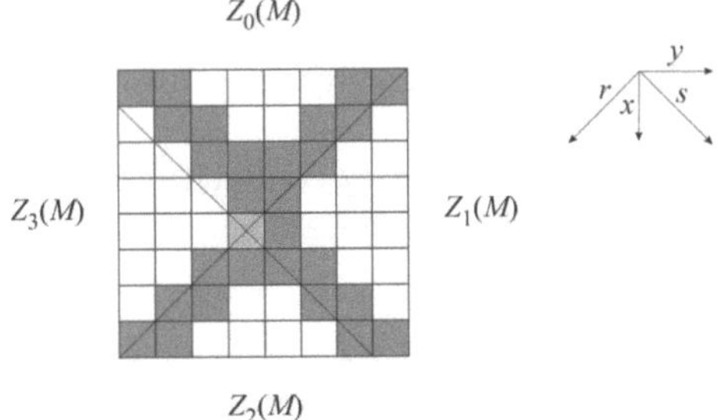

Fig. 2. Illustration of the concept of Q-convexity with respect to the $\mathbf{r} = (1, -1)$ and $\mathbf{s} = (1, 1)$ directions. Consider that the pixel M, shown in light grey, would belong to the foreground. In this case, no point violates the Q-convexity, and thus the image is Q-convex with respect to $\mathbf{r}$ and $\mathbf{s}$. If M were part of the background, however, the image would be non-Q-convex, since all four zones around M (when M is in the background) contain foreground pixels.

3   The Descriptors

In this section, we derive Q-concavity shape descriptors based on the Q-concave points and on the relative position of the rest of the points of the image [6]. Let F be a finite subset of $\mathbb{Z}^2$ and $\mathcal{B}$ be the bounding box of F of size $m \times n$. Consider any two orthogonal lattice directions $\mathbf{r} = (r_1, r_2)$ and $\mathbf{s} = (s_1, s_2)$. Let us denote the number of object points (foreground pixels) of F in $Z_k(M)$ by $n_k(M)$, for $k = 0, \ldots, 3$, i.e.,

$$n_k(M) = card(Z_k(M) \cap F\}) \ (k = 0, \ldots, 3) \ .$$

We say that F is $Q(\mathbf{r}, \mathbf{s})$-convex if for each $M \in \mathcal{B}$ $(n_0(M) > 0 \land n_1(M) > 0 \land n_2(M) > 0 \land n_3(M) > 0)$ implies $M \in F$. Otherwise we say F is non-Q-convex. In other words, if F is non-Q-convex, then there exists a point M violating the Q-convexity property, i.e. $n_k(M) > 0$ for all $k = 0, \ldots, 3$ and $M \notin F$. This gives rise to a quantitative representation of Q-concavity.

   We define the Q-concavity measure of F as the sum of the contributions of Q-concavity for each point in $\mathcal{B}$. Formally,

$$\varphi_F(M) = n_0(M)n_1(M)n_2(M)n_3(M)(1 - f(M)) \ ,$$

where M is an arbitrary point of $\mathcal{B}$ and $f(M) = 1$ if the point M belongs to the foreground, otherwise $f(M) = 0$. Moreover, let

$$\varphi_F = \sum_{M \in \mathcal{B}} \varphi_F(M) \ .$$

   For an arbitrary M, if $f(M) = 1$, then $\varphi_F(M) = 0$. Moreover, if $f(M) = 0$ and there exists a $p \in \{0, 1, 2, 3\}$ such that $n_p(M) = 0$ then also $\varphi_F(M) = 0$. Thus, F is Q-convex if and only if $\varphi_F = 0$. Furthermore, by definition, φ is invariant by reflection, point symmetry, and rotations of 90 degrees. In [6], the authors also showed that this Q-concavity measure φ extends the directional convexity in [4].

In order to measure the degree of Q-concavity, we normalize φ so that it ranges in $[0,1]$ based on [6]. There, the authors proved for lattice directions $\mathbf{r} = (1,0)$ and $\mathbf{s} = (0,1)$

Proposition 1. *Let F be a lattice set, $M = (i,j) \in \mathcal{B} \setminus F$ (i.e., $f(M) = 0$), and h_i^F and v_j^F denote the i-th row and j-th column sum of the representing matrix, respectively. Then, $\varphi_F(i,j) \leq ((card(F) + h_i^F + v_j^F)/4)^4$.*

The result can be easily generalized to hold for any pair of directions $\mathbf{r}$ and $\mathbf{s}$, replacing h_i^F and v_j^F by the number of points of F lying on straight lines parallel to $\mathbf{r}$ and $\mathbf{s}$ and passing through M, which we denote by r_i^F and s_j^F, respectively, to simplify notation. More precisely, (i,j) is the coordinate of M in the system $\mathbf{r}, \mathbf{s}$. As a consequence we may normalize each contribution as

$$\mathcal{E}_F^{\mathbf{r},\mathbf{s}}(i,j) = \frac{\varphi_F(i,j)}{((card(F) + r_i^F + s_j^F)/4)^4} \, . \tag{1}$$

To obtain a normalization for the global descriptor (1), we sum up each single contribution and then we divide by the number of non-zero contributions.

Definition 2. *Given a binary image F, its Q-concavity descriptor is defined by*

$$\mathcal{E}_F^{\mathbf{r},\mathbf{s}} = \sum_{(i,j)\in\bar{F}} \frac{\mathcal{E}_F^{\mathbf{r},\mathbf{s}}(i,j)}{card(\bar{F})} \, , \tag{2}$$

where $\bar{F}$ denotes the subset of background points in $\mathcal{B} \setminus F$ for which the contribution is not zero.

Figure 3 illustrates the values of the descriptors for two shapes according to (2) with $\mathbf{r} = (1,0)$ and $\mathbf{s} = (0,1)$. In case of the second shape, part of the background is entirely closed into the object, thus this shape receives a high value. The figure also shows the Q-concavity contribution of each pixel according to (1). We can see that pixels inside the concavities have lighter colors (representing higher values) than pixels outside.

We now introduce two multidirectional Q-concavity descriptors. The first one is based on the above extensions. Let $sl(\mathbf{v})$ denote the slope of vector $\mathbf{v}$, that is $\frac{v_2}{v_1}$, for $v_1 \neq 0$ and let $sl((0,1)) = \infty$. Furthermore, let $\mathbf{v}^\perp$ denote the orthogonal counterpart of $\mathbf{v}$ in the clockwise direction.

Definition 3. *Let $D = (\mathbf{d}_1, \mathbf{d}_2, \ldots, \mathbf{d}_k)$ be a set of lattice directions such that $0 < sl(\mathbf{d}_1) < sl(\mathbf{d}_2) < \cdots < sl(\mathbf{d}_k)$, and F be a binary image. The rotation-free Q-concavity signature of F with respect to D is defined as*

$$\mathcal{S}_F^D = (\mathcal{E}_F^{\mathbf{d}_1^\perp, \mathbf{d}_1}, \mathcal{E}_F^{\mathbf{d}_2^\perp, \mathbf{d}_2}, \ldots, \mathcal{E}_F^{\mathbf{d}_k^\perp, \mathbf{d}_k}).$$

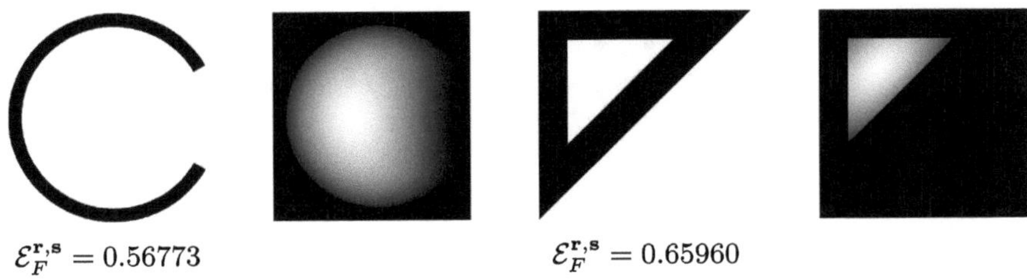

$$\mathcal{E}_F^{\mathbf{r},\mathbf{s}} = 0.56773 \qquad\qquad\qquad \mathcal{E}_F^{\mathbf{r},\mathbf{s}} = 0.65960$$

Fig. 3. Example shapes with their Q-concavity values with respect to the $\mathbf{r} = (1,0)$ and $\mathbf{s} = (0,1)$ directions, and the Q-concavity contribution of each pixel, the Q-concavity contribution of each pixel (lighter colors represent higher Q-concavity values).

The second one takes rotated versions of an image and calculates the Q-concavity on each of them along the $\mathbf{r} = (1,0)$ and $\mathbf{s} = (0,1)$ directions. Recall that the direction angle in degrees, $\theta_{\mathbf{v}}$, of an arbitrary vector $\mathbf{v} = (v_1, v_2)$ with $v_1 \neq 0$ is calculated as

$$\theta_{\mathbf{v}} = \frac{180^\circ}{\pi} \arctan \frac{v_2}{v_1} \ .$$

The direction $(0,1)$ will be associated to 90°.

Definition 4. *Let $D = (\mathbf{d}_1, \mathbf{d}_2, \ldots, \mathbf{d}_k)$ be a set of lattice directions such that $0 < sl(\mathbf{d}_1) < sl(\mathbf{d}_2) < \cdots < sl(\mathbf{d}_k)$, and F be a binary image. For each i $(i = 1, \ldots, k)$ let F_i denote the image that we obtain by rotating F at an angle $\theta_{\mathbf{d}_i}$. The* rotational Q-concavity signature *of F with respect to D is defined as*

$$\mathcal{R}_F^D = (\mathcal{E}_{F_1}{}^{(1,0),(0,1)}, \mathcal{E}_{F_2}{}^{(1,0),(0,1)}, \ldots, \mathcal{E}_{F_k}{}^{(1,0),(0,1)}).$$

4   Implementation

Let F be a binary image of size $m \times n$, $\mathbf{r} = (r_1, r_2)$ and $\mathbf{s} = (s_1, s_2)$ be orthogonal lattice directions. For every point $P \in \mathcal{B}$, to calculate $\varphi_F(P) = n_0(P)n_1(P)n_2(P)n_3(P)(1 - f(P))$, we calculate the number of points in each quadrant $Z_k(P)$ with $k = \{0, 1, 2, 3\}$. By the inclusion property, if $N \in Z_k(P)$, then $Z_k(N) \subseteq Z_k(P)$ $(k = 0, 1, 2, 3)$. This property allows to compute $n_k(P)$ $(k = 0, 1, 2, 3)$ by dynamic programming.

In particular, considering $k = 0$, we can find the two neighbors V and W of P such that $V = P - \mathbf{s}$ and $W = P - \mathbf{r}$, and a third point $U = P - (\mathbf{r} + \mathbf{s})$. The four points P, V, U, W determine a parallelogram whose area is given by $|det(\mathbf{r}, \mathbf{s})| = |r_1 s_2 - s_1 r_2|$. Moreover, if $|det(\mathbf{r}, \mathbf{s})| > 1$, by Pick's Theorem [10], there are exactly $|det(\mathbf{r}, \mathbf{s})| - 1$ points of $\mathcal{B}$ contained inside the parallelogram (see Fig. 4 for an illustration). We denote the set of such points by $I_0(P)$, i.e.,

$$I_0(P) = \{N \in \mathbb{Z}^2 : W_r < N_r < P_r, \ W_s < N_s < P_s\} \ ,$$

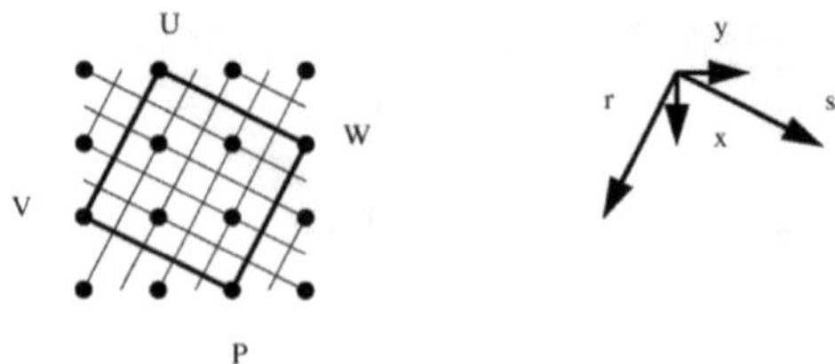

Fig. 4. (Left) The parallelogram determined by the point P and the lattice directions $\mathbf{r} = (2, -1)$, and $\mathbf{s} = (1, 2)$ and the coordinate system (right).

where we denote the coordinates of any point $M \in \mathcal{B}$ by (M_r, M_s) in the system defined by $\mathbf{r}$, $\mathbf{s}$ (see Sect. 2). Therefore, the following relation holds:

$$n_0(P) = n_0(V) + n_0(W) - n_0(U) + \sum_{N \in I_0(P)} f(N). \tag{3}$$

Analogously, we can define points U, V, W, and $I_k(P)$ for $k = \{1, 2, 3\}$ by renaming the vertices of the parallelogram following a clockwise rotation, and hence (3) becomes in general:

$$n_k(P) = n_k(V) + n_k(W) - n_k(U) + \sum_{N \in I_k(P)} f(N) \ .$$

In order to compute (3), we construct a mask g of size $(r_1 + s_1 + 1) \times (|r_2| + |s_2| + 1)$ that corresponds to the described parallelogram. For example, if $\mathbf{r} = (2, -1)$ and $\mathbf{s} = (1, 2)$, the corresponding mask of size 4×4 is

$$g = \begin{bmatrix} 0 & u & 0 & 0 \\ 0 & 1 & 1 & w \\ v & 1 & 1 & 0 \\ 0 & 0 & p & 0 \end{bmatrix}$$

where items labeled by (lowercase) letters correspond to the points defined above and are valued 0 (don't care). In this example, $det(\mathbf{r}, \mathbf{s}) = |2 \cdot 2 - (-1) \cdot 1| = 5$, and so there are four points in I_0. In order to compute n_0 for each point, we use zero padding of F to handle the starting and ending configurations of g with respect to F: we add $r_1 + s_1$ rows, and $|s_2|$ columns of 0's to the binary matrix F, so that $f(0, 0)$ aligns with item p of the mask g, when the upper-left corner of the padded matrix is aligned with the upper-left corner of the mask; and $|r_2|$ columns of 0's items at the end, so that $f(m - 1, n - 1)$ aligns with item p of the mask g, when the bottom-right corner of the padded matrix is aligned with the bottom-right corner of the mask. More precisely,

$$f(x, y) = \begin{cases} f(x, y), & 0 \leq x \leq m - 1, \ 0 \leq y \leq n - 1 \\ 0, & -(r_1 + s_1) \leq x < 0, \ -|s_2| \leq y < 0, \ n - 1 < y < |r_2| \end{cases}$$

Note that in relation (3) the abscissa of P is greater than the abscissa of the other points there. Therefore, we slide the mask g in the padded version of F

row by row from left to right, to select the items of F needed to compute (3), for each item:

$$g * f(x, y) = \sum_{i=0}^{a} \sum_{j=0}^{b} g(i, j) f(x - i, y - j) \, ,$$

where $a = r_1 + s_1$ and $b = |r_2| + |s_2|$.

By dynamic programming, n_0 values are computed following (3) by summing up values already computed. Therefore, for $0 \le x < m$ and $0 \le y < n$,

$$n_0(x, y) = n_0((x, y) - \mathbf{s}) + n_0((x, y) - \mathbf{r}) - n_0((x, y) - (\mathbf{s} + \mathbf{r})) + g * f(x, y),$$

with initial conditions $n_0(x, 0) = n_0(0, y) = 0$.

Analogously, n_k can be computed for all the items in F. Then, φ_F in each pixel of F is obtained by the product of n_k, $k = 0, 1, 2, 3$.

The time complexity is given by the mask sliding, which is upper bounded by $O(mn(r_1 + s_1)(|r_2| + |s_2|))$. Finally, to calculate $\mathcal{E}_F$, we have to add the computation of the projections in the given directions, which actually does not change the bound. Therefore:

Theorem 1. *Given two lattice directions* $\mathbf{r} = (r_1, r_2)$ *and* $\mathbf{s} = (s_1, s_2)$ *the Q-concavity descriptor* $\mathcal{E}_F^{\mathbf{r}, \mathbf{s}}$ *of an arbitrary* $m \times n$ *binary image* F *can be computed in* $O(mn(r_1 + s_1)(|r_2| + |s_2|))$ *time.*

5   Experiments

5.1   Choosing the Directions

Since the time complexity of calculating the Q-concavity depends on the length of lattice directions used (see Theorem 1), it is desirable to work with reasonably short direction vectors. Thus, in our first approach we searched for all integer angles in $[0°, 90°)$ the closest (in terms of direction angle) lattice direction $\mathbf{d} = (d_1, d_2)$ with $0 \le d_1, d_2 \le 10$. This determined a set of 20 directions (see Fig. 5) for which we identified the corresponding angles and rounded them to the nearest integer. The direction set formed this way and the corresponding angle set (denoted S_{int}) are given in Table 1. Note that in the signature definitions we allowed the direction $(0, 1)$ (with $90°$) instead of the direction $(1, 0)$ (with $0°$). Since the definition of Q-concavity is invariant by rotation of $90°$ and signatures are handled up to shifts and rotations, we decided to use here the latter direction and angle, for better interpretation.

Based on the above, we also defined different 'equiangular' sets $S_n = \{k \cdot (90/n)° \mid k = 0, 1, \ldots, n - 1\}$ with $n \in \{18, 9, 6, 3, 1\}$. For all these sets and for all angles $\alpha \in S_n \setminus S_{int}$, we used linear interpolation to calculate the Q-concavity values, with the data points corresponding to the two degrees in S_{int} that were closest to α in case $\alpha < 84°$. For $\alpha > 84°$, we used the Q-concavity values corresponding to $84°$ and $0°$ for interpolation. Note that S_1 corresponds to the original scalar Q-concavity descriptor of [6].

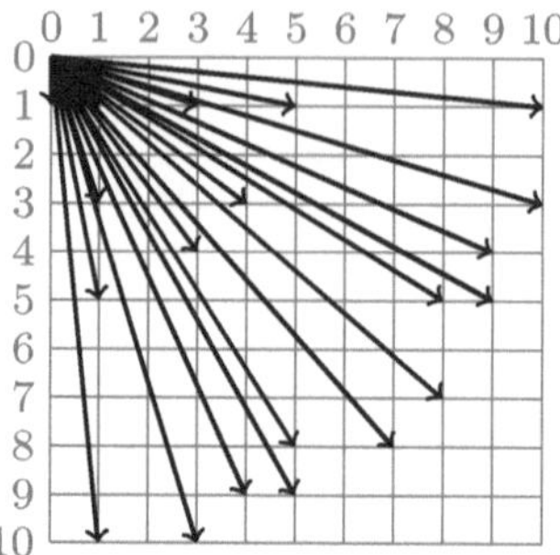

Fig. 5. Lattice directions used in the experiments

5.2 Calculating the Distance of Signatures

Let F and G be binary images, and let $\mathcal{S}_F^D$ and $\mathcal{S}_G^D$ their rotation-free Q-concavity signatures with D set of lattice directions for S_{int} or S_n. It is easy to see that if G is a reflected and/or rotated version of F, then – even assuming that resampling does not corrupt G – the two signatures may significantly differ (see Fig. 6 for two examples from the MPEG-7 dataset). The same holds for the rotational Q-concavity signatures $\mathcal{R}_F^D$ and $\mathcal{R}_G^D$.

To avoid this effect, we introduce a relation on $\mathbb{R}^k$. Let $\mathbf{u} = (u_1, u_2, \ldots, u_n) \in \mathbb{R}^k$ and let us introduce the notation

$$rev(\mathbf{u}) = (u_k, u_{k-1}, \ldots, u_1)$$

and

$$shift(\mathbf{u}, i) = (u_{k-i+1}, u_{k-i+2}, \ldots, u_k, u_1, u_2, \ldots, u_{k-i}) \quad i = 1, \ldots, k-1 .$$

Let

$$\mathcal{C}_{\mathbf{u}} = \{\mathbf{u}\} \cup rev(\mathbf{u}) \cup \bigcup_{i=1}^{k-1} shift(\mathbf{u}, i) \cup \bigcup_{i=1}^{k-1} rev(shift(\mathbf{u}, i)) ,$$

which gives a partitioning on $\mathbb{R}^k$. On this partitioning, let us define

$$d(\mathbf{u}, \mathbf{v}) = \min\{||\mathbf{u}' - \mathbf{v}'||_2 \mid \mathbf{u}' \in \mathcal{C}_{\mathbf{u}} \wedge \mathbf{v}' \in \mathcal{C}_{\mathbf{v}}\} = \min\{||\mathbf{u}' - \mathbf{v}||_2 \mid \mathbf{u}' \in \mathcal{C}_{\mathbf{u}}\} .$$

Finally, we normalize the vectors by their means to calculate the centroid distance: let $\bar{\mathbf{u}} = \frac{\sum_{i=1}^{k} u_i}{n}$, $\bar{\mathbf{v}} = \frac{\sum_{i=1}^{n} v_i}{n}$, and $d_C(\mathbf{u}, \mathbf{v}) = d(\mathbf{u} - \bar{\mathbf{u}}, \mathbf{v} - \bar{\mathbf{v}})$.

For our purpose, we consider vectors obtained by computing the rotation-free Q-concavity signature $\mathcal{S}_F^D$ (resp., rotational Q-concavity signature $\mathcal{R}_F^D$), where D is a set of lattice directions for S_{int} or for S_n.

We will use d_C to measure the distance of two signatures, i.e. $d_C(\mathcal{S}_F^D, \mathcal{S}_G^D)$ (resp., $d_C(\mathcal{R}_F^D, \mathcal{R}_G^D)$).

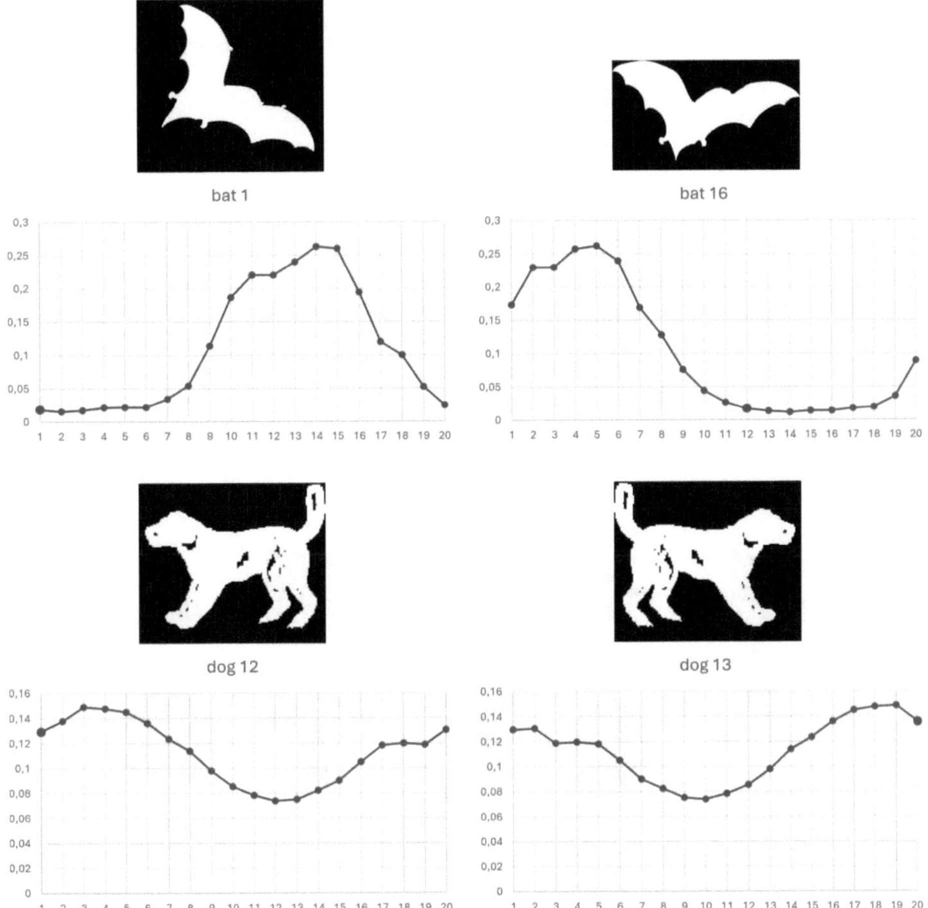

Fig. 6. Shapes and and their rotation-free Q-concavity signatures according to the angle set S_{int}. Since 'bat-1' is a rotated variant of 'bat-16', its signature is shifted with respect to the signature of 'bat-16'. Since 'dog-12' is a reflected variant of 'dog-13', its signature is reversed with respect to the signature of 'dog-13'. In both cases the red dots indicate the alignment point. Horizontal axis represents the order of directions listed in Table 1, while the vertical axis shows the Q-concavity values for the corresponding pairs of directions.

5.3   Hardware and Software Environment

The experimental tests were performed on an Intel Xeon CPU E5-2670 v2 (2.50GHz) processor under Ubuntu 16.04. The computation of the descriptors was implemented in Python. The code is available at https://github.com/bodnaar/convexity.

To produce the rotational descriptors, we used OpenCV functions: we first calculated the rotation matrix and then performed the affine transformation with linear interpolation.

5.4 Results and Discussion

We conducted experiments on the MPEG-7 Core Experiment CE-Shape-1 dataset designed to evaluate the performance of 2D shape descriptors. The dataset consists of 1,400 shapes grouped into 70 classes with 20 representatives of each class. 10 of the classes ('device0' to 'device9'; see one sample of each class in Fig. 7) are formed of geometrical shapes that we also investigated in more detail (we call this subset as **Device**). Since processing time for calculating the Q-convexity descriptors depends on the size of the image, for a fair comparison, we decided to rescale each image so that the size along the longest side became 128, by keeping aspect ratio.

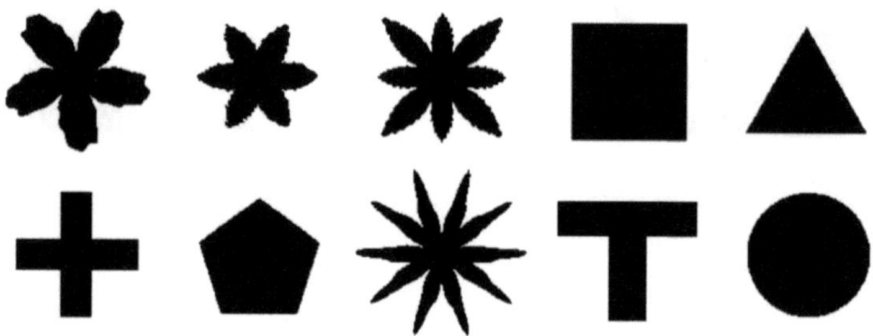

Fig. 7. Representatives of the **Device** classes ('device0' to 'device9' from left to right and top to bottom)

In the first experiment, our aim was to study the time required to calculate a single component of the Q-concavity descriptors for a particular direction defined by $\mathbf{r}$ and its perpendicular counterpart $\mathbf{s}$. Recall from Sect. 4 that for an $n \times n$ image this can be done in $O(n^2(r_1 + s_1)(|r_2| + |s_2|))$ time in the rotation-free and—independently of $\mathbf{r}$ and $\mathbf{s}$—in $O(n^2)$ time in the rotational case. We find that the average time for calculating an element of any rotational signature is $2.54s$. In addition, Table 1 reports the mean computational time on the entire MPEG-7 dataset for each element of the signature S_{int}. The results nicely reflect the fact that the time of computing the components of a rotation-free signature is highly influenced by the length of the chosen lattice direction.

Table 1. Lattice directions with the corresponding angles (rounded to the nearest integer) used in the experiments, and mean computational times in seconds needed to calculate Q-concavity values with different lattice directions for the rotation-free signatures on the MPEG-7 dataset.

degree	0	6	11	17	18	24	29	32	37	41
direction	(1, 0)	(10, 1)	(5, 1)	(10, 3)	(3, 1)	(9, 4)	(9, 5)	(8, 5)	(4, 3)	(8, 7)
mean	2.54	60.90	15.96	65.21	7.38	56.92	62.07	51.25	15.05	64.07
degree	45	49	53	58	61	66	72	73	79	84
direction	(1,1)	(7, 8)	(3, 4)	(5, 8)	(5, 9)	(4, 9)	(1, 3)	(3, 10)	(1, 5)	(1, 10)
mean	3.10	64.10	15.06	51.37	62.13	57.13	7.38	65.35	15.99	61.01

In the second test, we tried to classify the images with the 1-nearest-neighbor approach with the d_C distance of Sect. 5.2 and leave-one-out crossvalidation, using rotational and rotation-free Q-concavity descriptors. The results are presented in Table 2. It is clearly visible that using more directions yields better classification results. The descriptors corresponding to S_6, S_3, and especially the original scalar descriptor S_1 perform poorly on this dataset. It is also observable that there is no significant difference between the performance of S_{int}, S_{18}, and S_9 on the entire dataset (**All**), whereas on the less complex **Device** part S_{int} is significantly better both in the rotational and in the rotation-free cases. However, the most important observation is that for a sufficient number of directions (at least 9, in this experiment), the rotation-free signature outperforms the rotational one.

We point out that our aim was not to beat the cutting-edge classifiers on this dataset but to investigate the difference between the two presented approaches.

Table 2. Classification accuracy in percentage according to the different angle sets for the rotational ($\mathcal{R}$) and rotation-free ($\mathcal{S}$) Q-concavity signatures

	S_{int}	S_{18}	S_9	S_6	S_3	S_1
$\mathcal{R}$ – **All**	42.57	41.43	40	29.64	14.14	1.43
$\mathcal{R}$ – **Device**	71.5	67	61	54	33	10
$\mathcal{S}$ – **All**	46.86	48.86	47.21	33.5	15.71	1.43
$\mathcal{S}$ – **Device**	75	71	70	56.5	36.5	10

6 Conclusion

In this paper, we provided two possible extensions of the scalar Q-concavity shape descriptor presented in [6]. Whereas the original descriptor utilizes the horizontal and vertical directions, we now calculate Q-concavity of the shape given an arbitrary pair of orthogonal directions. In the first approach, we compute the scalar Q-concavity value with respect to the horizontal and vertical directions of any binary image rotated by a certain angle. However, since rotation is not a safe transformation on the 2D digital space, we also developed an alternative that works directly with an arbitrary pair of orthogonal lattice directions. We provided the algorithmic details of this approach as well as its computational complexity analysis.

In case of both approaches, the Q-concavity values measured along different directions can be gathered into a vector descriptor (signature). We conducted experiments on the MPEG-7 dataset with different sets of lattice directions. We concluded that even though the rotational-signature is much faster to compute than the rotation-free one, the latter one can ensure better classification accuracy. It depends on the given application whether computational time or accuracy is more important, thus both methods seem to be a viable option.

One question is, how to choose a set of lattice directions with which the rotation-free signature is relatively fast to calculate while accuracy can be kept sufficiently high. For this aim one could utilize, e.g., principal component analysis or floating search methods [12].

It is obvious that the rotation-free Q-concavity signature cannot distinguish between any two convex shapes, independently of the number of directions. Neglecting the side effect arising from rotating a discrete image, the same holds for the rotational signature.

In this paper, we restricted the study to two orthogonal dimensions in 2D. However, the definitions can be easily extended to higher dimensions as well as to two arbitrary directions, too.

Acknowledgements. This research was supported by project no. TKP2021-NVA-09 provided by the Ministry of Culture and Innovation of Hungary from the National Research, Development and Innovation Fund.

References

1. Balázs, P., Brunetti, S.: A q-convexity vector descriptor for image analysis. J. Math. Imaging Vision **61**(2), 193–203 (2019)
2. Balázs, P., Brunetti, S.: A measure of q-convexity for shape analysis. J. Math. Imaging Vision **62**, 1121–1135 (2020)
3. Balázs, P., Brunetti, S.: A comparative study of descriptors for quadrant-convexity. Mathematics **13**(13), 2114 (2025)
4. Balázs, P., Ozsvár, Z., Tasi, T.S., Nyúl, L.G.: A measure of directional convexity inspired by binary tomography. Fund. Inform. **141**(2–3), 151–167 (2015)
5. Boxter, L.: Computing deviations from convexity in polygons. Pattern Recogn. Lett. **14**, 163–167 (1993)
6. Brunetti, S., Balázs, P., Bodnár, P.: Extension of a one-dimensional convexity measure to two dimensions. In: Lecture Notes in Computer Science, Vol. 10256: 18th International Workshop on Combinatorial Image Analysis. pp. 105–116. Plovdiv, Bulgaria (2017)

7. Brunetti, S., Daurat, A.: An algorithm reconstructing convex lattice sets. Theoret. Comput. Sci. **304**(1–3), 35–57 (2003)
8. Brunetti, S., Daurat, A.: Reconstruction of q-convex lattice sets. In: Herman, G.T., Kuba, A. (eds.) Advances in Discrete Tomography and Its Applications, pp. 31–54. Birkhäuser Basel (2007). https://doi.org/10.1007/978-0-8176-4543-4
9. Gonzalez, R.C., Woods, R.E.: Digital Image Processing. Prentice Hall, 3rd edn. (2008)
10. Grünbaum, B., Shephard, G.C.: Pick's theorem. Am. Math. Mon. **100**(2), 150–161 (1993)
11. Latecki, L.J., Lakamper, R.: Convexity rule for shape decomposition based on discrete contour evolution. Comput. Vis. Image Underst. **73**(3), 441–454 (1999)
12. Pudil, P., Novovičová, J., Kittler, J.: Floating search methods in feature selection. Pattern Recogn. Lett. **15**(11), 1119–1125 (1994)
13. Rahtu, E., Salo, M., Heikkila, J.: A new convexity measure based on a probabilistic interpretation of images. IEEE Trans. Pattern Anal. Mach. Intell. **28**(9), 1501–1512 (2006)
14. Rosin, P.L., Zunic, J.: Probabilistic convexity measure. IET Image Proc. **1**(2), 182–188 (2007)
15. Sonka, M., Hlavac, V., Boyle, R.: Image Processing, Analysis, and Machine Vision, 3rd edn. Thomson Learning, Toronto (2008)
16. Stern, H.: Polygonal entropy: a convexity measure. Pattern Recogn. Lett. **10**, 229–235 (1989)
17. Zunic, J., Rosin, P.L.: A new convexity measure for polygons. IEEE Trans. Pattern Anal. Mach. Intell. **26**(7), 923–934 (2004)

Deep Learning-Based Domain-Specific Object Detection in a 3D Virtual Tour

Judit Szűcs$^{(\boxtimes)}$, Tibor Gaál, Krisztián Németh, and Tibor Guzsvinecz

Department of Information Technology and Its Applications, University of Pannonia,
Gasparich M. utca 18/A, 8900 Zalaegerszeg, Hungary
{szucs.judit,guzsvinecz.tibor}@zek.uni-pannon.hu,
nemeth.krisztian@uni-pannon.hu

Abstract. This paper presents a domain-specific object detection pipeline for a virtualized campus environment, fine-tuning a YOLOv8 model to recognize chairs, benches, cars, and bicycles. By combining targeted data collection, manual annotation, and iterative training on GPU hardware, we obtain an accurate and competitive detector using a moderate dataset. Evaluation on image sequences derived from the digital twin indicates advantages over a general-purpose baseline for the intended classes. This work highlights the effectiveness of specialization in object detection tasks and underscores the practical advantages of integrating computer vision with spatial digital twin frameworks.

Keywords: Object detection · Image processing · YOLOv8 · Digital twin · Deep learning · Computer vision

1   Introduction

Object detection in images (or videos) is a fundamental goal of computer vision in real-world applications. From autonomous vehicles and surveillance systems to e-commerce and augmented reality, the automatic recognition and localization of objects in visual data is becoming increasingly important.

Popular detection frameworks such as Faster R-CNN [7] and single-stage models like YOLO [13] have dramatically improved detection speed and accuracy over earlier methods. These models are typically trained on large, general-purpose datasets to detect a wide range of object categories.

However, deploying object detection in a specific context can benefit from customization. General models might not be optimally tuned to particular camera perspectives or the subset of object classes relevant to the application.

This paper addresses the detection of four classes of interest – chairs, benches, cars, and bicycles – in image sequences captured from a virtual 3D tour of the Zalaegerszeg University Centre of the University of Pannonia. By focusing on this well-defined setting, a specialized detector can be trained that takes advantage of context-specific features and expected object appearances. This approach is useful when only a few classes are important for analysis. Typical applications

P. Balázs et al. (Eds.): IWCIA 2025, LNCS 15985, pp. 214–229, 2026.
https://doi.org/10.1007/978-3-032-19347-6_14

include orientation and accessibility monitoring (e.g., density and location of benches), equipment inventory (e.g., number of classroom furniture), and mobility analysis near parking lots and bicycle racks. In these cases, a compact model specialized for local viewpoints, lighting, and object styles reduces misclassifications and improves responsiveness compared to wider sensors.

Our approach combines classical image processing with modern deep learning detection. A state-of-the-art YOLO-based convolutional neural network is then trained on a custom-collected dataset comprising images of the target object classes. We incrementally expanded the training data and refined the model through multiple versions, demonstrating improved performance with larger training sets. The final model is evaluated on the University Centre tour images and compared against a standard YOLOv8 detector to verify its competitiveness.

This paper is structured as follows: a literature review is conducted in Sect. 2. Methodology is detailed in Sect. 3, while the experiments and results are presented in Sect. 4. Finally, conclusions are made in Sect. 5.

2   Literature Review

Early digital image processing research established a pipeline of operations showing how to perform suitable analyses from raw images [3, 4, 8, 15].

Beyond general detectors, previous studies have emphasized customizing models to the visual statistics of specific scenes, such as built environments and digital twins. This type of specialization generally replaces broader class sets with greater accuracy and robust performance across recurring viewpoints. Our study follows this pattern, limiting the label space and aligning the training data with the digital twin of the University Centre.

Low-level image processing techniques – such as noise reduction, edge detection, etc. – are often applied to improve image quality and extract basic features. The purpose of these operations is to preserve essential information while removing artifacts, thus facilitating higher-level analysis.

High-level image analysis techniques are built on these foundations to effectively interpret image content. For example, image segmentation divides an image into meaningful regions (separating objects from the background), while object classification or recognition identifies what objects are present in the image being examined. Together, these techniques enable one to understand the scene by isolating and recognizing its components.

Classic digital image processing techniques for object recognition rely on manually designed features and exhaustive searches, which limit both their speed and generalizability. For instance, classical methods – such as Scale-Invariant Feature Transform (SIFT) or the Histogram of Oriented Gradients (HOG) – often use sliding-window classifiers to scan images for objects, which is computationally expensive and generally not suitable for real-time use with large amounts of data.

A major advancement in object detection emerged with deep convolutional neural networks (CNNs), which enabled automatic feature extraction through hierarchical layers, surpassing traditional methods in both precision and resilience to variation [12].

Based on this idea, the region-based convolutional neural network (R-CNN) model [7] and its successors (Fast R-CNN [6], Faster R-CNN [14]) were proposed. These approaches (called two-stage detectors) generate region proposals and then classify each region, achieving high accuracy but at the cost of speed.

A major milestone in real-time detection was the introduction of the You Only Look Once (YOLO) framework by Redmon et al. in 2016 [13]. The first YOLO model introduced a novel formulation of object detection, treating it as a unified regression problem where a single model simultaneously estimates bounding boxes and class scores across the image grid. Early single-stage models traded some accuracy for speed; however, continuous architectural advances have significantly closed this accuracy gap [5]. Subsequent YOLO versions and related one-stage models further improved accuracy while maintaining high speed [17]. For example, the YOLOv4 by Bochkovskiy et al. [2] incorporates numerous architectural enhancements and training proposals to achieve comparable performance on standard benchmarks.

YOLOv8 is one of the latest generations of the model [10]. It builds on previous versions and introduces a simplified architecture that improves detection accuracy and efficiency. Like its predecessors, YOLOv8 performs detection in a single stage. It normalizes the input image and then runs it through a deep CNN, which simultaneously outputs class predictions and the coordinates of the bounding boxes of multiple possible regions. YOLOv8 retains the simplicity and speed of YOLO while further improving accuracy and adaptability.

Recent enhancements allow YOLO-based systems to match or even surpass the precision of two-stage models, while retaining superior inference speed suitable for real-time applications [18].

3   Methodology

The object recognition solution was developed using a YOLO-based deep learning architecture, which was adapted to detect four specific object classes commonly found in the University Centre's environments: chairs, benches, cars, and bicycles. Fine-tuning pre-trained detectors on domain-specific or small-scale datasets is a well-established practice. We follow this approach by fine-tuning a general-purpose YOLO model with data aligned to our campus context.

The overall workflow is illustrated in Fig. 1. It consists of: (1) data collection and annotation, (2) model training using a YOLOv8 framework, and (3) iterative refinement through multiple model versions. Each component is described in the following subsections.

This section is split into five subsections. In Subsect. 3.1 the model architecture is detailed. The data collection process is shown in Subsect. 3.2, while the training procedure is presented in Subsect. 3.3. Subsection 3.4 contains the description of the digitalized virtual environment, and finally, Subsect. 3.5 presents the evaluation method.

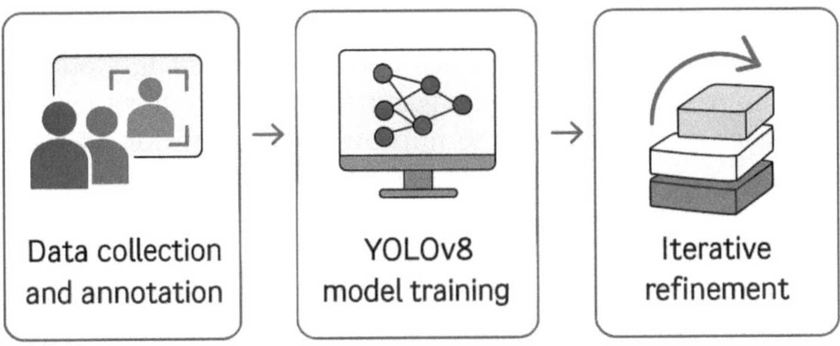

Fig. 1. Schematic workflow of the training pipeline and object detection algorithm.

3.1 Model Architecture

The model architecture employed in this study is based on the YOLOv8 framework, which offers a single-stage, real-time object detection approach. A pretrained YOLOv8 model was used as the initial base and fine-tuned to recognize the four target object classes: chairs, benches, cars, and bicycles. The architecture was modified to output only the specific classes of interest by adjusting the final classification layers.

To optimize detection performance, data augmentation techniques such as random flipping, scaling, and color jittering were applied during training. These methods were integrated using the Ultralytics training pipeline to ensure compatibility and reproducibility.

3.2 Data Collection and Annotation

Collecting a comprehensive dataset is a crucial first step. Thus, large annotated datasets are pivotal in training deep detectors. The Open Images Dataset (OID) is one of the largest, with millions of images and bounding-box annotations spanning hundreds of object classes [11]. OID v6, in particular, provides a rich source of training data for object detection [9].

Training relies on YOLO-format boxes aggregated from Open Images v6 and manually curated project images for the four classes. Validation uses a 10% split, stratified by class, from the same pool with no overlap with the training set. The evaluation uses image frames extracted from the virtual tour. These frames are not included in the training or validation sets. All splits are created once and kept fixed across versions to avoid leakage.

In our case, the target images for detection are frames extracted from a virtual 3D tour of the digital twins of the University Centre. These frames simulate a real-world walk-through, containing scenes of rooms, corridors, and outdoor areas. We focused on four object categories commonly found in this environment: chairs, benches, cars, and bicycles. To assemble a training set with sufficient variety, we augmented the images of the University Centre with additional examples of these objects from an external source.

Specifically, the Open Images v6 (OIDv6) dataset for images labeled as containing chairs, benches, cars, or bicycles were queried. Using Google Colab and the OID API, approximately 1000 images per class were downloaded, ensuring a mix of perspectives and settings to improve the adaptation to the University Centre environment. In total, the dataset comprised approximately 4000 images (across all classes), with 800–1000 images per object class in the later versions of our model.

For annotation, the Roboflow platform was used, a web-based tool that streamlines dataset management and labeling for computer vision projects. Roboflow supports converting annotations to various formats (including YOLO), dataset versioning, and even offers augmentation utilities. Although Roboflow provides an automatic annotation feature, we found it unsuitable for our dataset. Many images contained only partially visible objects or objects occluded by others, which the auto-annotator often failed to label correctly. To ensure high-quality training data, all images were manually annotated using bounding boxes around each instance of the four classes. This intensive process resulted in precise localization of the objects, at the cost of time.

After annotation, the dataset was reviewed and approximately 500 images were removed that were deemed unusable (e.g., excessive noise, ambiguous content, or irrelevant scenes). The final annotated dataset was split into training and validation sets (approximately 90% and 10% of the images respectively). A separate test set was not used for offline evaluation since the ultimate testing was conducted on the actual virtual tour frames in the deployment context.

The class distribution among the training images was roughly balanced, as we originally set a target of 1,000 images per class; the small decrease did not significantly distort the ratios. We then used Roboflow's export function to download the annotations in YOLOv8 format (text files associated with each image containing the coordinates of the bounding box) and the images, which were resized and divided into training and validation sets. These prepared images and labels formed the input data for our model training process.

3.3   Training Procedure

Training was carried out using the PyTorch-based implementation of YOLOv8, provided by the Ultralytics open-source package. A GPU-enabled environment (Google Colab) was utilized to significantly accelerate training, while CPU-based training was used only in the early experimentation phases due to its limited performance.

The dataset was divided into training and validation subsets in a 90:10 ratio. The model was trained over multiple epochs with early stopping based on validation loss stabilization. An adaptive learning rate schedule and standard optimization settings recommended by the YOLOv8 framework were adopted.

To evaluate the influence of dataset size, several versions of the model were created with increasing amounts of annotated data. Each new version was trained from scratch with the expanded dataset to measure improvements in detection accuracy. Mean average precision (mAP) and loss curves were recorded throughout the training to assess convergence and generalization.

3.4   Source of Visual Data: Digitized Virtual Environment

The dataset used for evaluation was derived from a previously developed digital twin of the Zalaegerszeg University Centre of the University of Pannonia. This virtual environment (VE) was generated through a detailed indoor and outdoor scanning process carried out between summer and autumn 2023 [16]. A combination of high-precision 3D laser scanning and high dynamic range (HDR) panoramic imaging was used to capture the architectural and environmental features of the site.

In total, roughly 80% of the University Centre was digitally captured. This effort yielded two types of digital twin models. One is a detailed point-cloud model (over 263 million points) that allows virtual inspection from arbitrary viewpoints, though its size makes real-time VR navigation impractical (Fig. 2).

Fig. 2. Upper view of the University Centre [16].

The other one is a Matterport-based [1] 360° virtual tour designed for easy online access (Fig. 3). The virtual tour connects interior and exterior scenes with clickable transitions and includes interactive information points in Hungarian. Each segment of this tour covers about 2,000 square meters, enabling seamless user exploration of the scanned University Centre.

Fig. 3. The 360° photo version [16].

The digital environment provided a valuable visual foundation for generating the image sequences used in the object detection pipeline. Screenshots and panoramic views were extracted from the Matterport-based tour and subsequently annotated for supervised training. This integration of high-fidelity spatial scanning and AI-powered visual recognition demonstrates a compelling use case for digital twins in academic and cognitive computing contexts.

3.5   Evaluation

The trained models were evaluated on the 2D image sequences extracted from the virtual University Centre tour, which were not included in the training or validation datasets. Detection results were visualized, and qualitative comparisons with a standard pre-trained YOLOv8 model were performed to validate the approach.

Object detection confidence scores and bounding box overlaps were visually inspected to identify true positives, false positives, and missed detections. Observations were recorded for each model version to document performance trends across different dataset scales.

4   Experiments and Results

To evaluate the effectiveness of the proposed domain-specific object detection pipeline, a series of experiments were conducted. The primary objective of these experiments was to assess the accuracy, robustness, and generalization capability of the custom YOLOv8-based detector, particularly under real-world conditions characterized by visual complexity, motion blur, and lighting variations.

The experimental design encompassed multiple stages: model training with progressive dataset sizes, performance benchmarking across different model versions, quantitative and qualitative validation on images, and real-world deployment tests using image sequences extracted from a 3D virtual tour of the University Centre. Additionally, the custom model was compared against a baseline YOLOv8 detector trained on the COCO dataset, to highlight the strengths and limitations of task-specific fine-tuning versus general-purpose training.

This section first describes the setup used for evaluating detection performance on image sequences representing various campus environments. We then present a qualitative comparison across different model versions, followed by a detailed analysis of results in comparison to a pre-trained baseline. Finally, we summarize key lessons learned regarding dataset construction, training strategy, and hardware requirements.

4.1   Test Setup on Image Sequences

The final object detection model was evaluated on 2D image sequences extracted from a digitized virtual tour of the digital twin of the Zalaegerszeg University Centre of the University of Pannonia. Three separate video sequences were used, each corresponding to a walking tour through distinct regions of the University Centre. The sequences featured a variety of indoor and outdoor settings: one traversed an interior hallway with benches and chairs, another passed through an outdoor parking area containing cars and bicycles, and the third followed a student gathering space with mixed furniture.

Testing was conducted by applying the detector frame-by-frame to the image sequences, which maintained the original resolution of approximately 1920×1080 pixels. A confidence threshold of 0.5 was applied to visualize only high-confidence predictions, thereby reducing clutter from low-certainty detections. The output of each run consisted of the processed frames with annotated bounding boxes and class labels, forming a visual sequence akin to a tracking-by-detection approach. Although explicit object tracking was not implemented, the consistent detection of visible objects across adjacent frames created the impression of temporal coherence.

Prior to video evaluation, model performance was validated on a held-out set of 200 static images, where high precision was achieved across all four target classes. However, testing on video frames introduced additional challenges such as motion blur, variable lighting, and partial occlusion, which were less prevalent in the training set. These factors are known to degrade performance in real-world applications.

4.2   Qualitative Results and Version Comparison

The final YOLOv8-based model (Version 21) demonstrated robust detection performance across the three video sequences. In most frames, visible chairs, benches, cars and bicycles were accurately identified, and their locations were annotated with bounding boxes. Figure 4 shows an example frame with multiple confident detections.

Fig. 4. Sample frame with multiple confident detections.

False detections were also observed but remained infrequent. These typically occurred under conditions of rapid camera movement or when background elements mimicked target object features. For example, in one case, a trash bin was misclassified as a bicycle, and in another, a wall shadow triggered a false chair detection. Figure 5 illustrates an example of a false positive detection: the algorithm drew a box where no actual target object was present. These errors were attributed to data limitations—such conditions were not well represented in the training set—and highlight the inherent complexity of visual generalization.

Fig. 5. Sample false detection.

Earlier model versions, such as Version 1 (trained on 107 images), failed to detect relevant objects under real-world conditions due to overfitting. Version 11, trained on 554 images, showed moderate improvement but still lacked consistent accuracy. In contrast, Version 21, trained on 3334 images, consistently identified objects with high confidence across video frames. This progression validates the necessity of both dataset scaling and GPU-accelerated iterative training.

4.3   Comparison with Baseline YOLOv8 Model

A comparative evaluation was performed using the baseline YOLOv8 model pre-trained on the COCO dataset. The same test sequences were processed, and the results were compared for the four shared object classes (car, bicycle, chair, bench). While the baseline model covers the object classes, class confusion still occurs in practice due to scene context and appearance differences.

Fig. 6. Comparison of the detection of the baseline YOLOv8 (left) and the custom (right) model.

Figure 6 presents a side-by-side detection output: green bounding boxes denote the baseline model's results, and blue boxes represent those of the custom model. In several cases, the custom model demonstrated superior class-specific accuracy and avoided class confusion due to its focused training.

Quantitative evaluation was complicated by differing class definitions; however, a summary on selected frames indicated that the custom model achieved slightly higher recall for cars, while the baseline model maintained marginally better precision on bicycles. Both models exhibited over 90% estimated precision and recall on clearly visible instances, reinforcing that the custom model performs competitively despite being trained on significantly fewer images.

Moreover, in certain scenarios, the custom algorithm outperforms its counterpart. This is illustrated in the following figures. Figures 7 and 8 present representative frames extracted from the comparative evaluation.

Fig. 7. Sample result of the custom model.

Fig. 8. Sample result of the baseline YOLOv8 model.

4.4   Impact of Dataset Size and Hardware Acceleration on Training Efficiency

This subsection summarizes how dataset scaling and the transition from CPU to GPU implementation affected convergence behavior and practical training efficiency.

An iterative development strategy was applied to trace how incremental data growth influenced convergence and stability. Instead of optimizing hyperparameters ad hoc, successive versions were trained under identical settings while gradually increasing the amount of annotated data and extending training duration. Early prototypes validated the workflow and confirmed the need for GPU acceleration; later iterations demonstrated that scaling the dataset steadily improved consistency of detections and reduced overfitting. This controlled progression established a transparent link between dataset size, training efficiency, and detection reliability without altering the architecture or introducing unreviewed experiments.

The training process was conducted in three main stages, each representing a progressively larger dataset and corresponding model version (V1, V11, and V21). The trainings followed the same configuration: confidence threshold of 0.25, Non-Maximum Suppression IoU (NMS) of 0.45, initial learning rate 1×10^{-3} with cosine decay scheduling, and on-the-fly data augmentation through flipping, rotation, and scale jittering. Early experiments involved training for 100 epochs.

Each training session produced built-in metrics and loss curves provided by YOLOv8. These were analyzed to guide further development. In particular, the progression from Version 1 to Version 21 illustrated improvements in loss stability, precision, and recall. While early versions exhibited erratic training curves and low confidence predictions, the final model showed well-converged metrics and consistent detection performance across the target object classes. This iterative refinement process confirmed that both dataset size and hardware acceleration were essential to achieving the desired accuracy and reliability of the object detection system.

Performance was monitored across epochs using mean average precision (mAP) and recall on the validation set, complemented by qualitative assessment on held-out sequences to compensate for the relatively small validation size.

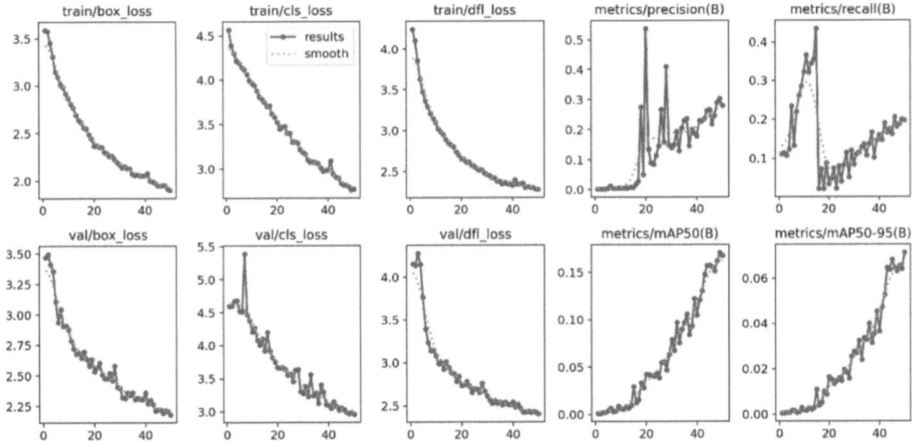

Fig. 9. Training and validation losses and validation-side metrics across epochs for Version 1.

The initial prototype (V1) was trained on only 107 manually annotated images using a CPU-based configuration (Intel Core i5-10400F). As shown in Fig. 9, the loss curves exhibit a clear downward trend on both training and validation sets, confirming that the network successfully learned the core patterns despite the very limited dataset. However, the corresponding precision and recall metrics fluctuate strongly, indicating unstable generalization. The model produced low-confidence detections and failed to recognize target objects in the test video sequences. These outcomes reflect a classic case of overfitting under data scarcity: the network minimizes training loss but lacks sufficient variability to generalize beyond the training samples.

To address this, the next versions were implemented by using larger datasets. The training initially ran on a CPU, which proved to be computationally expensive and time-consuming. After minor modifications to the configuration, the process was successfully migrated to GPU-based execution, resulting in a significant reduction in training time and enabling more efficient experimentation.

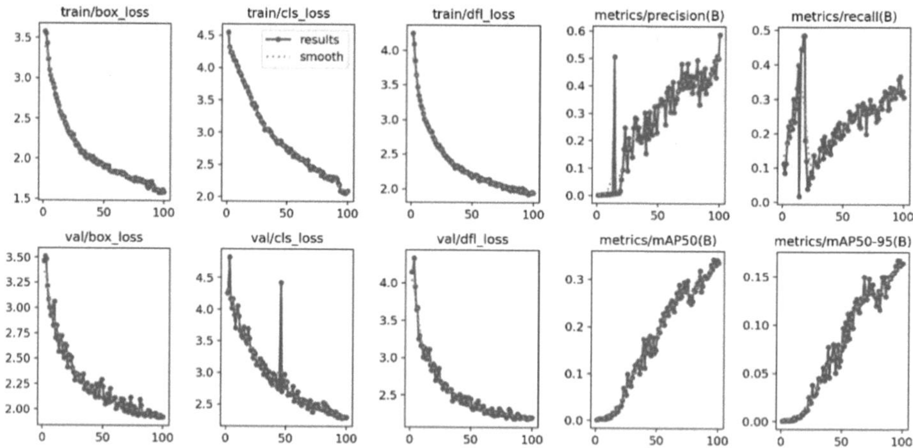

Fig. 10. Training and validation losses and validation-side metrics across epochs for Version 11.

For the second important milestone (V11) the training set was extended to 554 images. Figure 10 displays the corresponding training curves of V11, showing smoother convergence and reduced oscillation compared to V1, though the overall precision remained modest. These findings motivated the creation of a substantially larger dataset and longer training schedule.

The performance difference between the CPU and GPU configurations is quantified in Table 1. For the same dataset and hyperparameters, a single epoch took 223.74 s on the CPU but only 14.16 s on the GPU, representing a more than fifteenfold acceleration. The resulting 50- and 100-epoch runs demonstrated that GPU acceleration not only shortened training time but also made iterative experimentation feasible.

Table 1. Training time comparison of YOLOv8 on CPU vs GPU using 554 images.

Device	1 Epoch (s)	50 Epochs (s)	100 Epochs (s)
CPU	223.74	11,186.8 (3.1 h)	22,374 (6.21 h)
GPU	14.16	708.2 (11.8 min)	1,416.4 (23.6 min)

Building on these insights, the final version (V21) was trained with 3334 annotated images using the GPU exclusively. To ensure full convergence, the process ran for 200 epochs. The complete training time profile, summarized in Table 2, confirms the scalability of the configuration. Even with more than triple the data volume, the GPU maintained reasonable turnaround times suitable for experimental iteration.

Table 2. Training time of the final version on GPU.

Device	1 Epoch (s)	50 Epochs (s)	100 Epochs (s)	200 Epochs (s)
GPU	78.22	3,935.33	7,857.49	15,643.7

Figure 11 visualizes the evolution of loss and evaluation metrics during the 200-epoch training of V21. The smooth, monotonic reduction in box, classification, and distribution focal losses—paired with consistent growth in precision, recall, and mAP—demonstrates stable optimization and robust generalization. The narrow gap between training and validation curves suggests balanced capacity and effective regularization. Importantly, mAP@50 saturates earlier than mAP@50–95, revealing that later epochs mainly refine localization accuracy rather than classification confidence. This pattern supports that a larger dataset and an extended schedule improve convergence without evident overfitting on the validation split.

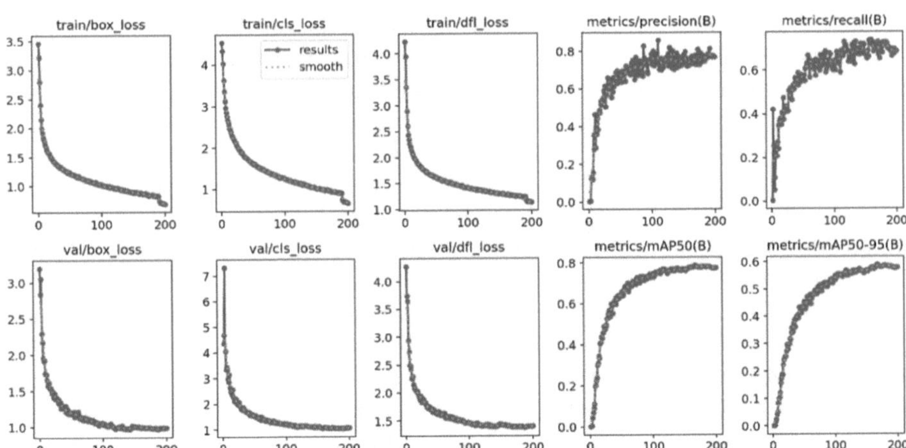

Fig. 11. Training and validation losses and validation-side metrics across epochs for Version 21.

From a methodological perspective, these three iterations collectively confirm that scaling annotated data, supported by GPU acceleration, directly improves training stability and predictive reliability. Early models validated the pipeline but lacked sufficient variability to sustain accuracy. Intermediate configurations demonstrated the feasibility of GPU-based acceleration and incremental dataset growth. The final model converged smoothly, maintained high precision and recall, and eliminated erratic training behavior. In summary, Figs. 9, 10 and 11 and Tables 1–2 jointly establish that the combination of dataset expansion and hardware optimization was decisive in achieving the target performance level for such scenarios.

4.5   Limitations and Scope

Our evaluation uses frames from the same virtual campus ecosystem that was used to define the task. Consequently, domain shift beyond this environment is only partially covered. We do not report an external, out-of-domain test at camera-ready time, and our results are limited to per-class metrics on held-out, in-domain frames. Therefore, our claims are competitive rather than state-of-the-art, and generalization to unrelated scenes remains future work.

5   Conclusions

This paper has demonstrated that domain-adapted object detection models can achieve high accuracy and practical reliability using a focused training strategy, even when operating on a dataset of limited scale. By fine-tuning a YOLOv8 detector for the recognition of chairs, benches, cars, and bicycles in a virtualized campus environment, the study achieved results that meet or exceed those of a generalized baseline model trained on significantly larger datasets.

The success of the approach relied on multiple factors: the use of a GPU-accelerated training framework that enabled efficient iteration; the manual curation and annotation of high-quality image data; and the iterative refinement of the model across 21 versions. Evaluation on realistic image sequences extracted from a digital twin of the University Centre confirmed the robustness of the detector in dynamic and partially occluded settings. In particular, the final model showed improved temporal consistency, higher detection confidence, and fewer false detections.

Overall, this work affirms the viability of building high-performing, application-specific detection models through targeted training. It also underscores the potential of integrating computer vision with digital twin technologies to support intelligent spatial analytics in smart campus and urban planning contexts. Future work may explore the integration of object tracking, multi-class extensions, and deployment in augmented or virtual reality interfaces.

Acknowledgments. This work has been implemented by the TKP2021-NVA-10 project with the support provided by the Ministry of Culture and Innovation of Hungary from the National Research, Development and Innovation Fund, financed under the 2021 Thematic Excellence Programme funding scheme.

References

1. Explore Pannon Egyetem Zalaegerszegi Egyetemi Központ in 3D (2024). https://my.matterport.com/show/?m=45j6qV29Yjt&ts=1. Accessed 20 July 2025
2. Bochkovskiy, A., Wang, C.Y., Liao, H.Y.M.: YOLOv4: Optimal speed and accuracy of object detection (2020)
3. Burger, W., Burge, M.J.: Digital Image Processing: An Algorithmic Introduction Using Java. Springer (2016)
4. Burger, W., Burge, M.J.: Digital image processing: an algorithmic introduction using Java. Springer (2008). https://doi.org/10.1007/978-1-84628-968-2_12
5. Diwan, T., Anirudh, G., Tembhurne, J.V.: Object detection using YOLO: challenges, architectural successors, datasets and applications. Multimedia Tools Appl. **82**(6), 9243–9275 (2023)
6. Girshick, R.: Fast r-cnn. In: Proceedings of the IEEE International Conference on Computer Vision (ICCV) (2015)
7. Girshick, R., Donahue, J., Darrell, T., Malik, J.: Region-based convolutional networks for accurate object detection and segmentation. IEEE Trans. Pattern Anal. Mach. Intell. **38**(1), 142–158 (2015)
8. Gonzalez, R.C., Woods, R.E.: Digital Image Processing. Pearson, 4th edn. (2018)
9. Google AI: Open images dataset v6 (2020). https://storage.googleapis.com/openimages/web/index.html, Accessed 13 May 2024
10. Jocher, G., Chaurasia, A., Qiu, J., Stoken, T.: Yolov8: Ultralytics' official implementation of yolo models (2023). https://github.com/ultralytics/ultralytics, Accessed 13 May 2024
11. Kuznetsova, A., et al.: The open images dataset v4: unified image classification, object detection, and visual relationship detection at scale. Int. J. Comput. Vision **128**(7), 1956–1981 (2020)
12. Mohammed, A., Ibrahim, H.M., Omar, N.M.: Optimizing retinanet anchors using differential evolution for improved object detection. Sci. Rep. **15**(1), 20101 (2025)
13. Redmon, J., Divvala, S., Girshick, R., Farhadi, A.: You only look once: Unified, real-time object detection. In: Proceedings of the IEEE Conference on Computer Vision and Pattern Recognition (CVPR). pp. 779–788 (2016)
14. Ren, S., He, K., Girshick, R., Sun, J.: Faster R-CNN: Towards real-time object detection with region proposal networks. In: Advances in Neural Information Processing Systems (NeurIPS) (2015)
15. Szeliski, R.: Computer Vision: Algorithms and Applications. Springer, 2nd edn. (2022). https://szeliski.org/Book/
16. Szűcs, J., Guzsvinecz, T.: Creation of digital twins using spatial artificial intelligence-a pilot study. Acta Polytechn. Hung. **22**(6) (2025)
17. Terven, J., Córdova-Esparza, D.M., Romero-González, J.A.: A comprehensive review of yolo architectures in computer vision: from yolov1 to yolov8 and yolo-nas. Mach. Learn. Knowl. Extr. **5**(4), 1680–1716 (2023)
18. Zhao, Z., Zheng, P., Xu, S.T., Wu, X.: Object detection with deep learning: a review. IEEE Trans. Neural Netw. Learn. Syst. **30**(11), 3212–3232 (2019)

Binary Tomography on the Cairo Pattern

Benedek Nagy[1,2]([✉]) [iD] and Tibor Lukić[3] [iD]

[1] Department of Mathematics, Eastern Mediterranean University, via Mersin-10,
Famagusta, North Cyprus, Turkey
[2] Institute of Mathematics and Informatics, Eszterházy Károly Catholic University,
3300 Eger, Hungary
nbenedek.inf@gmail.com
[3] Faculty of Technical Sciences, Chair of Mathematics,
University of Novi Sad, Novi Sad, Serbia
tibor@uns.ac.rs

Abstract. In this paper, we investigate a tomography reconstruction approach for binary images on the Cairo pattern. In discrete tomography the underlying grid plays an essential role, as the projection data and the structure of the grid together give the information that we can use for the reconstruction purpose. The Cairo pattern is one of the most known pentagonal tessellations of the plane and it has a nice symmetric structure for orthogonal projections giving the information in a somewhat different way as similar direction projections on the traditional square grid. We present how gradient based reconstruction method can be used to reconstruct images based on pentagonal pixels in the Cairo pattern. A short experimental evaluation is also presented.

Keywords: Tomography reconstruction · Nontraditional grids ·
Energy minimization · Gradient based optimization · Regularization ·
Pentagonal grids

1 Introduction

Tomography is a branch of image processing focused on reconstructing unknown images from given projection data [9]. Mathematically, an image is represented as a function whose range can be either continuous or discrete. In *Computerized Tomography*, this function typically has a continuous range. A specialized area within this field, known as *Discrete Tomography* (DT) [10,11], studies cases where the image function takes values from a finite, discrete set. When this set consists of a limited number of predefined intensity levels, the technique is referred to as *Multi-Level Tomography* [14]. A notable example is *Binary Tomography* (BT), where the image is composed of only two intensity values-commonly zero and one. The range of applications for tomography methods is extensive. Tomographic image reconstruction techniques are commonly employed across various industrial investigations, particularly in nondestructive material testing [4]. Another major area of application is in medical diagnostics, such as CT

P. Balázs et al. (Eds.): IWCIA 2025, LNCS 15985, pp. 230–244, 2026.
https://doi.org/10.1007/978-3-032-19347-6_15

(computed tomography) scans used in radiology. Tomography also plays a significant role in security screening – for instance, X-ray computed tomography is widely used at airports for baggage inspection [12]. BT methods are especially valuable when detecting the presence or absence of specific materials within a homogeneous structure, such as identifying atoms in crystalline materials [11].

There are various tilings/tessellations of the plane by various tiles, (usually, finitely many types are used in a tessellation). The three regular grids are built up by sole regular tiles: the square, the hexagonal and the triangular grids are having same side-length regular shapes: squares, hexagons and triangles, respectively, in a tessellation by a side-to-side manner. The graph theoretical dual of a regular grid is also a regular grid as the square grid is self-dual, while the other two regular grids are dual of each other. The eight semi-regular grids are built up by still only regular polygons, but more than one type is used and the tesselation is still side-to-side with same side-length polygons such that at each corner the the meeting polygons can be written in the same order. These grids are denoted based on the order of these polygons using their face code [5], e.g., the grid T(3,3,4,3,4) has three triangles and two squares at each corner in the given order. Actually, this grid is called isosnub quadrille or snub square tiling. The dual grids of the semi-regular grids are called Catalan grids [5], they build up by a sole but not regular tile. Moreover, this tile appear in various (at least two different) orientations. No regular pentagons tesselate the plane, but some pentagonal tessellations belong to the Catalan grids. Actually, there are 15 known tesselation families by sole pentagons [34]. In each pentagonal tilings, the used pentagon tile occur in various orientations. In this paper, we focus on the Cairo pattern which is the dual of the aforementioned T(3,3,4,3,4) grid [25,39]. In this tiling the used pentagon appears in four different orientations, thus it is also called as 4-fold Pentille Tessellation. In the next section, we give a mathematical description of the grid.

There are a number of successful tomographic image reconstruction methods proposed in literature starting with the seminal woks of Ryser [35] and Gale [6] in the 1950's, like *Algebraic Reconstruction Technique* (ART) [8], *Simultaneous Iterative Reconstruction Technique* (SIRT) [7], *Discrete Algebraic Reconstruction Technique* (DART) [1,2], *Spectral Projected Gradient* (SPG) [19], *Difference of Convex functions* (DC) [32,36]. However, most of these methods suppose a conventional image grid, such as grids with quadratic or rectangular pixel shapes. Only a few attempts were made to apply nonconventional grids in tomographic reconstructions. For the hexagonal grid, in [23] the Gibbs priors were used. Energy minimization based method has also been applied on this grid in [21]. We mention methods proposed for triangular grid as well: in [24,29,30] evolutionary, memetic and genetic approaches were used. While in [20,22] simulated annealing and energy minimization were used. Further, in [26,28] these techniques were complemented by some special tricks based on some properties of the grid. Moreover, also a technique is developed when the relation of the hexagonal and triangular grids is used for efficient reconstruction [27].

In the reconstruction process the basic criteria is the matching of the reconstructed image with the measured projection data. The acquired data is often the most reliable information about the unknown image.

The paper is organized as follows. Section 2 provides a description of the Cairo pattern along with some of its properties. Section 3 presents a detailed description of the proposed tomographic reconstruction method on the Cairo pattern image grid. Section 4 offers a brief experimental evaluation of the proposed method. Finally, the conclusion is presented in Sect. 5.

2   Mathematical Description of the Cairo Pattern

The Cairo pattern is one of the side-to-side pentagonal mono-tilings. It is also known with other, more mathematical names, e.g., 4-fold pentille grid [5]. Digital geometry, by giving a coordinate system and computing digital (path-based) distances were done on it in [13,37] and in [31]. Here, in this section, we briefly recall this grid based on the latter paper, and give a mathematical description addressing its pentagonal pixels (tiles) for later use. Figure 1 presents a coordinate system for the Cairo pattern addressing the pentagonal pixels by coordinate pairs. The grid is built up by a sole pentagon shape, but it is not a regular pentagon and it appears in 4 different orientations: the orientations are named after the direction to where they are directed as a kind of arrows: D, U, L and R (shown also for some of the pixels in the figure).

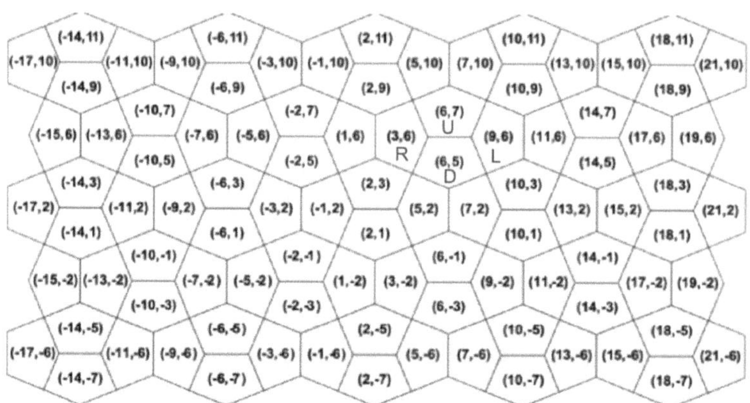

Fig. 1. The Cairo pattern image grid and coordinate system.

Observe that to keep the description symmetric, no tile with $(0,0)$ appears. Instead there are corners (gridpoints) where one of each of the four types meet, e.g., where type D $(2,1)$, type R $(-1,2)$, type U $(-2,-1)$ and type L $(1,-2)$. Based on the coordinates assigned we have the following:

- The x and y coordinates of a type D pixel have values
  $\equiv 2 \pmod 4$ and $\equiv 1 \pmod 4$, respectively, with sum $\equiv 3 \pmod 8$.
- The x and y coordinates of a type R pixel have values
  $\equiv 3 \pmod 4$ and $\equiv 2 \pmod 4$, respectively, with sum $\equiv 1 \pmod 8$.
- The x and y coordinates of a type U pixel have values
  $\equiv 2 \pmod 4$ and $\equiv 3 \pmod 4$, respectively, with sum $\equiv 5 \pmod 8$.
- The x and y coordinates of a type L pixel have values
  $\equiv 1 \pmod 4$ and $\equiv 2 \pmod 4$, respectively, with sum $\equiv 7 \pmod 8$.

Further, any pixel and any of its five side neighbor pixels have the sum of their absolute valued coordinate difference at most 4.

3 Tomography Reconstruction Problem on the Cairo Pattern

In this paper, we examine a reconstruction model for transmission tomography. A key feature of transmission tomography is that both the source and the detector are positioned outside the object being studied. The key information is obtained from the difference in ray intensity between the detector and the source, see Fig. 2.

Based on the description above, the orthogonal projections (horizontal and vertical) are very natural directions for the Cairo pattern, see Fig. 2. For horizontal projections, the y coordinate can have only two values: y and $y-1$. Similarly, for vertical projections, the x coordinate can have only two values: x and $x-1$. Actually, the number of the object pixels intersected by the projection lines are counted. Since this is not a regular grid and the pixels appear in various orientations, by using our discrete coordinate system, some of the pixels intersected by the same projection line have either the same coordinate value or it may differ with 1, as we have described above. In each projection line, there are tiles with exactly 3 of the 4 orientations. In horizontal projections, there are tiles with two of the orientations such that two of the neighbor projection lines go through on them (see blue and red projection lines in Fig. 2). In these two types, R and L, of tiles the value of the y coordinate is 2 (mod 4), then, in the upper projection (blue) the tiles have y coordinate 1 more than this, in the lower projection (red), the third type (orientation) tiles have y coordinate value that is 1 less than the given 2 (mod 4) value.

The reconstruction problem is approached mathematically. Let $u(x,y)$ denote the discrete image function not known, where (x,y) represents the coordinates of pixels (see Fig. 1) belonging to the image domain Ω. The image function u may be reshaped into a large column vector. The values of $u(x,y)$ belonging to the same column of the image, formed by a single vertical line (see an example in Fig. 3), represent one segment in the vector representation, with these segments stacked one below the other.

The problem of tomographic reconstruction can be expressed mathematically as the following system of linear equations:

$$A\,u = b, \quad A \in \mathbb{R}^{m \times n}, \ u \in \mathbb{R}^n, \ b \in \mathbb{R}^m, \tag{1}$$

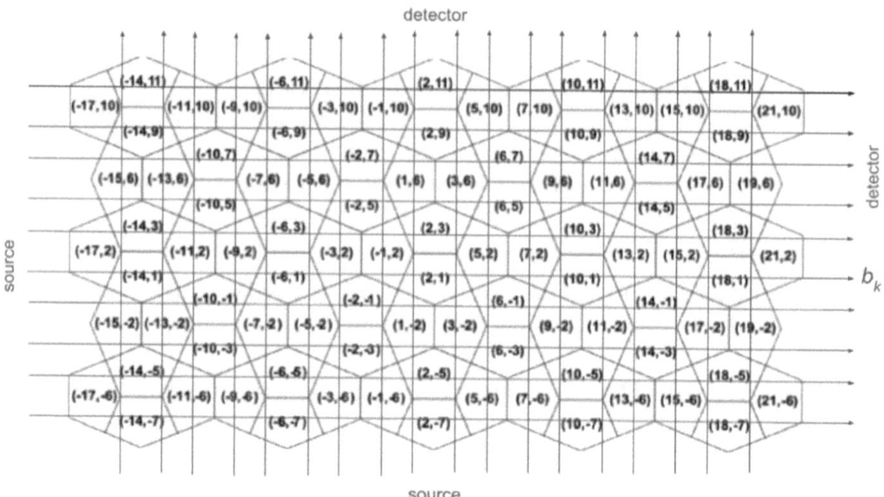

Fig. 2. The transmission parallel beam projection geometry on the Cairo pattern image grid. The calculation of the projection values is modeled on k-th projection ray (shown in green): $b_k = u(-17, 2) + u(-14, 1) + u(-11, 2) + \ldots + u(18, 1) + u(21, 2)$.

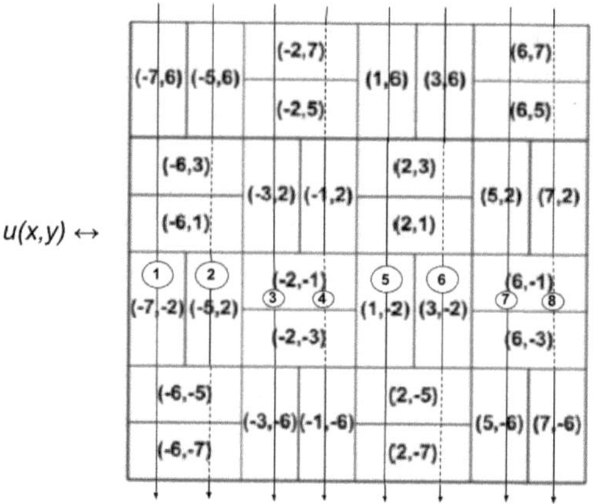

Fig. 3. Simplified and rectangular shape representation of a Cairo pattern image grid. The vector representation of u is formed as follows $[u(-7, 6), u(-6, 3), u(-6, 1), u(-7, -2), u(-6, -5), u(-6, -7), -2-, -3-, \ldots, -8-]^T$.

where u denotes the unknown image, containing n pixels (pentagons), to be reconstructed, in vector form. The vector b consists of projection values obtained through measurement or calculation. The matrix A is the so-called *projection matrix* and its rows hold information about pixels (pentagons) through which a corresponding projection ray passes. For example, the k-th row of A, denoted by A_k, contains ones at the pixel positions that are intersected by the k-th projection ray, and zeros elsewhere. In this way, it is ensured that the inner product $\langle A_k, u \rangle$ is equal to the projection value b_k, see Fig. 2. Therefore, each individual projection ray contributes a single linear equation to the system of Eqs. (1). The total number of projection rays taken is m. Although the projection directions shown in Fig. 2 are just horizontal and vertical, in a general case, the source-detector system can be rotated around the image centre for an arbitrary angle. Each such projection direction contributes a new set of parallel projection rays to the reconstruction problem (1).

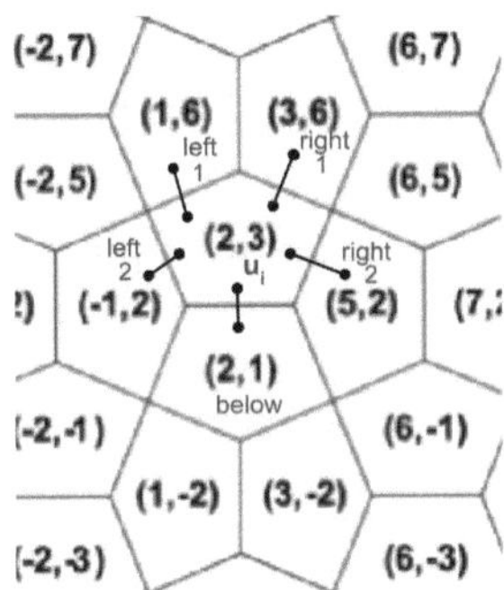

Fig. 4. Side neighbors of the pixel u_i which are located to its right and below.

In the reconstruction problem, as we explained above, both the projection matrix A and the projection vector b are known. The objective is to determine (reconstruct) the unknown image u. The system (1) is often underdetermined ($m \ll n$), because there are many more pixels (unknowns) than projection rays (equations). Therefore, in the general case, we can expect very large number of possible solutions (with growing image size, infinitely many in the limit). On the other hand, the system of Eqs. (1) is a large-scale problem, and consequently its solution requires an optimization approach. We address these problems by formulating a specially designed optimization problem, rather than solving problem (1) directly. We propose the following optimization problem

$$\min_{u} \; EG(u) := w_P \|Au - b\|_2^2 + w_H \sum_{i=1}^{n} [(u_i - u_{r1})^2 + (u_i - u_{r2})^2 + (u_i - u_b)^2] . \quad (2)$$

The objective (or energy) function (2) is a sum of two quadratic terms. The first one, also called the *data fitting* term, expresses consent of a solution with the given projection data. The role of the second term, also called *smooth regularization* term, is to ensure the smoothness or homogeneity of the solution. The application of this regularization is based on the assumption (a priori knowledge) about the compactness of the original image, in the sense that it consists of relatively large hole-less regions (which is usually a valid assumption for images of real objects). Indices r and b mark pixels which are side-adjacent to u_i, to the right and to the bottom, respectively, see Fig. 4. In a coordinate form, the pixel (x', y') is a side neighbor of (x, y) if the condition $|x - x'| + |y - y'| \leq 4$ holds. A pixel (x', y') in an image is considered to be positioned to the right and below its side neighboring pixel (x, y) if the following condition holds: $(x' > x)$ or $(x' = x$ and $y' < y)$. This method of neighbor selection ensures that each intensity difference between two side neighbors is counted exactly once in the calculation of the regularization term. This is the closest analogy to the well-known quadratic Total Variation (TV) regularization used in the case of conventional (quadratic or rectangular) grids. The quadratic TV type regularization, which in continuous case has a form

$$\iint_\Omega \|\nabla u(x, y)\|^2 \, dx dy \,, \tag{3}$$

has an isotropic diffusion type effect on an applied image $u(x, y)$, and it is often used in tomography reconstruction methods [20, 21, 38]. The integrand function in the TV regularization term (3) relies on the image gradient $\nabla u(x, y)$, calculated at each point. The discrete version of the gradient is calculated as the intensity difference between the considered pixel and its neighboring pixels to the right and below. Therefore, TV regularization on conventional grids takes into consideration each side difference in the entire image exactly once. We captured this feature and transferred it to our regularization design on the Cairo pattern grid. The parameters $w_P > 0$ and $w_H > 0$ control the influence of the data fitting and regularization terms, respectively, in the reconstruction process. We note that other types of regularization can also be considered, see [15, 16, 18].

A great feature of the proposed model (2) is that the gradient of the energy function can be derived analytically. Specifically, its gradient can be expressed as follows:

$$\nabla EG(u) = \left[\frac{\partial EG(u)}{\partial u_1}, \frac{\partial EG(u)}{\partial u_2}, \dots, \frac{\partial EG(u)}{\partial u_n}\right]^T,$$

where

$$\frac{\partial EG(u)}{\partial u_i} = 2w_P(A^T(Au - b))_i + 2w_H[(u_i - u_{r1}) + (u_i - u_{r2}) + (u_i - u_b)$$
$$+ (u_i - u_{l1}) + (u_i - u_{l2})],$$

where u_{l1} and u_{l2} are the remaining two side-neighbors of u_i that are not considered in the regularization term of the formula (2). This means that the solution of the model (2) can be found using a gradient-based iterative optimization algorithm. This is a good property because gradient-based algorithms are deterministic, relatively fast and accurate.

The model (2) does not include any binary-enforcing term; therefore, its solutions will be gray-level, particularly in cases where the amount of projection data is small. To ensure that the final solution is binary, we have to add a binary-enforcing term to the energy function. We decided to add the following concave term $\langle u, e - u \rangle$, where vector $e = [1, 1, \dots, 1]^T$. Its minimum value (zero) on the domain $[0, 1]^n$ is achieved when all pixel intensities of u are either zero or one, i.e., u is a binary image. Accordingly, we propose the following model for binary tomography:

$$\min_{u \in [0,1]^n} EB(u) := EG(u) + \mu \cdot \langle u, e - u \rangle , \tag{4}$$

where parameter $\mu > 0$ controls the magnitude of the influence of the term $\langle u, e - u \rangle$ in the reconstruction process. The gradient of the binary model (4) can also be derived analytically:

$$\nabla EB(u) = \nabla EG(u) + \mu(e - 2u).$$

In addition, the gradient $\nabla EB(u)$ can be calculated in a computationally very efficient manner. Therefore, the binary model (4) is also suitable for solving using a gradient-based algorithm. The optimization problem defined by model (4) is a constrained one, with the feasible set given by $\Gamma = [0, 1]^n$. We use the Spectral Projected Gradient (SPG) optimization algorithm for this task. This choice is motivated by earlier successful applications of SPG in similar problems, see [15–17,21]. According to the results of Birgin et al. [3], the SPG algorithm converges to the constrained stationary point for any arbitrary initial solution $u^0 \in \Gamma$, provided the following conditions are satisfied: 1) Γ is a closed and convex set, 2) the objective function is continuously differentiable, and 3) the projection P_Γ of an arbitrary point $u \in R^N$ onto the set Γ is given. It is easy to see that the first two conditions are satisfied. The third condition is also satisfied, since we can define the projection P_Γ function as follows:

$$[P_\Gamma(u)]_i = \begin{cases} 0, & u_i \leq 0 \\ 1, & u_i \geq 1 \\ u_i, & \text{elsewhere} \end{cases}, \qquad \text{where } i = 1, \ldots, n .$$

P_Γ is a projection with respect to the Euclidean distance. Therefore, for every fixed parameter μ, the SPG method is guaranteed to converge to at least a local minimum when solving problem (4). The design of the energy function follows the concept of convex-concave regularization, where the quadratic data fitting term is regularized by both convex (smoothing) and concave (binarization) terms, for more details see [36].

The reconstruction process implies successive solving of the problem (4), increasing the effect of binarization at each step. The process starts with a small binarization parameter $\mu > 0$ and solve the (convex) problem (4) using the SPG algorithm. After that, the parameter μ is gradually increased and the problem (4) is solved again, using the previous solution as the starting point. This process is repeated until the final binary solution is obtained. It is important that the parameter μ is increased properly during the process. If the increase rate is too steep, then the binarization force will overpower the data term, and the final solution will not fit the projection data. On the other hand, an excessively slow increase rate can lead to an unnecessarily slow reconstruction process. The proposed binary reconstruction method is detailed in the pseudocode of Algorithm 1.

Algorithm 1: Binary reconstruction

Parameters: $\epsilon_{out} = 10^{-3}$; $\mu_\Delta = 0.2$;
$u^{init} = [0.5, 0.5, \ldots 0.5]^T$; $\mu = 0$.
while $\max\limits_{i}\{\min\{u_i^{init}, 1 - u_i^{init}\}\} > \epsilon_{out}$

do

    `/* Solve by SPG algorithm: */`
    $u^{new} \leftarrow \arg\min\limits_{u\in[0,1]^n} EB(u)$;

    $u^{init} \leftarrow u^{new}$;
    $\mu \leftarrow \mu + \mu_\Delta$;
end

---

4   Experimental Results

This section brings presentation of experimental results provided by the proposed approach. Three original test images are used in the experimental work as they are shown in Fig. 5. They are synthetic (phantom) and binary images of the same size. The size is such that substantially corresponds to the conventional square grid image of size 32×32, however, the pentagons are double in area than squares and there are only 512 pentagon-shaped pixels on each of the image. In the following, we present reconstruction simulations that utilize only vertical and horizontal projection directions, based on the proposed binary reconstruction model (4). To the best of our knowledge, no other reconstruction method has been proposed in the literature based on the Cairo pattern image grid; therefore, we present and discuss only our own results.

Figure 5 shows the vertical and horizontal projection values (i.e., the number of white pentagons per row and column) for each test image. The task of the reconstruction process is to reconstruct the original images using only these data.

In tomographic image reconstruction, the two most significant measures of reconstruction quality can be expressed as follows:

$$E_R(u^r) = \sum_{i=1}^{n} |u_i^r - u_i^*|,$$

$$rE_R(u^r) = \frac{E_R(u^r)}{n_O} \cdot 100\,\%,$$

where u^r is the reconstructed image, while u^* denotes the original image and n_O is the number of object pixels in u^*. The reconstruction error E_R quantifies the difference between the reconstructed image u^r and the original image u^*. In the case of binary reconstructions, the function E_R expresses the *number of misclassified pixels*, while rE_R expresses the *relative number of misclassified pixels*, normalized by the size of the object, that is, the number of white pentagons.

In Fig. 6, the reconstructions of the original test images are shown. The reconstructions are obtained using only two natural directions: vertical and horizontal.

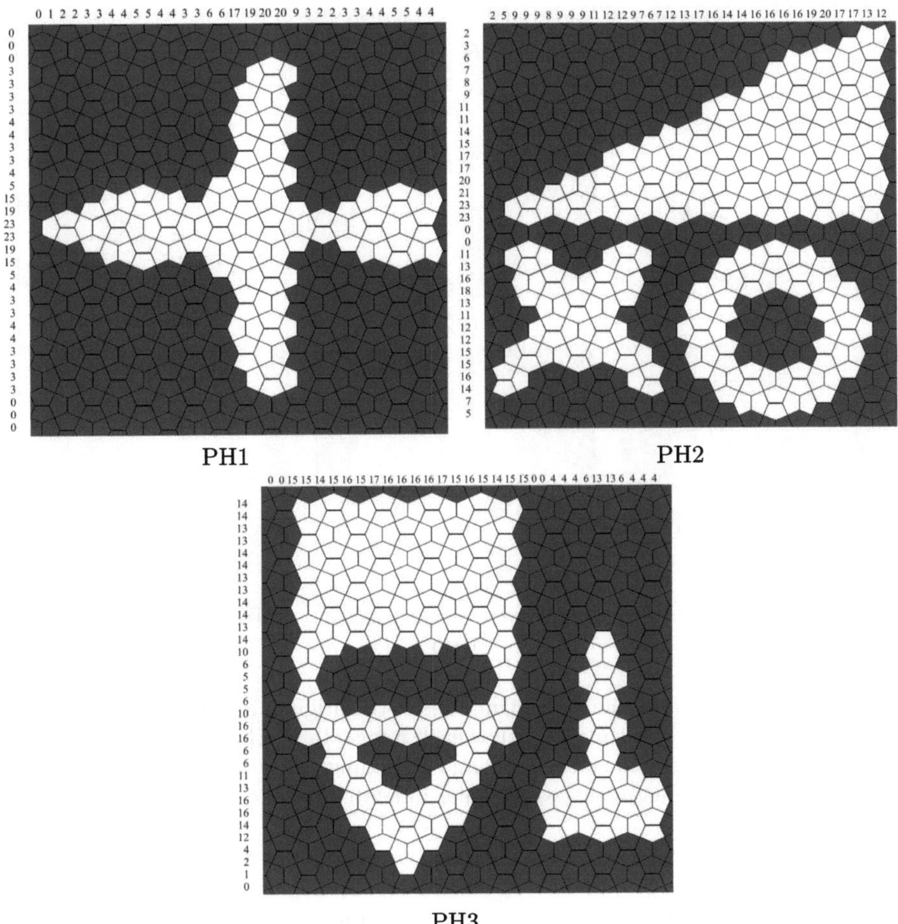

Fig. 5. Original test images used in experiments. Horizontal and vertical projection values of the test images are shown to the left and above of each image, respectively.

The most complex figure appears in the PH2 test image, which contains three separate regions, as well as a hole and cracks. Therefore, this image presented the most challenging task for the reconstruction process. As expected, the largest reconstruction error in quality (72.83%) occurred with this image. Additionally, a large switching component is also responsible for this large pixel error. The benefit of utilizing smooth regularization in the energy function can be observed in the presented reconstructions. Specifically, this regularization prevents the appearance of separate or "island-type" pentagons. This is fully consistent with the test images, which mostly exhibit compact or homogeneous structures.

In Fig. 7, counterparts of the original Cairo pattern test images are shown. These images are digitized on a conventional square grid of size 32×32. For the reconstruction, the SPG algorithm is used, proposed in [19]. This algorithm

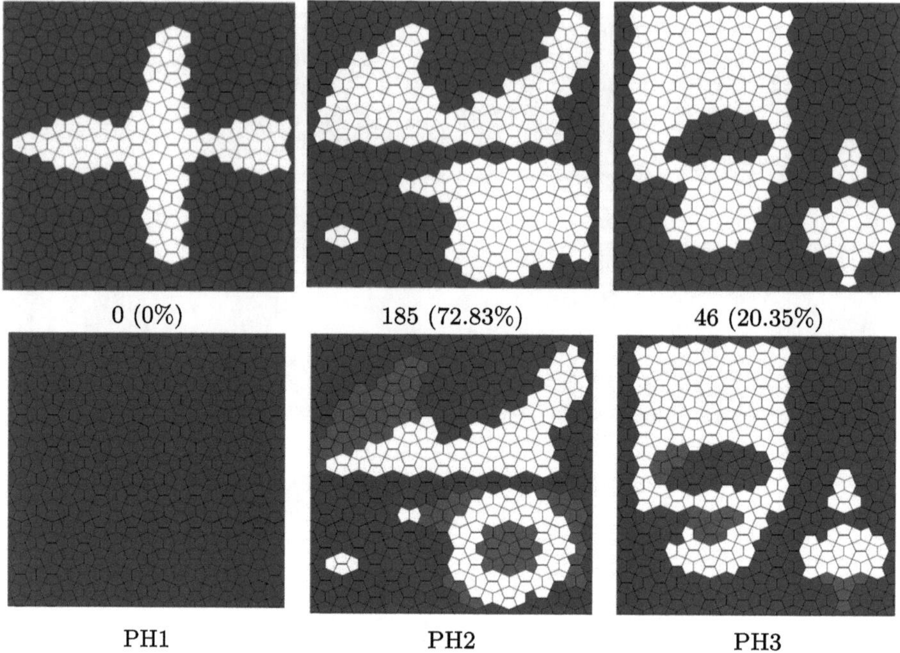

<table>
<tr><td align="center">0 (0%)</td><td align="center">185 (72.83%)</td><td align="center">46 (20.35%)</td></tr>
<tr><td align="center">PH1</td><td align="center">PH2</td><td align="center">PH3</td></tr>
</table>

Fig. 6. The first row shows the reconstructions of the original images, with the number of misclassified pentagons displayed below each image, in brackets the relative number of misclassified pixels are shown. The second row highlights the misclassified pentagons: red indicates missing pentagons, while blue indicates redundant ones. (Color figure online)

is gradient-based and applies an energy function that is very similar to the proposed energy function EG, except for the smoothing terms, which differ due to the specific characteristics of the image grids. These facts motivate us to choose this algorithm for the comparison. The projection directions are vertical and horizontal, as in Fig. 6, and the number of applied projection rays is the same: 32 per direction. The worst pixel error results on the square grid can be explained by the fact that the number of pixels on the square grid is double compared to the Cairo grid images, while the amount of projective data remains the same.

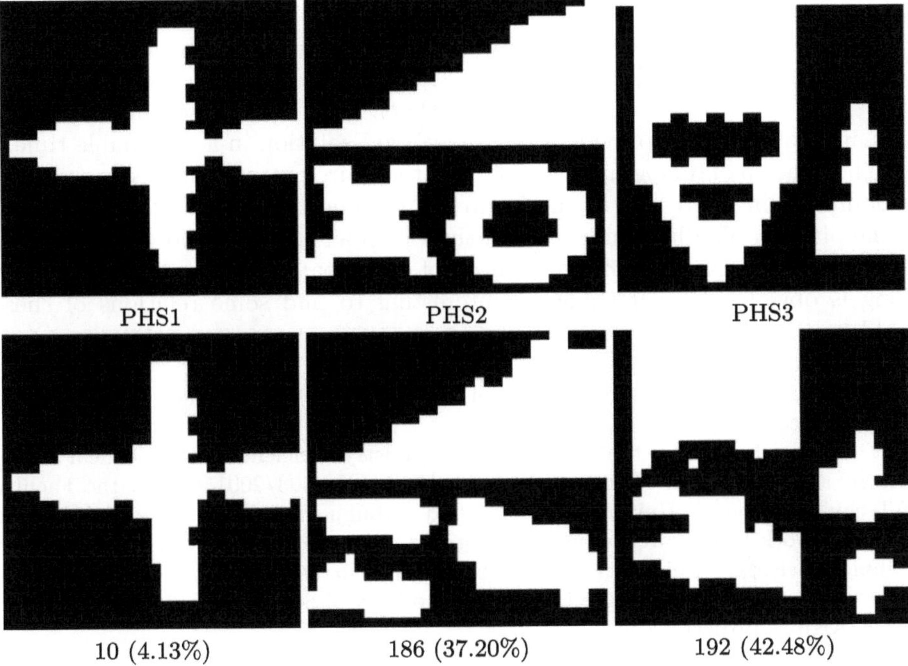

Fig. 7. The first row shows the original images on the conventional square grid. These images correspond to the images on the Cairo pattern grid PH1, PH2, and PH3, respectively. The second row shows the reconstructions obtained by the SPG algorithm. The number of misclassified pixels is displayed below each image, with the relative number of misclassified pixels shown in brackets.

5   Conclusions

In this paper, we consider the application of a nonconventional Cairo pattern image grid in the tomographic image reconstruction process. The Cairo pattern image grid belongs to the class of side-to-side pentagonal mono-tilings. Utilizing this specific image grid, we introduce a new deterministic and gradient-based tomographic image reconstruction method. The energy function of the proposed reconstruction model is tailored to the characteristics of the Cairo pattern grid. It is fully analytical, and its gradient is explicitly determined, enabling the efficient application of gradient-based algorithms for energy minimization. The proposed reconstruction method employs the Spectral Projected Gradient (SPG) optimization algorithm. The presented experimental results, based on a limited set of test images, demonstrate the promising performance of the proposed method. Future work could involve a deeper investigation into the properties of the Cairo pattern grid in the context of reconstruction, such as the characterization of switching components. Tomography reconstruction problems are usually either hard, or there could be exponentially many solutions. In the case of Cairo pattern it seems that the number of solutions can be exponential on the size of the image. To find a solution that could be close to the original image, we have used the assumption that the objects built up by relatively large connected components (our smooth regularization term is about that). As in many cases, our algorithm uses approximation to provide a relatively good solution in a reasonable time.

The relation of tomography problem on the Cairo pattern with reconstruction of domino tilings left also to the future. In domino tilings a (rectangular) part of the plane is tiled by dominoes and their numbers are given for each row and column [33]. Looking Fig. 3, the relation of Cairo pattern and a kind of domino tiling is obvious, thus it could be interesting to find some relations of these problems.

Acknowledgement. The comments of the anonymous reviewers are gratefully acknowledged.

T. Lukić acknowledge the support by the Ministry of Science, Technological Development and Innovation (Contract No. 451-03-137/2025-03/200156) and the Faculty of Technical Sciences, University of Novi Sad through project Scientific and Artistic Research Work of Researchers in Teaching and Associate Positions at the Faculty of Technical Sciences, University of Novi Sad 2025 (No. 01-50/295). He also acknowledges support received from the Hungarian Academy of Sciences through the DOMUS HUNGARICA project.

References

1. Batenburg, K.J., Sijbers, J.: Dart: a practical reconstruction algorithm for discrete tomography. IEEE Trans. Image Process. **20**(9), 2542–2553 (2011)
2. Batenburg, K.J., Sijbers, J.: DART: a fast heuristic algebraic reconstruction algorithm for discrete tomography. In: Proceedings of International Conference on Image Processing (ICIP). pp. 133–136 (2007)
3. Birgin, E.G., Martínez, J.M., Raydan, M.: Algorithm: 813: SPG - software for convex-constrained optimization. ACM Trans. Math. Softw. **27**, 340–349 (2001)
4. Carmignato, S., Dewulf, W., Leach, R.: Industrial X-Ray Computed Tomography. Vol. 10, pp. 978-3. Springer, Cham (2018)
5. Conway, J., Burgiel, H., Goodman-Strauss, C.: The Symmetries of Things. AK Peters (2008)
6. Gale, D.: A theorem on flows in networks. Pac. J. Math. **7**(2), 1073–1082 (1957)
7. Gilbert, P.: Iterative methods for the three-dimensional reconstruction of an object from projections. J. Theor. Biol. **36**(1), 105–117 (1972). https://doi.org/10.1016/0022-5193(72)90180-4
8. Gordon, R., Bender, R., Herman, G.T.: Algebraic reconstruction techniques (ART) for three-dimensional electron microscopy and x-ray photography. J. Theor. Biol. **29**(3), 471–481 (1970). https://doi.org/10.1016/0022-5193(70)90109-8
9. Herman, G.T.: Image reconstruction from Projections. Real-Time Imaging, **1**(1), 3–18, Springer-Verlag (1980)
10. Herman, G.T., Kuba, A.: Advances in Discrete Tomography and Its Applications. Birkhäuser (2007)

11. Herman, G.T., Kuba, A.: Discrete tomography: Foundations, algorithms, and applications. original edition: Birkhäuser (1999); new edition: Springer Science & Business Media (2012)
12. Kisner, S.J.: Image Reconstruction for X-Ray Computed Tomography in Security Screening Applications. Ph.D. thesis, USA (2013)
13. Kovács, G., Nagy, B., Turgay, N.D.: Distance on the Cairo pattern. Pattern Recognit. Lett. **145**, 141–146 (2021). https://doi.org/10.1016/J.PATREC.2021.02.002
14. Lukić, T.: Discrete Tomography Reconstruction Based on the Multi-well Potential. In: Aggarwal, J.K., Barneva, R.P., Brimkov, V.E., Koroutchev, K.N., Korutcheva, E.R. (eds.) IWCIA 2011. LNCS, vol. 6636, pp. 335–345. Springer, Heidelberg (2011). https://doi.org/10.1007/978-3-642-21073-0_30
15. Lukić, T., Balázs, P.: Binary tomography reconstruction based on shape orientation. Pattern Recogn. Lett. **79**, 18–24 (2016)
16. Lukić, T., Balázs, P.: Limited-view binary tomography reconstruction assisted by shape centroid. Visual Comput. **38**, 695–705 (2022)
17. Lukić, T., Balázs, P.: Moment preserving tomographic image reconstruction model. Image Vis. Comput. **146**, 105036 (2024). https://doi.org/10.1016/j.imavis.2024.105036
18. Lukić, T., Balázs, P.: Shape circularity assisted tomography reconstruction. Phys. Scr. **95**(10), 105211 (2020). https://doi.org/10.1088/1402-4896/abb633
19. Lukić, T., Lukity, A.: A spectral projected gradient optimization for binary tomography. In: Computational Intelligence in Engineering, SCI, vol. 313, pp. 263–272. Springer-Verlag (2010)
20. Lukić, T., Nagy, B.: Deterministic discrete tomography reconstruction method for images on triangular grid. Pattern Recogn. Lett. **49**, 11–16 (2014)
21. Lukić, T., Nagy, B.: Regularized binary tomography on the hexagonal grid. Phys. Scripta **94**(2), 025201 (2019)
22. Lukić, T., Nagy, B.: Energy-Minimization Based Discrete Tomography Reconstruction Method for Images on Triangular Grid. In: Barneva, R.P., Brimkov, V.E., Aggarwal, J.K. (eds.) IWCIA 2012. LNCS, vol. 7655, pp. 274–284. Springer, Heidelberg (2012). https://doi.org/10.1007/978-3-642-34732-0_21
23. Matej, S., Vardi, A., Herman, G., Vardi, E.: Discrete tomography: Foundations, algorithms, and applications, chap. Binary Tomography Using Gibbs Priors. In: [11] (2012)
24. Moisi, E., Nagy, B.: Discrete tomography on the triangular grid: a memetic approach. In: Proceedings of 7th International Symposium on Image and Signal Processing and Analysis (ISPA 2011). pp. 579–584. Dubrovnik, Croatia (2011)
25. Nagy, B.: Non-traditional 2d grids in combinatorial imaging - advances and challenges. In: Barneva, R.P., Brimkov, V.E., Nordo, G. (eds.) Combinatorial Image Analysis - 21st International Workshop, IWCIA 2022, Messina, Italy, July 13–15, 2022, Proceedings. Lecture Notes in Computer Science, vol. 13348, pp. 3–27. Springer (2022). https://doi.org/10.1007/978-3-031-23612-9_1
26. Nagy, B., Lukić, T.: Dense projection tomography on the triangular tiling. Fund. Inform. **145**(2), 125–141 (2016). https://doi.org/10.3233/FI-2016-1350
27. Nagy, B., Lukić, T.: Binary tomography on triangular grid involving hexagonal grid approach. In: Barneva, R.P., Brimkov, V.E., Tavares, J.M.R.S. (eds.) Combinatorial Image Analysis - 19th International Workshop, IWCIA 2018, Porto, Portugal, November 22–24, 2018, Proceedings. Lecture Notes in Computer Science, vol. 11255, pp. 68–81. Springer, Cham (2018). https://doi.org/10.1007/978-3-030-05288-1_6

28. Nagy, B., Lukić, T.: Binary tomography on the isometric tessellation involving pixel shape orientation. IET Image Process **14**(1), 25–30 (2020). https://doi.org/10.1049/IET-IPR.2019.0099

29. Nagy, B., Moisi, E.V.: Binary tomography on the triangular grid with 3 alternative directions - A genetic approach. In: 22nd International Conference on Pattern Recognition, ICPR 2014, Stockholm, Sweden, August 24–28, 2014. pp. 1079–1084. IEEE Computer Society (2014). https://doi.org/10.1109/ICPR.2014.195

30. Nagy, B., Moisi, E.V.: Memetic algorithms for reconstruction of binary images on triangular grids with 3 and 6 projections. Appl. Soft Comput. **52**, 549–565 (2017). https://doi.org/10.1016/J.ASOC.2016.10.014

31. Nagy, B., Saadat, M.: Digital distances on the 4-fold pentille tessellation. Sci. Rep. **15**(1), 35601 (2025). https://doi.org/10.1038/s41598-025-17010-4

32. Pham Dinh, T., Hoai An, L.T.: A D.C. Optimization Algorithm for solving the trust-region subproblem. SIAM J. Optim. **8**(2), 476–505 (1998)

33. Picouleau, C.: Reconstruction of domino tiling from its two orthogonal projections. Theor. Comput. Sci. **255**(1–2), 437–447 (2001). https://doi.org/10.1016/S0304-3975(99)00312-6

34. Rao, M.: Exhaustive search of convex pentagons which tile the plane. Tech. rep., arXiv preprint arXiv:1708.00274 (2017)

35. Ryser, H.: Combinatorial properties of matrices of zeros and ones. Can. J. Math. **9**, 371–377 (1957)

36. Schüle, T., Schnörr, C., Weber, S., Hornegger, J.: Discrete tomography by convex-concave regularization and D.C. programming. Discrete Appl. Math. **151**(1–3), 229–243 (2005)

37. Turgay, N.D., Nagy, B., Kovács, G., Vizvári, B.: Weighted distances in the Cairo pattern. Pattern Recognit. Lett. **166**, 105–111 (2023). https://doi.org/10.1016/J.PATREC.2023.01.004

38. Weber, S., Nagy, A., Schüle, T., Schnörr, C., Kuba, A.: A benchmark evaluation of large-scale optimization approaches to binary tomography. In: Kuba, A., Nyúl, L.G., Palágyi, K. (eds.) Proceedings of 13th International Conference on Discrete Geometry for Computer Imagery (DGCI 2006). LNCS, vol. 4245, pp. 146–156. pringer, Berlin, Heidelberg (2006). https://doi.org/10.1007/11907350_13

39. Williams, R.: The Geometrical Foundation of Natural Structure: A Source Book of Design. Dover, New York (1979)

Author Index

P. Balázs et al. (Eds.): IWCIA 2025, LNCS 15985, p. 245, 2026.
https://doi.org/10.1007/978-3-032-19347-6